Lecture Notes in Mathematics

Edited by A. Dold and B. Eckm

755

Global Analysis

Proceedings of the Biennial Seminar of the Canadian Mathematical Congress, Calgary, Alberta, June 12 – 27, 1978

Edited by
M. Grmela and J. E. Marsden

Springer-Verlag
Berlin Heidelberg New York 1979

Editors

Miroslav Grmela
Centre de Recherche de Mathématiques Appliquées
Université de Montréal
Montreal, Québec/Canada H3C 3J7

Jerrold Eldon Marsden
Department of Mathematics
University of California
Berkeley, CA 94720/USA

AMS Subject Classifications (1970): 10 H xx, 58 F xx, 58 G xx

ISBN 3-540-09703-1 Springer-Verlag Berlin Heidelberg New York
ISBN 0-387-09703-1 Springer-Verlag New York Heidelberg Berlin

Printed in Germany

Printing and binding: Beltz Offsetdruck, Hemsbach/Bergstr.
2141/3140-543210

Preface

This volume represents the invited papers for the seminar on global analysis held at the University of Calgary, June 12-27, 1978. Not all the lecturers were able to provide notes and their work is being, or has been, published elsewhere.

We thank all the participants for a lively and very successful conference, and the Canadian Mathematical Congress for its support.

M. Grmela

J. Marsden

Contents

Conference Participants

ADLER, Mark
Department of Mathematics
University of Minnesota
Minneapolis, Minnesota 55455

ARMS, Judy
Department of Mathematics
University of Utah
233 Widtsoe Building
Salt Lake City, Utah 84112

BAXTER, John
Department of Mathematics
The University of Calgary
Calgary, Alberta T2N 1N4

BERTHIER, Anne Marie
Department of Mathematics
Rice University
Box 1892
Houston, Texas 77001

BLATTNER, Robert J.
Department of Mathematics
University of California
Los Angeles, California 90024

BROOKE, Jim
Department of Mathematics
University of Alberta
Edmonton, Alberta T6G 2G1

CHURCHILL, Rick
Department of Mathematics
Hunter College
State University of New York
New York, New York 10021

COUCH, Eugene
Department of Mathematics
The University of Calgary
Calgary, Alberta T2N 1N4

DUISTERMAAT, J.J.
Mathematisch Instituut
University of Utrecht
Budapestlaan 6, De Uithof
Utrecht, The Netherlands

GEBA, K.
Department of Mathematics
University of Gdansk
Gdansk Oliwa, Poland

GRMELA, M.
Mathematics Institut
Universite de Montreal
Montreal, Quebec

GUTKIN, Eugene
Department of Mathematics
The University of Utah
Salt Lake City, Utah 84112

IHRIG, Ed
Department of Applied Mathematics
McMaster University
Hamilton, Ontario

ISCOE, Ian
Department of Mathematics
Colonel by Drive
Carleton University
Ottawa, Ontario K1S 5B6

JOEL, Jeffrey S.
Mathematical Reviews
University of Michigan
611 Church Street
Ann Arbor, Michigan 48109

JONKER, Leo
Department of Mathematics
Queen's University
Kingston, Ontario K7L 3N6

KAMINKER, Jerry
IUPUI
Department of Mathematics
Indianapolis, Indiana 46205

KLEMOLA, Tapio
Department of Mathematics
University of Montreal
P. O. 6128, Station A
Montreal, Quebec

KOKOSKI, Richard
Department of Physics
University of Toronto
Toronto, Ontario M5S 1W4

KOSTANT, Bertram
Department of Mathematics
Massachusetts Institute of Technology
Cambridge, Massachusetts 02139

KUNZLE, H.P.
Department of Mathematics
The University of Alberta
Edmonton, Alberta T6G 2G1

KUPKA, I.
Department of Mathematics
The University of Toronto
Toronto, Ontario

LAUE, Hans
Department of Physics
The University of Calgary
Calgary, Alberta T2N 1N4

LORENZ, Edward
Department of Meterology
Massachusetts Institute of Technology
Cambridge, Massachusetts 02139

MALLET-PARET, John
Department of Mathematics
Brown University
Providence, Rhode Island 02912

MARSDEN, J.
Department of Mathematics
University of California
Berkeley, California 94720

McKEAN, Henry
Department of Mathematics
Courant Institute
251 Mercer Street
New York, New York 10012

NORMAN, Dan
Department of Mathematics and Statistics
Queen's University
Kingston, Ontario K7L 3N6

RABINOWITZ, P.
Department of Mathematics
University of Wisconsin
Madison, Wisconsin 53706

RATIU, Tudor
Department of Mathematics
University of California
Berkeley, California 94720

RAWNSLEY, John
School of Theoretical Physics
Dublin Institute for Advanced Studies
Dublin 4, Ireland

ROWLEY, Brian
Department of Mathematics
McGill University
805 Sherbrooke Street West
Montreal, Quebec H3A 2K6

ROD, David L.
Department of Mathematics
The University of Calgary
Calgary, Alberta T2N 1N4

SATTINGER, David
Department of Mathematics
University of Minnesota
Minneasplis, Minnesota 55455

SAVAGE, Jim
Department of Physics
The University of Alberta
Edmonton, Alberta T6G 2G1

SIMMS, David
Department of Mathematics
Trinity College
University of Dublin
Dublin, Ireland

SNIATYCKI, J.
Department of Mathematics
The University of Calgary
Calgary, Alberta T2N 1N4

SJOSTRAND, Johannes
Department of Mathematiques
Universite de Paris Sud
Centre d'Orsay
Orsay, France F91405

SYMES, William
Mathematics Research Center
University of Wisconsin
Madison, Wisconsin 53706

TERRIER, J. M.
Department of Mathematiques
Universite de Montreal
Case Postale 6128
Montreal, Quebec

TORRENCE, R. J.
Department of Mathematics
The University of Calgary
Calgary, Alberta T2N 1N4

VAN MOERBEKE, P.
Department of Mathematics
University of California
Berkeley, California 94720

WAN, Yieh-Hei
Department of Mathematics
State University of New York
Buffalo, New York

ON A TRACE FUNCTIONAL FOR FORMAL PSEUDO-DIFFERENTIAL OPERATORS AND THE HAMILTONIAN STRUCTURE OF KORTEWEG-DEVRIES TYPE EQUATIONS

M. Adler

Abstract We study the Lie geometric structure behind the Hamiltonian structure of the Korteweg deVries type equations.

I. Introduction The Korteweg-deVries equation for $q \in C_0^\infty(R)$,

(1.1) $$q_t = 6qq_x - 2q_{xxx} ,$$

has been intensively studied over recent years. I want to focus attention on its inherent Hamiltonian structure and status as a completely integrable system, and moreover on generalizations of it discussed by Gel'fand-Dikii [1] . Gardner discovered that (1.1) can be written as a Hamiltonian system in the following form:

(1.2) $$\frac{d}{dt} q = X_H \equiv \mathcal{J} \frac{DH}{Dq} ,$$

where $\mathcal{J} = \frac{\partial}{\partial x}$, $H = H[q] = \int_R (q^3 + \frac{1}{2} q_x^2) dx$, and $\frac{DH}{Dq}$ is the directional derivative of H with respect to q . The vector field X_H is said to be Hamiltonian precisely because we can use $\mathcal{J}$ to define the following Poisson bracket:

$$\{H(q),F(q)\} = \int_R (\frac{DH}{Dq}) \mathcal{J} (\frac{DP}{Dq}) dx .$$

The above $\{\cdot,\cdot\}$ is a Poisson bracket because it is a skew-symmetric bilinear derivative in its arguments which satisfies the Jacobi identity. This is easily verified, assuming that H,F are integrals of polynomials in q and its derivatives, which shall always be assumed. The system (1.1,2) is said to be completely integrable because there exist a denumerable sequence of independent functionals $H_j = \int_R P_j dx$, $j = 1,\dots$, $H_1 = \int_R q\, dx$, the P_j's polynomials in q and its derivatives, which are in involution with respect to $\{\cdot,\cdot\}$, i.e. ,

$$\{H_j, H_k\} = 0 \text{ , for all } j,k \in Z .$$

In addition, Lax [2] discovered that (1.1) can be expressed in the following form:

$$(1.3) \qquad \frac{dL}{dt} = [B,L] \text{ , } L = -\partial_x^2 + q(x,t) \text{ ,}$$
$$B = -4\partial_x^3 + \partial(q\partial_x + \partial_x q\cdot) .$$

In general one has an equivalence between

$$q_t = X_{H_j}(q) \quad \text{and} \quad \frac{dL}{dt} = [B_j, L] \text{ ,}$$

with B_j a formally skew-symmetric $2j-1$ order differential operator, with coefficients polynomials in q and its derivatives. Recently, Gel'fand-Dikii [1] discovered a generalization of the above situation, namely, if one takes

$$L = (-i\partial_x)^n + \sum_{j=0}^{n-2} q_j(-i\partial_x)^j \text{ , } q_i \in C_0^\infty(R) \text{ , } i = 0,..,n-2 \text{ ,}$$

then the Lax-equations $\frac{dL}{dt} = [L,B_j]$, $j = 1,\dots$, with appropriate differential operators B_j , are equivalent to the Hamiltonian equations

$$(1.4) \qquad q_t = X_{H_j}(q) \equiv \mathcal{J}\frac{DH}{Dq} \text{ ,}$$

with $q = (q_0, q_1, q_{n-2})^T$, $\frac{DH}{Dq} = (\frac{DH}{Dq_0}, \frac{DH}{Dq_1}, \ldots, \frac{DH}{Dq_{n-2}})^T$,

and $H = \int_R P\,dx$, P being a polynomial in the q_i and their derivatives with respect to x . In the above $\mathcal{J}$ is a (n-1) x (n-1) matrix differential operator with coefficients polynomials in q and its derivatives, while as before, $\mathcal{J}$ defines a Poisson bracket { , } in the previously specified sense, via $\{H,F\} = \int_R (\frac{DH}{Dq}, \mathcal{J}(\frac{DF}{Dq}))dx$, with (,) the R^{n-1} scalar product. Here also the coefficients of the B_j's are polynomials in q and its derivatives, and in fact Gel'fand-Dikii [1] gives an algebraic construction for them. Once again we have the H_j's form an involutive system, i.e., $\{H_j, H_k\} = 0$, $j,k \in Z$. The case n = 2 , with slight modification leads to the Boussinesq equation. We now give a natural geometric interpretation to the above situation.

The Lax-equation (1.3) description of the dynamical equations of motion (1.1,2) suggest a group theoretical basis for the underlying Hamiltonian structure and integrability phenomena. In fact this is the case. We will describe in the next section this structure with the aid of the formal calculus of variations. We shall see that in a sense to be made specific, the symplectic structure is the co-adjoint structure of Kostant-Kirillov [3], and the integrability a consequence of a simple geometric argument found in [4], whose formal abstraction is due independently to B. Kostant and B. Symes. In fact, the same mechanism is also behind the nonperiodic Toda systems, as was reported in [4], and also in [5,6]. In a joint work with P. Moerbeke and T. Ratiu, to appear, it will be shown that this mechanism is also crucial in the periodic Toda systems and their generalizations [7], and the Euler-Arnold top. [8]. In the Calogero-Moser systems, group theoretic constructions are also seem to play a crucial role [9,10,11], and especially Lie algebra decompositions. The crucial geometrical construction for the Gel'fand-Dikii equations is that of a trace functional for formal pseudo-differential operators. In the last section we prove Theorem 2, that the bracket to be defined in the next section is in fact a Poisson bracket.

II. The Geometric Description

In this section we give the geometric description of the Hamiltonian structure and complete integrability of the KdV type equations, which involves the definition of the trace functional alluded to in the introduction. We give no proofs, but in the next section we prove the basic theorem of this section. We need some terminology, discussed and motivated in [4] in some detail.

Let R be a commutative ring over the complex numbers, equipped with a derivation D, i.e. R is a differential ring. We define the indefinite integrals I as R modulo DR, i.e. $I = R/DR$. The map which takes $r \in R$ to its equivalence class in I is denoted $r \rightarrow \overline{r}$, and equality in I is denoted by $\doteq$. The ring of polynomials in $a_1,\dots,a_n \in R$ and their derivatives shall be denoted $R[a_1,\dots,a_n]$, similarly for $I[a_1,\dots,a_n]$. Define Φ, the ring of formal pseudo-differential operators, to be the formal Laurant series in the variable ξ over the differential ring R, i.e.

$$\Phi = \{ \sum_{-\infty<i\leq N} a_i \xi^i \,|\, a_i \in R,\ N \text{ arbitrary},\ N < \infty\} .$$

The rule of multiplication in Φ is the rule for pseudo-differential operators:

(2.1) $\phi_1 \circ \phi_2 \equiv \sum_{\nu \geq 0} \frac{1}{\nu!}(\frac{\partial}{\partial\xi})^\nu \phi_1 \cdot (-iD)^\nu \phi_2$.

In the above , $\frac{\partial}{\partial\xi}$ denotes formal term by term differentiation of the formal ξ series, and $\cdot$ denotes formal series multiplication. We define the planes $\mathcal{a}_{\ell,j} = \{ \sum_{\ell \leq i \leq j} a_i \xi^i | a_i \in R\}$ and projections $P_{\ell,j}$ onto $\mathcal{a}_{\ell,j}$ by $P_{\ell,j}(\Sigma a_i \xi^i) = \sum_{\ell \leq i \leq j} a_i \xi^i$.

In order to do geometry, we shall employ a trace, and we define the map $\Phi \to I$ by

(2.2) $tr(\Sigma a_i \xi^i) \equiv \langle \Sigma a_i \xi^i \rangle \doteq \overline{a_{-1}} \in I$.

The important fact about the trace is

Theorem 1

$$tr[a,b] \doteq 0 ,$$

where $[a,b] = a \circ b - b \circ a$. This is easily proved from (2.1,2) using formal integration by parts and the formal power series version of the Cauchy Residue Theorem. Using the trace, we can define, by Theorem 1,a symmetric inner product

(2.3) $\langle a,b \rangle \equiv \langle a \circ b \rangle = \langle b \circ a \rangle = \langle b,a \rangle$.

We note that if we define, by the binomial theorem,

$$(\xi - iD)^{-j-1} r \equiv \sum_{k \geq 0} \xi^{-j-1-k} \binom{j+k}{k} (iD)^k r \ , \quad r \in R \ ,$$

then if $a = \sum_{i\geq 0} a_i \xi^i$, $b = \sum_{j\geq 0} (\xi - iD)^{-j-1} b_j$ (the latter a form by which any element in $\mathcal{O}_{-\infty,-1}$ can be represented) then it is easily shown [4] that

(2.4)(a) $\langle a,b\rangle \doteq \overline{\Sigma\, a_i b_i}$.

If one defines $K = \mathcal{O}_{0,\infty}$, $N = \mathcal{O}_{-\infty,-1}$, then from (2.2,3) it immediately follows

(2.4)(b) $K^{\perp} = K$, $N^{\perp} = N$, and $\Phi = K + N$,

the $\perp$ being with respect to $\langle\cdot,\cdot\rangle$, (2.3) . Moreover (2.4) shows the maps $K \hookrightarrow \mathrm{Hom}(N,I)$, $N \hookrightarrow \mathrm{Hom}(K,I)$ defined by $K \to \langle K,\cdot\rangle$, etc., are injections and we may think of K,N as dual with respect to $\langle\cdot,\cdot\rangle$, and Φ as being self-dual, i.e. $K^* = N$, $N^* = K$, $\Phi^* = \Phi$.

So by the above construction we have $\Phi = K + N$, K and N dually paired, Φ self-dual, and $K^{\perp} = K$, $N^{\perp} = N$. The formal Lie algebra Φ acts on $\Phi^* \sim \Phi$ by the (linearization of the) co-adjoint action, and so we may speak of Φ invariant functions on $\Phi^* \sim \Phi$. (Such functions must satisfy an infinitesimal condition.) Similarly, the formal affine group $1 + N$, with formal Lie algebra N , acts on K , now identified with N^*, by the co-adjoint action. The orbit θ_A , through $A \in K$, under the action of $1 + N$ is easily seen to be, using (2.3),

(2.5) $$\theta_A = \{[g^{-1}Ag]_+ \mid g \in 1 + N\} \ , \quad [\ \ ]_+ = P_{0,\infty}[\ \ ] \ ,$$

and similarly the 'tangent space' of θ_A at A is

(2.6) $$T\theta_A(A) = \{[n,A]_+ \mid n \in N\} \ .$$

We make the important observation that $\xi^n + \mathcal{C}_{0,n-1} \equiv \Phi_n \subset K$ is preserved by the co-adjoint action of $1+N$ on K, and the coefficient of ξ^{n-1} is an orbit invariant. These concepts and what follows are made precise in [4].

As is well known, at least in the case of genuine finite dimensional Lie groups, there is a natural orbit symplectic structure on the co-adjoint orbits, the Kostant-Kirillov orbit structure [3]. This is also true in our case. Lets parametrize Φ by $\Phi \ni A = \Sigma\, a_i \xi^i$, and suppose $H \in I[a_0,\ldots,a_{n-1}\}$, then thinking of $\Phi_n \subset K \subset \Phi^* = \Phi$, its natural to define the gradient of H, $\nabla_K H \in (\Phi^*)^* = \Phi$, as

$$(2.7) \qquad \nabla_K H = \sum_{j=0}^{n-1} (\xi - iD)^{-j-1} \frac{DH}{Da_j}, \quad \frac{D}{Da_j} = \Sigma(-1)^k \frac{\partial}{\partial a_j^{(k)}}, \quad a^{(k)} = \partial_x^k a .$$

Note by (2.4), $\langle \nabla_K H, \delta A \rangle \doteq \Sigma \overline{\frac{DH}{Da_j} \delta a_j} \equiv dH(\delta A)$, which is just the usual formula from the (formal) calculus of variations. We are really thinking of H as a function just on Φ_n, or even K, but it's sometimes more convenient (as in the above) to interpret H as a function on $\Phi^* = \Phi$. Since $\Phi_n \subset K \subset \Phi^*$ is an invariant manifold of the $1+N$ action, and H may also be interpreted as a function on K, it follows from (2.6) (see [4]), that the Hamiltonian vector field of the orbit structure, X_H, generated by H at the point $A = \xi^n + \sum_{0 \le i \le n-1} a_i \xi^i$, is given by

$$(2.8) \qquad X_H(A) = [\nabla_K H, A]_+ = [\underline{\tilde{P}}_{(n-1),-1}(\nabla H), A)^+ .$$

In the above, the projector $\tilde{P}_{j,k}(\Sigma(\xi-iD)^{\ell}a_{\ell}) \equiv \sum_{j\leq\ell\leq k}(\xi-iD)^{\ell}a_{\ell}$, and so X_H doesn't depend on $\frac{DH}{Da_{n-1}}$, even though X_H depends on the orbit invariant a_{n-1} as a parameter. In addition we immediately have from (2.8) that the formula for the orbit Poisson structure is given by:

(2.9) $\{H,P\}(A) = \langle A,[\nabla_K H, \nabla_K F]\rangle$,

and more importantly we have:

Theorem 2 $\{\cdot,\cdot\}$ is indeed a Poisson bracket.

(See Section 3 for the proof.)

Before changing the subject, we stress one point. If we consider functions $H\in I[a_{-\ell},\ldots,a_0,\ldots,a_K]$, $A = \Sigma\, a_i\xi^i \in \Phi^* = \Phi$, then we must distinguish between three gradients, ∇_K , ∇_N , and the total gradient $\nabla = \nabla_K + \nabla_N$, where since $K^{\perp} = K$, $N^{\perp} = N$, we have $\nabla_K = P_{-\infty,-1}\circ\nabla$, $\nabla_N = P_{0,\infty}\circ\nabla$, (see [4]) . In [4] we showed that upon restriction to each orbit, we had in some appropriate sense, a nondegenerate two form, and as a consequence, with the aid of some technical machinery, (2.9) defined a Poisson structure upon restriction to an orbit. In the next section we give a simple, direct proof that (2.9) defines a Poisson structure on the entire K , which was only indicated in [4] .

We now briefly digress, as we need another construction. Given $A = \xi^n + \sum_{i<n} a_i\xi^i$, we can construct $A^{\frac{1}{n}}$, i.e. a solution to $x^n = A$, in the form $\tau = \xi + \sum_{i\geq 0} b_i\xi^{-i}$, determining the b_i recursively (see [4]) as polynomials in the a_j and their derivatives. We can adjoint to R , without changing notation, the formal element σ (which we intuitively think of as $\mathrm{sgn}(\xi)$) such that $\sigma^2 = 1$, $\frac{\partial\sigma}{\partial\xi} = 1$, $D\sigma = 0$, and obtain in addition the "branched" root of A, $\sigma^{n+1}\cdot\tau$. We then let

$A^{\frac{m}{n}} \equiv \tau^m$ (with either choice of τ) and in fact $(\tau^m)^n = (\tau^n)^m = A^m$, and so indeed τ^m is an nth root of A^m, i.e. satisifes $y^n = A^m$. In particular if $A = \xi^n + \sum_{0 \leq i \leq n-1} a_i \xi^i$, the $b_j \in R[a_0, a_1, \dots, a_{n-1}]$, and $\langle A^{\frac{m}{n}} \rangle \in I[a_0, a_1, \dots, a_{n-1}]$.

We now give, in a suitable form, an abstract finite dimensional geometrical theorem due independently to B. Kostant and B. Symes, which is an abstraction of a geometrical argument in [4] which proved the complete integrability of the Gelfand-Dikii systems. Unfortunately, it doesn't quite apply technically to our situation, but it gives the geometrical essence of whats going on, and so we state it anyway, and refer the reader to [4] for the technically complete and more direct argument mentioned above in [4].

Theorem 3 (Kostant-Symes) Let L be a Lie algebra paired through a non-degenerate bilinear form, $\langle \cdot, \cdot \rangle$, with a vector space which we may as well identify with L^*. Let L support the direct sum vector space decomposition $L = K + N$, with K, N Lie subalgebras. This, of course, induces the splitting $L^* = K^\perp + N^\perp$, with respect to $\langle\ ,\ \rangle$. Associated with these splittings are the projections $P_K, P_N, P_{K^\perp}, P_{N^\perp}$ onto $K, N, K^\perp, N^\perp$ along $N, K, N^\perp, K^\perp$ respectively. We may identify, by the nondegeneracy of $\langle \cdot, \cdot \rangle$, $K^\perp \sim N^*$, $N^\perp \sim K^*$. Along with the breakup of $L^* = K^\perp + N^\perp$, we have for functions H, F, G respectively on $L^*, K^\perp, N^\perp$ the respective gradients ∇H, $\nabla_{K^\perp} P$, $\nabla_{N^\perp} G$, contained in L, N, K respectively. We automatically have $L = (L^*)^* \ni \nabla H = \nabla_{K^\perp} H|_{K^\perp} + \nabla_{N^\perp} H|_{N^\perp}$ where in addition

$\nabla_{K^\perp} H|_{K^\perp} (\in (K^\perp)^* \sim N) = P_N \nabla H$, $\nabla_{N^\perp} H|_{N^\perp} = P_K \nabla H$. By the above identifications, the co-adjoint action of the connected group with Lie algebra N induces the Kostant-Kirillov orbit symplectic structure on $K^\perp \sim N^*$. We may thus speak of N invariant manifolds in $K^\perp$, which are just unions of orbits. In addition, we have the co-adjoint action of the connected group associated with L on L^* , and thus we have the notion of L invariant functions on L^* , which by calculus can be thought of as an infinitesimal notion. Having described the setup we now give the conclusion: Given an invariant manifold Γ of $K^\perp$, consider the algebra $\mathfrak{A}(\Gamma)$ of functions on L^* which are L invariant on Γ . These are precisely the functions H on L^* such that $[\nabla H(A), A] = 0$ for $A \in \Gamma$. These functions, thought of as functions on Γ by restriction to Γ , form an involutive system of Γ functions with respect to the orbit symplectic structure induced on $\Gamma \subset K$ by the N action. Moreover, for such functions the Hamiltonian vector field $X_H(A) = (ad)_B^*(A)$, $B = -P_K \nabla H(A) = -\nabla_{N_\perp} H|_{N^\perp}$ $A \in \Gamma$, the $*$ being with respect to $\langle \cdot,\cdot \rangle$, and hence we have the Lax-equations of motion

$$\dot{A} = X_H(A) = (ad)_B^*(A) \quad .$$

In the event that $L = L^*$, $(ad)^* = -(ad)$, which occurs if $\langle \, , \, \rangle$ is symmetric, as it is if defined by a trace operation like the Killing form, we have the usual form of the Lax-equations, $\dot{A} = X_H = [A,B]$, $B = -P_K \nabla H$, $A \in \Gamma \subset K^\perp$, $B \in K$.

In the form in which we've stated the theorem, it applies to the Arnold-Euler-Dubrovin equations [8,12], and the periodic symmetric Toda systems. This will be reported elsewhere, in a joint work with P. Moerbeke and T. Ratiu. In [4,5,6], the above theorem is applied to the nonperiodic Toda systems. In order to apply the above theorem to the Gel'fand-Dikii systems in the imprecise fashion indicated previously, we let (see [4]):

$$\Phi = \Phi^* = L \ , \ K = \mathcal{a}_{0,\infty} \ , \ N = \mathcal{a}_{-\infty,-1} \ ,$$

$\langle\ ,\ \rangle$ be as defined in (2.3), and so $K^\perp = K$, $N^\perp = N$, $(\mathrm{ad})^* = -(\mathrm{ad})$, while $\Gamma = \Phi_n$, and $H^\nu = \mathrm{tr} L^{\frac{\nu}{n}} \in \mathcal{a}(\Gamma)$, $\nu = 0,1,\ldots$. We thus conclude: (in a formulation appropriate to this paper)

<u>Theorem 4</u> (Gel'fand-Dikii) Let $A \in \Phi_n$, and let $H^\nu = \mathrm{tr}\, A^{\frac{\nu}{n}}$, then

$$\dot{A} = X_H(A) = [B_\nu, A],\ B_\nu = \frac{\nu}{n}[A^\mu]_+ = P_K \nabla H \ , \quad \mu = \frac{\nu}{n} - 1 \ ,$$

where the X_H refers to (2.7) .

<u>Corollary</u> The H^ν , $\nu = 1,2,..,n-1$ are constants on Φ_n , i.e. are orbit invariants on Φ_n , and so in particular Φ_n contains more than one orbit. For the KdV equation, where $A = -D^2 + a_0$, the orbit invariant is $\overline{a_0}$, which is intuitively $\int a_0\, dx$. It is well known that this 'first' integral of the motion for the KdV is not on the same footing as the 'higher' integrals.

<u>Remark 1</u> All the above easily generalizes to operators with matrix coefficients as was also considered by Gel'fand-Dikii [13] . Just define the differential ring to be the (noncommutative differential) ring of $m \times m$ matrices with coefficients in the former differential ring R , $ML(m,R)$, and define $\mathrm{tr}\, \Sigma a_i \xi^i \doteq \overline{(\text{matrix trace } a_{-1})}$. Then as before $\mathrm{tr}([A,B]) \doteq 0$,

and all the previous definitions, arguments and theorems apply to this case with almost no modification (see [4]) .

Remark 2 The case of formally self-adjoint operators is taken care of by merely representing Φ_n in the form $\Phi_n = \{\xi^n + \sum_{j=0}^{n-1} (b_j \xi^j + (\xi - iD)^j b_j) \mid b_j \in R$ (or $ML(m,R))\}$, which comes down to reparametrizing Φ_n . Again all the previous considerations apply to this case, but of course the formulas are different, as we are using different coordinates. This is the proper setting for the Boussinesq equation.

III. The Poisson Bracket

In this section we prove Theorem 2 that indeed formula (2.8) defines a Poisson bracket.

Proof of Theorem 2. We must show $\{\ ,\ \}$, (2.9), satisfies the Jacobi identity

(3.1) $\{G,\{H,F\}\} + \{F,\{G,H\}\} + \{H,\{F,G\}\} \equiv \lambda(G,H,F) \doteq 0$.

(Note Remarks 1,2 shall also apply to this proof.) Here remember $H \in I(a_0,..,a_{n-1})$, so we may think of H as a function on Φ_n . We must first compute the gradient $\nabla_{K^\perp}$, which for this section we let be denoted by ∇ , of

(3.2) $\{H,F\} = \langle A, [\nabla H, \nabla F] \rangle$,

which has two components. Note that in general (see [4]) we have the geometric definition $dG \equiv \langle \nabla G, dA \rangle$, $\nabla G \in \Phi_n^*$, which uniquely defines ∇G as $\nabla G = \sum_{j=0}^{n-1} (\xi - i))^{-j-1} \frac{DG}{Da_j}$. From (3.2) we have

$$\delta\{H,P\} \doteq \langle\delta A,[\nabla H,\nabla F]\rangle + \langle A,\delta[\nabla H,\nabla P]\rangle \quad ,$$

and so we have

$$(3.3) \qquad \nabla\{H,F\} = \pi[\nabla H,\nabla F] + \sigma(H,P) \quad .$$

In the above $\pi = \tilde{P}_{-(n-1),-1}$ (see (2.7)) as a consequence of $\delta A \in \mathcal{a}_{0,n-1}$, and (2.4). The second term $\sigma(H,F)$ comes from the contribution from the second term of (3.3)

When one substitutes this σ term and the two others into $\lambda(G,H,F)$, (3.1), it automatically will make no contribution. This is for the simple reason that any bracket of the form (see (1.4) for notation) $(C(\frac{DH}{Dq}),\frac{DP}{Dq})$, with C a formally skew-symmetric, constant coefficient, differential operator, defines a Poisson bracket from general principles, see (13,14). In the appropriate setting, this is just the fact that $d\omega = 0$, if ω is a constant coefficient form. The three contributions of the form $\sigma(H,F)$, and only these would occur in (3.1) if A had constant coefficients. Hence upon substituting only these three terms into (3.1), one would get exactly the same form term, $\hat{\lambda}(G,H,F)$, as one gets in computing $\lambda(G,H,F)$ for a bracket of the above mentioned form, $(C(\frac{DH}{Dq}),\frac{DP}{Dq})$. But since the previous bracket is a Poisson bracket, such a $\hat{\lambda}(G,H,F)$ is identically zero. Hence we shall forget about the σ term in (3.3) and only work with the other term, which we indicate by writing

$$(3.4) \qquad \nabla\{H,F\} \cong \pi[\nabla H,\nabla P] \quad .$$

We compute from (3.2,4),

$$\{G,\{H,F\}\} \cong \langle A,[\nabla G,(\pi[\nabla H,\nabla F])]\rangle$$

(by Theorem 1) $$\cong \langle [A,\nabla G],\ \pi[\nabla H,\nabla P]\rangle$$

$$\cong \langle [A,\nabla G],\ [\nabla H,\nabla P]\rangle = \langle A,[\nabla G,[\nabla H,\nabla P]]\rangle \ .$$

We must justify the second from the last step. Note $[A,\nabla G] \in \mathcal{O}_{-\infty,n-1}$, and $[\nabla H,\nabla F] \in \mathcal{O}_{-\infty,-2}$, hence by formula (2.4), and the definitions of π ;

$$\langle [A,\nabla G],\ [\nabla H,\nabla F] - \pi[\nabla H,\nabla F]\rangle = 0 \quad ,$$

thus justifying the above step. From (3.1), and the above calculation we have

$$\lambda(G,H,P) = \langle A,\tau(\nabla G,\nabla H,\nabla P)\rangle \quad ,$$

where $\tau(L,M,N)$ is the cyclic sum $[L[M,N]] + \dots$, which vanishes identically since $[\cdot,\cdot]$ is a Lie bracket, and so $\lambda(G,H,F) = 0$, thus concluding the proof of Theorem 2.

<u>Remark 3</u> One conjectures that the Benni equations considered in [15] can be put in the framework of this paper.

References

1. I.M. Gel'fand and L.A. Dikii, Fractional Powers of Operators and Hamiltonian Systems, Funkcional'nyi i ego Prilozenija, Vol. 10, No. 4, 1976.

2. P.D. Lax, Integrals of Nonlinear Equations of Evolution and Solitary Waves, Comm. Pure Appl. Math., Vol. 21, 1968, 467-490.

3. J. Marsden, Applications of Global Analysis in Mathematical Physics, Chapter 6, Publish or Perish, Inc., 1974.

4. M. Adler, On a Trace Functional for Formal Pseudo-Differential Operators and the Symplectic Structure of the Korteweg-deVries Type Equations, to appear in Inventiones, (also obtainable in slightly different form as an MRC Tech. Report, University of Wisconsin at Madison, 1978).

5. B. Kostant, to appear.

6. B. Symes, to appear.

7. O.I. Boboyovlensky, On Perturbations of the Periodic Toda Lattices, Comm. Math. Physics, 51, 1976, 201-209.

8. L. Dikii, Hamiltonian Systems Connected with the Rotation Group, Funkcional'nyi Analiz i ego Prilozenija, Vol. 6, No. 4, 83-84.

9. J. Moser, Three Integrable Hamiltonian Systems Connected with Isospectral Deformations, Adv. in Math., Vol. 16, No. 2, May 1975.

10. D. Kazhdan , B. Kostant, and S. Sternberg, Hamiltonian Group Actions and Dynamical Systems of Calogero Type, C.P.A.M., Vol. 23, 1-27, (1978).

11. M. Adler, Completely Integrable Systems and Symplectic Actions, MRC Tech. Report #1830, University of Wisconsin-Madison, 1978, an augmented version to appear in J.M.P., 1979.

12. S.V. Manakov, Note on the Integration of Euler's Equations of the Dynamics of an n-Dimensional Rigid Body, Funkcional'nyi Analiz i ego Prilozenija, Vol. 10, No. 4, 93-94, October-December (1976).

13. I.M. Gel'fand and L.A. Dikii, The Resolvant and Hamiltonian Systems, Funkcional'nyi Analiz i ego Prilozenija, Vol. 11, No. 2, 11-27, April-June, 1977.

14. P. Lax, Almost Periodic Solutions of the KdV Equation, SIAMS Reviews, 1976.

15. I.M. Gel'fand, Yu I. Manin, and M.A. Shubin, Poisson Brackets and the Kernel of a Variational Derivative in the Formal Variational Calculus, Funkcional'nyi Analiz i ego Prilozenija, Vol. 10, No. 4, (1976).

16. B.A. Kupershmidt and Yu I. Manin, Equations of Long Waves with a Free Surface. II. Hamiltonian Structure and Higher Equations, Funkcional'nyi Analiz i ego Prilozenija, Vol. 12, No. 1, 25-37, January-March (1978).

A Remark on a Generalized Uncertainty Principle

Anne Marie Berthier

A) I would like to talk about work done with W.O. Amrein. It concerns certain properties of Schrödinger equations. Although this work describes particular operators, it is possible, I think, to obtain from our research more general results of interest. First I will discuss generalities and afterwords I will indicate a solution for the case of Schrödinger operators.

The first question is to know what happens to the "uncertainty principle." The answer to this question is given at different levels of precision:

1) In Quantum Mechanics one calls quantities simultaneously measurable if the corresponding self-adjoint operators commute . The classical situation concerns the measurement of the momentum P and the position Q. In the one-dimensional case, the operators P and Q can be represented in $\mathcal{H} = L^2(\mathbb{R})$ by

$$(Qf)(x) = xf(x)$$

$$(Pf)(x) = -i\hbar \frac{\partial f}{\partial x} \quad .$$

Let Δx and Δp be the "dispersion" of $f(x)$ and $\tilde{f}(p)$, then we have

$$\tilde{f}(p) = \frac{1}{\sqrt{2\pi}} \int dx\, e^{-ipx}\, dx \quad ,$$

and $\Delta x \cdot \Delta p \simeq \hbar$ where $\hbar$ is Planck's constant. Note that the greater the precision x-coordinate the greater the uncertainty Δp of the momentum and conversely [1].

2) Now a more rigourous argument: The principle is essentially identified as the commutation rule [2]. Let A and B be self-adjoint operators and $[A,B] = iC$ and let ρ be a density measure (called in physics: state). A state of the system is represented by a positive trace class operator ρ in the Hilbert space $\mathcal{H}$, called the density operator (or density matrix when a specific basis is referred to) which also satisfies $\mathrm{Tr}\,\rho = 1$. Let a and b be the mean values of A and B in the state ρ.

$$a = \mathrm{Tr}\,\rho A \qquad b = \mathrm{Tr}\,\rho B$$

The dispersion of A and B in the state ρ is defined by

$$(\Delta a)^2 = \mathrm{Tr}\,\rho\,(A-a)^2 \qquad (\Delta b)^2 = \mathrm{Tr}\,\rho\,(B-b)^2$$

where $(\Delta a)^2$ and $(\Delta b)^2$ are the mean quadratic states. It can be shown that

$$(\Delta a)(\Delta b) \geq \tfrac{1}{2}|\mathrm{Tr}\,\rho\,C| \quad . \qquad (1)$$

If $C = \alpha I$ we obtain

$$(\Delta a)(\Delta b) \geq \tfrac{1}{2}\alpha \;;$$

in particular we have $(\Delta p)(\Delta q) \geq \frac{1}{2}\hbar$. Δp and Δq now have a precise meaning. For the physicist the problem of uncertainty is finished with the equation (1). The mathematician may not be satisfied for the following reason: the operators used by the physicist are not in general bounded and it is difficult (perhaps not possible) to define the commutation between two unbounded operators on a sufficiently large domain of vectors. For example the equation (1) has only meaning for a particular case, namely only if the state ρ belongs to $D(A^2) \cap D(B^2) \cap D(C)$. Therefore it is necessary to find a formulation of the uncertainty principle using bounded operators only.

The problem of unbounded operators can be avoided by either considering their unitary groups

$$U_\alpha = e^{-i\alpha A} \qquad V_\beta = e^{-i\beta B}$$

or by considering the spectral families $\{E_\lambda\}$ and $\{F_\mu\}$ associated with A and B respectively.

In this talk, I shall consider only the second possibility [2]. By studying the commutation relation between the spectral families $\{E_\lambda\}$ and $\{F_\mu\}$ we obtain more detailed results by considering only $[A,B]$. It can be shown, for instance, that two different operators $A_1 \neq A_2$ satisfying the same commutation relation

$[A_1,B] = [A_2,B]$ can sometimes be distinguished in terms of their spectral families.

Let E and F be two spectral projections of A and B. Let us consider the relation between E and F. Given E and F we can find a decomposition of the Hilbert space $\mathcal{H}$ into mutually orthogonal subspaces

$$\mathcal{H} = \mathcal{H}_{00} \oplus \mathcal{H}_{01} \oplus \mathcal{H}_{10} \oplus \mathcal{H}_{11} \oplus \mathcal{H}_g \equiv \mathcal{H}_c \oplus \mathcal{H}_g \tag{1'}$$

where

$$\mathcal{H}_{mn} = \{f \in \mathcal{H} \mid Ef = mf,\ Ff = nf;\ m,n = 0,1\} \ .$$

$\mathcal{H}_{11}$ is the range of $E \cap F$; $\mathcal{H}_{00}$ is the range of $E^{\perp} \cap F^{\perp}$ where $E^{\perp} = I - E$. $\mathcal{H}_c$ is the orthogonal sum of these four subspaces and $\mathcal{H}_g$ is the orthogonal complement of $\mathcal{H}_c$.

$$\left.\begin{aligned} E &= 0 \oplus 0 \oplus I \oplus I \oplus E_g \equiv E_c \oplus E_g \\ E^{\perp} &= I \oplus I \oplus 0 \oplus 0 \oplus E_g^{\perp} \equiv E_c^{\perp} \oplus E_g^{\perp} \\ F &= 0 \oplus I \oplus 0 \oplus I \oplus F_g \equiv F_c \oplus F_g \\ F^{\perp} &= I \oplus 0 \oplus I \oplus 0 \oplus F_g^{\perp} \equiv F_c^{\perp} \oplus F_g^{\perp} \end{aligned}\right\} \tag{2}$$

We have

$$\left.\begin{aligned} E_c F_c &= F_c E_c = E \cap F \\ E_c^{\perp} F_c^{\perp} &= F_c^{\perp} E_c^{\perp} = E^{\perp} \cap F^{\perp} \end{aligned}\right\} \qquad (3)$$

E_g and F_g are assumed to be in "generic position" in $\mathcal{H}_g$ [4] i.e.

$$E_g \cap F_g = E_g \cap F_g^{\perp} = E_g^{\perp} \cap F_g^{\perp} = E_g^{\perp} \cap F_g = 0 \quad .$$

We can proceed in two ways:

a) study the five subspaces

b) find the numerical values from the decomposition of $\mathcal{H}$.

a) If E and F commute, then $\mathcal{H}_g = \{\theta\}$ and none of the four subspaces are nonzero except in special cases. Hence in the general case $E \cap F \neq 0$, i.e. there exist states which have at the same time the property E and the property F. If E and F do not commute, the dimensional study of these four subspaces is more difficult, as we shall see later.

b) One possible numerical measure was given by Lenard [5].

.Define the numerical range of E and F as

$$N_{EF} = \{(x,y) \mid x = (f,Ef),\ y = (f,Ff),\ \|y\| = 1\} \quad .$$

.If E and F commute, N_{EF} is the square $\{(x,y) : 0 \le x;\ y \le 1\}$ except in the case where $E = 0, I$ and $F = 0, I$.

.If E and F are in generic position. Then N_{EF} is a convex subset of the unit square and its symmetric with respect to reflection about diagonals $x = y$ and $x+y = 1$. Convexity follows from the general Theorem of Toeplitz and Hausdorff [6] upon identifying N_{EF} with $N_{(E+iF)}$ in the complex plane $x+iy$. Let $\mathcal{S} = N_{EF}$ in this case, choose the center of the unit square $(\frac{1}{2},\frac{1}{2})$. We find an ellipse $\epsilon(\theta) = \{x,y\} : x = \frac{1}{2} + \frac{1}{2} (\cos t + \theta)$, $y = \frac{1}{2} + \frac{1}{2} \cos (t-\theta) \quad t \in \mathbb{R}\}$, $\hat{\epsilon}(\theta)$ is the convex hull i.e. the solid ellipse.

.In the general case, N_{EF} is the convex envelope of the five following sets which correspond to a non-zero summand in (1'):

$$\{(0,0)\},\ \{(0,1)\},\ \{(1,0)\},\ \{(1,1\},\ N_{E_g F_g} \quad ,$$

which corresponds to a nonzero subspace in (2).

Let us give an example from a paper by A. M. Berthier and J. M. Jauch in [7].

<u>Example</u>: Let Δ,Σ be finite intervals and let $A = Q$ and $B = P$. We have

$$E = E_{\Delta}(Q) \qquad F = F_{\Sigma}(P) \quad \text{in} \quad L^2(\mathbb{R}) \quad .$$

If $f = E_{\Delta}(Q)f$, $f(x)$ is localized in Δ. Hence $\tilde{f}(p)$ is an entire function in the p-complex plane and cannot vanish on a set of positive measure. Thus

$$\mathcal{H}_{11} = 0 \qquad \mathcal{H}_{10} = 0 \qquad \mathcal{H}_{01} = 0 \ .$$

Lenard in [5] proved that in this case $\dim \mathcal{H}_{00} = \infty$. Thus N_{EF} is a degenerate ellipse. Therefore N_{EF} is an ellipse which degenerates into a straight line.

The numerical range is interesting but not very useful for experimental physics because it involves contribution from all states. It would be interesting to look for other quantities.

B) Now let us consider the determination of the subspaces $\mathcal{H}_{mn}$. In all cases that we have considered it was easier to show that a dimension is infinite than to show that it is zero. In our first paper [8] we took two operators P and Q in $L^2(R^n)$ but for Δ and Σ arbitrary sets of finite Lebesgue measure. In this case it is not possible to use analytic functions.

Still it is possible to show that $\dim \mathcal{H}_{00} = \infty$ and $\dim \mathcal{H}_{11} = 0$. This last result means that a function of $L^2(R^n)$ (or more generally of $L^p(R^n)$) has the following properties:

$$f(x) \neq 0 \quad \text{only for} \quad x \in \Delta, \ |\Delta| < \infty$$

$$\tilde{f}(p) \neq 0 \quad \text{only for} \quad p \in \Sigma, \ |\Sigma| < \infty \ ;$$

is null almost everywhere.

Another method is closer to physics [3]. We take $A = Q$ the multiplication operator in $L^2(\mathbb{R}^n)$ and $B = H$ a hamiltonian; Δ is compact in $\mathbb{R}^n$ and E_Δ is a multiplication operator by the characteristic function χ_Δ of Δ. Hence the state verifying $E_\Delta f = f$ have the property to be localized in Δ.

Let Σ be a finite interval in $\mathbb{R}$, F_Σ be the spectral projector of H associated with Σ. The states verifying $F_\Sigma f = f$ have the property that their energy is localized in the interval Σ. From the uncertainty relations we expect that $E_\Delta \cap F_\Sigma = 0$, i.e. it is not possible to localize a state at the same time, in energy and in space.

If H describes a free particle (relativistic or not relativistic) i.e. if $H = (P^2 + m^2)^{\frac{1}{2}}$ where $H = P^2$, it is easy to see that $E_\Delta \cap F_\Sigma = 0$. The reason being that a function of compact support cannot have a Fourier transform of compact support. By using the analyticity as above we find that $f = 0$ if $E_\Delta f = F_\Sigma f = f$. If we consider a non free particle i.e.

$$H = (P^2 + m^2)^{\frac{1}{2}} + V ,$$

the problem of showing that $E_\Delta \cap f_\Sigma = 0$ is more difficult. It is not possible to use perturbation techniques which are useful to show the self-adjointness of H, describing the spectrum or the scattering. In fact we can show that $E_\Delta f = f$ and $F_\Sigma f = f$ implies $f = 0$. A knowledge of the result for the unperturbed operator H_0 gives nothing because the relation between F_Σ^0 and F_Σ is very difficult. We find a proof in the non relativistic case by using the fact that H is a differential operator. I believe that the result is also true for the relativistic case but with our method it is not possible to obtain the result.

In the non-relativistic case, we consider the following operator

$$\hat{H} = \sum_{j,k=1}^{n} a_{jk}\left(-i \frac{\partial}{\partial x_j} + b_j(x)\right)\left(-i \frac{\partial}{\partial x_k} + b_k(x)\right) + V(x) \quad ,$$

with domain $D(\hat{H}) = C_0^\infty(\mathbb{R}^n \backslash N)$ under the following conditions

- .a_{jk} is real symmetric
- .b_j are real and $C^1_{Loc}(\mathbb{R}\backslash N)$
- .N is closed set of measure zero
- .$\hat{H}$ is symmetric
- .H is self-adjoint
- .V is real and belongs to $L^2_{Loc}(\mathbb{R}\backslash N)$
- .$V_t = \exp(-iHt)$, the strongly unitary parameter group associated with H
- .$\{F_\lambda\}$ is the spectral family of H.

<u>Lemma 1</u>: <u>Let</u> Δ <u>be closed in</u> R^n. <u>Suppose that</u> $f \in D(H)$ <u>and</u> $E_\Delta f = f$ <u>then</u> $E_\Delta Hf = Hf$.

<u>Proof</u>: Let $g \in C_0^\infty(R^n \backslash (\Delta \cup N))$. We have

$$(Hf,g) = (f, E_\Delta \hat{H} g) \quad \text{and} \quad (Hf,g) = (\hat{H}^* E_\Delta f, g) \quad .$$

Now $(\hat{H}g)(x) = 0$ if $x \in \Delta$. This implies that $E_\Delta \hat{H} g = 0$. Hence $(Hf,g) = 0$ and $Hf \perp L^2(R^n \backslash (\Delta \cup N)) = L^2(R^n)$. Therefore $Hf \in L^2(\Delta)$ and $E_\Delta Hf = Hf$.

<u>Lemma 2</u>: <u>Let</u> Δ <u>be closed in</u> R^n, <u>and</u> $\Sigma \subset R$ <u>bounded</u>. <u>Then</u> V_t <u>commutes with the projection</u> $E_\Delta \cap F_\Sigma$, <u>i.e.</u>

$$[V^t, E_\Delta \cap F_\Sigma] = 0 \; .$$

<u>Proof</u>: Let $S = E_\Delta \cap F_\Sigma$. Let $Sf = f$. Since $F_\Sigma f = f$; we have for $f \in D(H^k)$ for all $k = 0,1,2$

$$V_t f = \sum_{k=0}^{\infty} \frac{1}{k!} (-iHt)^k f \quad .$$

By the lemma 1, since $E_\Delta f = f$ we have $E_\Delta H^k f = H^k f$ by recursion. This implies

$$E_\Delta V_t f = V_t f = V_t E_\Delta f \quad .$$

Since $F_\Sigma V_t = V_t F_\Sigma$, $V_t f \in (E_\Delta \cap F_\Sigma)\mathcal{H}$ and V_t leaves invariant the subspace $(E_\Delta \cap F_\Sigma)\mathcal{H}$.

Proposition 1: Let Δ be closed in $\mathbb{R}^n$, $\Sigma \subset \mathbb{R}$ bounded.

a) if $E_\Delta \cap F_\Sigma$ is compact, then $E_\Delta \cap F_\Sigma \cap F_c = 0$

b) if $\lim\limits_{T\to\infty} \frac{1}{T} \int_0^T \|E_\Delta V_t g\|^2 dt \to 0 \quad \forall f \in \mathcal{H}_c(H)$ then

$E_\Delta \cap F_\Sigma \cap F_c = 0$.

c) Assume in addition $E_\Delta f = f$, $Hf = \mu f$ which implies $f = 0$. Then $E_\Delta \cap F_\Sigma = 0$.

Proof:

i) Let $S = E_\Delta \cap F_\Sigma$. The condition a) and lemma 2 give that $[V_t, S] = 0$ and $H|_{S\mathcal{H}}$ is self-adjoint in finite dimensional space in i.e. $H|_{S\mathcal{H}}$ has a discrete spectrum. Denote by $\mathcal{H}_p(H)$ be the closed subspace spanned by all the eigenvectors of H. Then we have

$$S\mathcal{H} \subseteq \mathcal{H}_p(H) \quad .$$

Since S reduces H, there exist a basis of $S\mathcal{H}$ composed by eigenvectors of H. They verify that $E_\Delta f = f$. Under the hypothesis c), they are all equal to zero and $E_\Delta \cap F_\Sigma = 0$.

ii) $|(f,g)|^2 = \frac{1}{T}\int_0^T dt|(V_tf,V_tg)|^2$

$= \frac{1}{T}\int_0^T dt\ |(E_\Delta SV_tf,V_tg)|^2 = \frac{1}{T}\int_0^T dt\ |(SV_tf,E_\Delta V_tg)|^2$

$\le \|f\|\ \frac{1}{T}\int_0^T dt\ \|E_\Delta V_tg\|^2 \to 0$ as $T \to \infty$ for all $g \in \mathcal{H}_c(H)$.

Therefore $f \perp \mathcal{H}_c(H)$ this follows from b). The proof of c) is similar to the above.

Proposition 2: Let Δ be compact in $R^n\backslash N$ and $\Sigma \subset R^n$ be bounded. Let $H_0 = -\Delta$. Assume that there exists a number M_Δ such that

$$\|VE_\Delta(H_0+M_\Delta)^{-1}\| < a_0$$

where a_0 is the smallest eigenvalue of $\{a_{jk}\}$. Then $E_\Delta \cap F_\Sigma$ is compact.

Proof:

α) Let $K_0 = -\sum_{j,k=1} a_{jk}(\frac{\partial}{\partial x_j})(\frac{\partial}{\partial x_k})$ be a self adjoint operator. We have

$$E_\Delta \cap F_\Sigma = E_\Delta(K_0+1)^{-1}(K_0+1)E_\Delta \cap F_\Sigma \quad .$$

The first part $E_\Delta(K_0+1)^{-1}$ is compact. It suffices to show that the last part is bounded or defined everywhere by the closed graph theorem.

For simplification, suppose that $b_j \equiv 0$.

β) Clearly $K_0 \geq a H_0$. This implies

$$\|VE_\Delta(K_0+a_0M_\Delta)^{-1}\| \leq \|VE_\Delta(H_0+M_\Delta)^{-1}\| \cdot$$

$$\cdot \|(H_0+M_\Delta)(K_0+a_0M_\Delta)^{-1}\| < a_0 \frac{1}{a_0} < 1 \quad .$$

Then VE_Δ is relatively bounded with respect to K_0 and $K_0 + VE_\Delta = H_\Delta$ is self-adjoint, $D(H_\Delta) = D(K_0) = D(H_0)$.

γ) Let $E_\Delta \cap F_\Sigma f = f$; choose a function $\psi \in C_0^\infty(\mathbb{R}^n \backslash N)$ such that $\psi(x) = 1$ for all $x \in \Delta$ [9, Thm. 1,2.2] and $g \in \mathcal{S}(\mathbb{R}^n)$. We have by lemma 1

$$(Hf,g) = (E_\Delta Hf,g) = (E_\Delta Hf,\psi g) = (Hf,\psi g) = (f,E_\Delta \hat{H}\psi g)$$

$$= (f,E_\Delta H_\Delta \psi g) = (f,E_\Delta H_\Delta g) = (f,H_\Delta g) \quad .$$

H_Δ is essentially slef-adjoint on $\mathcal{S}(\mathbb{R}^n)$. Therefore $f \in D(H_\Delta) = D(K_0)$ and $K_0 E_\Delta \cap F_\Sigma$ is defined everywhere.

C) Study of $E_\Delta^\perp \cap F_\Sigma^\perp$.

<u>Lemma 3</u>: (<u>Halmos</u>) [10] <u>$\mathcal{H}_g$ may be represented by $\mathcal{H}_g = \mathcal{K} \oplus \mathcal{K}$ such that</u>

$$E_g = \begin{pmatrix} I & 0 \\ 0 & 0 \end{pmatrix} \qquad F_g = \begin{pmatrix} C^2 & CS \\ CS & S^2 \end{pmatrix}$$

where C *and* S *are positive contractions on* $\mathcal{H}$ *with* $C^2 + S^2 = I$.

Lemma 4: *There exists a unitary operator* U *in* $\mathcal{H}$ *such that*

$$E^{\perp}F^{\perp} = UEFU^{*} + E^{\perp} \cap F^{\perp} - E \cap F \quad .$$

Proof: Since $\mathcal{H} = \mathcal{H}_c \oplus \mathcal{H}_g$, $U = I \oplus U_g$ where U_g is a unitary operator defined by

$$U_g = \begin{pmatrix} 0 & -I \\ I & 0 \end{pmatrix} .$$

Lemma 5: *Let* E *and* F *be such that* EF *is a compact operator. Then* $E^{\perp}F^{\perp}$ *is compact if and only if* $\dim(E^{\perp} \cap F^{\perp})\mathcal{H} < \infty$.

Proof: Denote by $\mathcal{B}_0$ the set of all compact operators. $EF \in \mathcal{B}_0$ implies that $E \cap F \in \mathcal{B}_0$. Therefore $E^{\perp}F^{\perp} \in \mathcal{B}_0$ if and only if $E^{\perp} \cap F^{\perp} \in \mathcal{B}_0$.

Application: Let $b_k \in C^1(\mathbb{R}^n)$ and $V \in L^2_{Loc}(\mathbb{R}^n)$. In this case $E_{\Delta}F_{\Sigma}$ is compact where Δ is compact. Let ψ be a function belongs to $C_0^{\infty}(\mathbb{R}^n \setminus N)$ such that $\psi(x) = 1$ for all $x \in \Delta$ [9, Thm 1.2.2] and $\psi\Delta = E_{\Delta}$. We have

$$E_{\Delta}F_{\Sigma} = E_{\Delta}(H_0+1)^{-1}(H_0+1)\psi F(\Sigma) \quad .$$

Then $E_{\Delta}(H_0+1)^{-1} \in \mathcal{B}_0$ and $(H_0+1)\psi F_{\Sigma} \in \mathcal{B}(\mathcal{H})$ where $\mathcal{B}(\mathcal{H})$ denotes the set of all bounded operators and everywhere defined linear operators in $\mathcal{H}$.

$E_\Delta^\perp F_\Sigma^\perp \in \mathcal{B}_0$

If $E_\Delta^\perp F_\Sigma^\perp \in \mathcal{B}_0$, let us consider that Δ' is a compact such that $|\Delta'| > |\Delta|$ and $\Delta \subset \Delta'$. Then we have $E_{\Delta'} E_\Delta^\perp F_\Sigma^\perp \in \mathcal{B}_0$.

$$E'_\Delta E_\Delta^\perp F_\Sigma^\perp = E'_\Delta(I-E_\Delta)(I-F_\Sigma) = E_{\Delta'} - E_\Delta - E_{\Delta'}F_\Sigma + E_\Delta F_\Sigma \quad .$$

We see that $E'_\Delta - E_\Delta \in \mathcal{B}_0$. This is a contradiction since $L^2(\Delta\backslash\Delta')$ is infinite dimensional space. Therefore

$$\dim(E_\Delta^\perp \cap F_\Sigma^\perp) = \infty \quad .$$

D) Examples

1. We use only local conditions on the scalar potential $V(x)$ and on the vector potential $b_j(x)$. The results are true for every self-adjoint extension of $\hat{H}$.

2. Let us now look at some particular classes of Hamiltonians. Let us consider a N-body system.

$$H = \sum_{i=1}^{N} \frac{1}{2m_i} \Delta x_i^2 + \sum_{i<k} V_{ik}(\underline{x}_i - \underline{x}_k) \quad \text{in} \quad L^2(\mathbb{R}^{3N}) \quad .$$

If V_{ik} belongs to $L^2_{Loc}(\mathbb{R}^3)$ and Δ is bounded set in $\mathbb{R}^{3(N-1)}$ where $\mathbb{R}^{3(N-1)}$ is relative coordinates spaces, we obtain

$$\| V_{ik} E_\Delta (H_{0,rel} + M)^{-1} \| \to 0 \quad \text{as} \quad M \to \infty \ .$$

Therefore $E_\Delta \cap F_\Sigma$ is compact where Σ is a bounded set in $\mathbb{R}$.

3. We next consider potentials that are strongly singular at $x = 0$. The hamiltonian is then defined as lemma 1. The impossibility of simultaneous localization in energy and in configuration space follows provided that the following conditions are verified:

a) every vector in $\mathcal{H}_c(H)$ is evanescent as $t \to +\infty$ or as $t \to -\infty$

b) one of the wave operators $\Omega_\pm$ exists and it is asymptotically complete.

The evanescence of all states in $\mathcal{H}_c(H)$ for a class of spherically symmetric singular potentials in $L^2(\mathbb{R}^3)$ follows from [11]. This class includes both attractive and repulsive potentials e.g. all potentials of the form $V(x) = \alpha|x|^{-\beta}$ with $\alpha \in \mathbb{R}$ and $\beta > 1$.

4. Some examples of potentials that are singular at infinity and covered by our theorem are the Stark Potential, the harmonic oscillator and $V(x) = \gamma|x|^m$ for any $\gamma \in \mathbb{R}$ and $m \geq 0$. In all these examples there are no wave functions that can be simultaneously localized in energy and in position. For the harmonic oscillator the spectrum is discrete and the eigenfunctions are explicitely known, and one can directly see that no finite linear combination of eigenfunctions can have compact support in configuration space.

5. Let us now consider non singular potentials, i.e. the case where $N = 0$. If $V \in L^2_{Loc}(\mathbb{R}^3)$, we have $E_\Delta \cap F_\Sigma = 0$ for any bounded set Δ and Σ, i.e. no state vectors can at the same time be localized in a bounded region of space and have bounded support in the spectral representation of the hamiltonian.

References:

[1] L. D. Landau and E. M. Lifschitz. Quantum Mechanics non relativistic Theory, vol 3 of course of Theoretical Physics, Pergamon Press, p. 47-48.

[2] J. M. Jauch. "The quantum probability theory," Synthese, 29, (3), (1974).

[3] W. O. Amrein and A. M. Berthier. "Impossibility of simultaneous localization of wave packets in energy and in configuration space," Reports in Math. Physics. vol. 11, 1, (1977), 123-132.

[4] J. Dixmier. Rev. Sci. 86, 387 (1969).

[5] A. Lenard. "The numerical range of a pair of projections," J. Functional Analysis, 10, (1972), 410-423.

[6] M. H. Stone. Linear transformations in Hilbert space, Am. Math. Society Colloquium Publications, New-York, 1932, p. 130-132.

[7] A. M. Berthier and J. M. Jauch. "A theorem on the support of functions in $L^2(\mathbb{R})$ and of Fourier transforms," Letters in Math. Physics 1, (1976) 93-97.

[8] W. O. Amrein and A. M. Berthier. "On support properties of L^p-functions and their Fourier transforms," J. of Functional Analysis, vol. 24, (1977).

[9] L. Hörmander. Linear partial differential operators, Springer.

[10] P. R. Halmos. "Two subspaces," Trans. American Math. Soc., 144, 381.

[11] W. O. Amrein and V. Georgescu. Helv. Acta, 47 (1974), 249-264.

Rice University
Mathematics Department
Houston, Texas 77001

Permanent Address

University Paris VI
UER 47, Mathematics
4 Place Jussieu
75230 Paris Cedex 05
FRANCE

DISSIPATIVE DYNAMICAL SYSTEMS OF MACROSCOPIC PHYSICS

by

Miroslav Grmela

Centre de Recherches Mathématiques
Université de Montréal
Montréal, Québec, Canada

1. Introduction

Vis à vis macroscopic systems (for example air, water) physics has developed the following strategy for their investigation: (an excellent historical account is available in [1], [2]).

1. A class of macroscopic systems S, (the elements of S are the systems to be investigated) and another class of macroscopic systems J, (the elements of J are measurement instruments) are chosen. The results of the interaction of the systems in S with the measurement instruments in J is recorded and collected. Experience shows that there exist pairs (J,S) and conceptual structures, theories, so that the results of the measurements are in good agreement with consequences of the corresponding theories. A well established dynamical theory of macroscopic systems can be formally represented by $DT = (J,S,DS_Q,Q,p)$ where S and J were introduced above, DS_Q is a family of dynamical systems parametrized by $q \in Q$, p is a map from Q to S', the elements of S' are subsets of S. The parameters q are called phenomenological (or fundamental) quantities introduced by the theory. The individuality of the systems inside S is expressed through q. The macroscopic systems corresponding to the same q are indistinguishable in the theory DT. Examples of the well established dynamical theories of macroscopic systems are: classical mechanics (CM), the Boltzmann kinetic theory (KT), the Navier-Stokes-Fourier fluid mechanics (FM) and thermodynamics (TH). The intersection $S_{CM} \cap S_{KT} \cap S_{FM} \cap S_{TH}$ is certainly not empty. The fundamental differences in the above theories originate in differences in J_{CM}, J_{KT}, J_{FM} and J_{TH}. For example, a macroscopic system in CM is regarded as a collection of small subsystems, particles. The time evolution of the particles is governed by the Hamiltonian dynamics, J_{CM} is composed of the instruments measuring positions and velocities of the particles. In TH, only specially prepared systems are observed. The process of preparation consists for example in leaving the system isolated for sufficiently long time. J_{TH} is composed of the instruments measuring M+1 thermodynamic parameters like, for example, the temperature and mass densities of the M components. A state of a prepared system is called a thermodynamic equilibrium state.

2. An attempt is made to find certain relationship among the well established dynamical theories. The creation of a theory (step 1) and the search for its relations to other theories (step 2) are often related one to the other [see [1],[2]]. The relationship, if found, provides a foundation of one theory in terms of the others. For example, it is believed that it should be possible to obtain KT, FM, TH from a theoretical investigation of the orbit space corresponding to CM. Similarly, it is believed that FM and TH can be obtained from KT and TH from FM. Intuitively, the problem of deriving one theory DT_2 from another theory DT_1 consists in recognizing a "quasi" invariant submanifold (invariant up to an equivalence — coarse graining — relation) in the phase space associated with $DS_{1\tilde{Q}_1}$, where $\tilde{Q}_1 \subset Q$. In other words, the problem consists in recognizing a pattern in the orbit space corresponding to DT_1 for $\tilde{Q}_1 \subset Q$. The problem of the relation between DT_1 and DT_2, or as we also say the problem of a compatibility of DT_1 and DT_2, can therefore be regarded as the problem of global analysis. The relation CM → TH provided a motivation for ergodic theory (the problem CM → TH remains however largely open). Analysis and a general formulation of the Gibbs postulates, also referring to CM → TH, is known as thermodynamic formalism [3].

The author's objective is to focus the attention on KT, FM, TH and on the compatibility relations among them. An attempt is made to suggest an appropriate setting for this study. Some applications are introduced in section 4. Section 5 contains a few remarks about similarities and differences between CM and KT, FM, TH.

2. Structure of DDS

The compatibility among the theories KT, FM, TH is first searched for in the unity of a structure in their mathematical formulations. A detailed study of the Boltzmann kinetic theory and of the Navier-Stokes-Fourier fluid mechanics reveals indeed a common mathematical structure that is shared by these theories.

Let Q denote phenomenological quantities through which the individuality of the macroscopic systems (say classical fluids) is expressed. To each $q \in Q$ a pair (H,R) is attached; H denotes the set of all admissible states (the elements of H denoted f, characterize completely the admissible states of fluids), the time evolution of f is governed by $\frac{\partial f}{\partial t} = R(f)$. Complete and admissible is meant with respect to the set of the observations and the measurements that form the empirical basis for the dynamical theory considered. The family of pairs (H,R) parametrized by $q \in Q$ is called the family of dissipative dynamical systems of macroscopic physics (abbreviated DDS) provided (DDS1)-(DDS7) introduced below are satisfied.

(DDS1) A *state* $f \in H$ is a mapping $D\times R \to I$, $(x,t) \to f(x,t)$, where I is a finite dimensional manifold, D is a cartesian product of the set $\{1,2,..,M\}$ with a finite dimensional manifold $D' = \Omega\times D''$, where Ω is the subset of $\mathbb{R}^3$ in which the macroscopic system is confined, M denotes the number of components of the macroscopic system. The integral $\int_D dx$ is understood to be the sum over the M integers and a well defined integral over D'. By t we denote the time.

(DDS2) *Thermodynamic equilibrium states.* An involution $J:H \to H$ is defined in H, ($J\cdot J$ is the identity operator in H). Thermodynamic equilibrium states E are defined by $E = \{f \in H | R(f)=0, Jf=f\}$ or equivalently $E = \{f \in H | R^+(f)=0, R^-(f)=0, Jf=f\}$, where $R^{\pm}(f) = \frac{1}{2}(R(f) \pm JR(Jf))$. A class of boundary conditions for which E is not empty is denoted B.

(DDS3) *Conservation laws.* There exist $c_1,\dots,c_{M_0}$, $c_i:H \to G$ (the elements of G are mappings $D\times\mathbb{R}\to\mathbb{R}$), such that $c_i(Jf) = c_i(f)$ and $\frac{\partial c_i}{\partial t} = C_i(f)$, where $\int_D dx\, C_i(f)$ depends only on $f|_{\partial D}$; ∂D denotes the boundary of D, $i=1,\dots,M_0$; $M_0 = M+1-N_c$, N_c is the number of different chemical reactions in the macroscopic systems considered.

(DDS4) *Dissipation law.* There exists $s:H \to G$ such that $s(Jf) = s(f)$ and $\frac{\partial s}{\partial t} = C_s(f)+\sigma$, where $\int_D dx\, C_s(f)$ depends only on $f|_{\partial D}$ and $\int_D dx\sigma \le 0$ for all $f \in H$.

(DDS5) *Dissipative equilibrium states* defined as $\tilde{E}_+ = \{f \in H | \int_D dx\sigma = 0\}$ are identical with $\{f \in H | R^+(f) = 0\}$.

(DDS6) *Thermodynamics.* Let $w = s + \Sigma_{i=1}^{M_0} \tau_i c_i$, where $s,c_1,\dots,c_{M_0}$, were introduced in (DDS4), (DDS5); $\tau_i \in \mathbb{R}$, $i=1,\dots,M_0$ are parameters parametrizing the class of admissible boundary conditions B (see (DDS2)). The equilibrium states E defined in (DDS2) are identical with $\{f \in H \,|\, \frac{\delta W}{\delta f} = 0\}$, where $W = \int_D dx\, w$, $\frac{\delta W}{\delta f}$ is defined by $\frac{d}{d\eta}W(f+\eta\varphi)\big|_{\eta=0} = \int_D dx \frac{\delta W}{\delta f(x)}\varphi(x)$. The quantities $\tau_1,\tau_2,\dots,\tau_{M_0}$, $\tau = -\frac{1}{\mathrm{vol}\Omega} W\big|_E$ ($\mathrm{vol}\Omega$ denotes the volume of Ω) determine thermodynamics implied by $\frac{\partial f}{\partial t} = R(f)$. It means that there is one to one relation between $(p,T,\mu_1,\dots,\mu_{M_0-1})$ and $(\tau,\tau_1,\dots,\tau_{M_0})$, p is the thermodynamic pressure, T is the thermodynamic temperature, μ_i is the thermodynamic chemical potential of the i-th component, $i=1,\dots,M_0-1$. The quantity W will be called the non equilibrium thermodynamic potential. The definition of τ implies immediately that $\frac{\partial \tau}{\partial \tau_i} = -(\mathrm{vol}\Omega)^{-1}\int_D dx c_i\big|_E$, $i=1,\dots,M_0$. If $\int_D dx c_i\big|_E$ can be related to thermodynamic densities (i.e. the mol numbers of the components and the inner energy, then the transformation $(\tau,\tau_1,\tau_2,\dots,\tau_{M_0}) \rightleftarrows (p,T,\mu_1,\dots,\mu_{M_0-1})$ is obtained explicitly . Note that the quantity τ introduced above is not in general a single valued function of $(\tau_1,\dots,\tau_{M_0})$. We shall consider in this paper only the situations in which τ is a single valued function. Physically, we exclude phase transitions and critical phenomena.

(DDS7) *Onsager-Casimir symmetry.* Let $F_0 \in E$ be a regular thermodynamic equilibrium state. Linear operators $A = D_f^2 W|_{F_0}$ (the second derivative of W with respect to f evaluated at F_0), P^+ (the linear part of R^+ at F_0) and P^- (the linear part of R^- at F_0) are defined on the Hilbert space H_0; the elements of the H_0 are f; the inner product in H_0 is $\langle f,g\rangle = \int_D dx\, f\cdot g$. We say that $F_0 \in E$ is regular if F_0 is independent of $\underline{r} \in \Omega$ and A is everywhere defined, bounded, positive definite operator. The operators P^+ and P^- are densely defined and closed. Moreover, P^+ is selfadjoint and dissipative with respect to $\langle\cdot,A\cdot\rangle$ and P^- is skewadjoint with respect to $\langle\cdot,A\cdot\rangle$. If in addition P^+ and P^- or their appropriate extensions or reductions have identical domains, then the above properties of P^+ and P^- imply that $P = P^+ + P^-$ is dissipative with respect to $\langle\cdot,A\cdot\rangle$ and selfadjoint with respect to the indefinite inner product $\langle .,AJ.\rangle$. (Note that $AJ = JA$ since $W(Jf) = W(f)$) In the particular context of nonequilibrium thermodynamics these properties of P coincide with the symmetry introduced by Onsager and Casimir [4]. The Onsager-Casimir symmetry implies (DDS4) in a small neighborhood of a regular equilibrium state. In many cases we are able to verify only (DDS7) and not (DDS4).

3. Examples

1. *The Boltzmann kinetic theory.* A state of a gas is completely described in the Boltzmann kinetic theory by one particle distribution function $f:\Omega\times\{1,\dots,M\}\times \mathbb{R}^3\times\mathbb{R} \to \mathbb{R}_+$, $(\underline{r},i,\underline{v},t) \mapsto f(\underline{r},i,\underline{v},t)$; $f(\underline{r},i,\underline{v},t)d^3\underline{r}\, d^3\underline{v}$ is the number of the gas particles at $\underline{r}+d\underline{r}$ with velocities $\underline{v}+d\underline{v}$ at time t. The involution J is defined by $f(\underline{r},i,\underline{v},t) \overset{J}{\mapsto} f(\underline{r},i,-\underline{v},t)$ (see more about the meaning of J in section 5). The time evolution of f is governed by $\frac{\partial f}{\partial t} = R_B(f)$, where $R_B(f) = R_B^-(f)+R_B^+(f)$; $R_B^-(f) = -v_\alpha \partial_\alpha f$ ($\partial_\alpha \equiv \frac{\partial}{\partial r_\alpha}$, we use the summation convention i.e. $a_\alpha b_\alpha = \Sigma_{\alpha=1}^3 a_\alpha b_\alpha$), $R_B^+(f) = \Sigma_{j=1}^M \int d^3\underline{\nu}\, d^2\underline{\varkappa}\, k[f(\underline{r},i,\underline{v}',t)f(\underline{r},j,\underline{\nu}',t)-f(\underline{r},i,\underline{v},t)f(\underline{r},j,\underline{\nu},t)]$, where $(\underline{v},\underline{\nu}) \xrightarrow{T_\varkappa^{(i,j)}} (\underline{v}',\underline{\nu}')$ is the two parameter family (the parameters are $\underline{\varkappa} \in \mathbb{R}^3$, $|\underline{\varkappa}| = 1$) of transformations satisfying the following properties: (i) $T_\varkappa^{(i,j)}$ is one to one for all $\underline{\varkappa}$; (ii) determinant of the Jacobian of $T_\varkappa^{(i,j)}$ equals to one for all $\underline{\varkappa}$; (iii) $m_i\underline{v}+m_j\underline{\nu} = m_i\underline{v}'+m_j\underline{\nu}'$, $m_i v^2+m_j \nu^2 = m_i v'^2+m_j \nu'^2$; (iv) $T_\varkappa^{(i,j)}(-\underline{v},-\underline{\nu}) = -T_\varkappa^{(i,j)}(\underline{v},\underline{\nu})$; (v) $T_{-\varkappa}^{(i,j)} \equiv T_\varkappa^{(i,j)}$; (vi) k depends on $i,j,|\underline{v}-\underline{\nu}|$, $(v_\alpha-\nu_\alpha)\varkappa_\alpha$, k is invariant with respect to $T_\varkappa^{(i,j)}$, $\underline{\varkappa} \to -\underline{\varkappa}$. The kernel k is positive and symmetric with respect to $i \rightleftarrows j$, $\underline{\varkappa} \to -\underline{\varkappa}$. Physically, $(\underline{v},\underline{\nu})$ are velocities of the particles of i-th resp. j-th component before collision, $(\underline{v}',\underline{\nu}')$ are the velocities after collision, k is proportional to the cross section of the collision. We assume that there is no chemical reaction.

The equation $\frac{\partial f}{\partial t} = R_B(f)$ is obtained as a balance of particles; $R^-(f)$ is the change due to the free flow of particles, $R^+(f)$ is the change due to the binary collisions among the particles. The phenomenological quantities entering the Boltzmann theory are $q_B = (k,m_1,\dots,m_M)$. The conserved quantities are $c_i = f(\underline{r},j,\underline{v},t)\cdot\delta_{ij}$, $i,j=1,\dots,M$, $c_{M+1} = \frac{1}{2}m_i v^2 f(\underline{r},\underline{v},i,t)$, the dissipative quantity is $s=f(\log f-1)$. Straightforward calculation shows that

$$\sum_{i=1}^{M} \int_\Omega d^3\underline{r}\, d^3\underline{v}\sigma = \frac{1}{4}\sum_{i=1}^{M}\sum_{j=1}^{M} \int d^3\underline{v}\int d^3\underline{\nu}\int d^2\underline{x}k)\log\frac{f(\underline{r},i,\underline{v},t)f(\underline{r},j,\underline{\nu},t)}{f(\underline{r},i,\underline{v}',t)f(\underline{r},j,\underline{\nu}',t)}\cdot$$

$$\cdot[f(\underline{r},i,\underline{v}',t)f(\underline{r},j,\underline{\nu}',t)-f(\underline{r},i,\underline{v},t)f(\underline{r},j,\underline{\nu},t)].$$

Since $\log\frac{Y}{X}(X-Y) \le 0$, $X \in \mathbb{R}_+$, $Y \in \mathbb{R}_+$ and the equality holds if and only if $X=Y$, we have $\Sigma_{i=1}^{M} \int_\Omega d^3\underline{r}\int d^3\underline{v}\sigma \le 0$ and $\tilde{E}_+ = \{f=n(\underline{r},i,t)\exp(N-\frac{1}{2}bm_i v^2+c_\alpha m_i v_\alpha)$, N is determined by $\int_\Omega d^3\underline{r}\int d^3\underline{v}f = n(\underline{r},i,t)$, n is an arbitrary function of $(\underline{r},i,t)$, $b,\underline{c}$ are arbitrary functions of $\underline{r},t\}$. Direct calculation gives

$$R^-(f)\Big|_{E_+} = \exp(N-\frac{1}{2}m_i b(\underline{r})v^2)\begin{pmatrix} n(\underline{r},i)(\frac{1}{2}m_i v^2-\frac{\partial N}{\partial b})v_\alpha\partial_\alpha b(\underline{r}) \\ -n(\underline{r},i)v_\alpha\partial_\alpha \log n(\underline{r},i)\end{pmatrix},$$

where $E_+ = \{f \in \tilde{E}_+ | Jf=f\}$, thus (DDS6) is verified. Thermodynamic mass density of the i-th component is $\frac{1}{\mathrm{vol}\Omega}(\int_D dx\, c_i)\big|_E$, thermodynamic energy density is $\frac{1}{\mathrm{vol}\Omega}\cdot(\int_D dx\, c_{M+1})\big|_E$, therefore $\tau_i = -\frac{\mu_i}{T}$, $i=1,\dots,M$, $\tau_{M+1} = \frac{1}{T}$, $\tau = \frac{p}{T}$. If $f=F_0(1+\varphi)$, $\varphi \in H_0$, $F_0 = n(i)\exp(N-\frac{1}{2}bm_i v^2)$, where $n(i)$ is independent of $\underline{r}$ and $b > 0$ is a constant, then $\langle\varphi,A\psi\rangle = \Sigma_{i=1}^{M}\int_\Omega d^3\underline{r}\, d^3\underline{v}\, n(i)\, e^{-\frac{1}{2}bv^2}\varphi(\underline{r},i,\underline{v},t)\psi(\underline{r},i,\underline{v},t)$. The Onsager-Casimir symmetry of P can be verified directly.

2. *The Navier-Stokes-Fourier fluid mechanics.* A state of a fluid is described by $f:\Omega\times\mathbb{R} \to \mathbb{R}_+\times\mathbb{R}^3\times\mathbb{R}$ (we assume $M=1$), $(\underline{r},t) \mapsto (n(\underline{r},t)\underline{u}(\underline{r},t),e(\underline{r},t))$, n is the mass density, e the density of the inner energy per one mole, $\underline{u}$ is the density of the impulse per one mole. The involution J is defined by $(n,\underline{u},e) \overset{J}{\mapsto} (n,-\underline{u},e)$. The equation $\frac{\partial f}{\partial t} = R(f)$ is obtained as follows. Let $c_1 = n$, $c_{3\alpha} = nu_\alpha$, $c_2 = \frac{1}{2}u_\gamma c_{3\gamma}+ne$ (c_3 is the total impulse density, c_2 is the density of the total energy). The time evolution of $c_1,c_2,\underline{c}_3$ is governed by the conservation laws $\partial c_1/\partial t = -\partial_\alpha\overset{\circ}{c}_{1\alpha}$, $\partial c_2/\partial t = -\partial_\alpha\overset{\circ}{c}_{2\alpha}$, $\partial c_{3\beta}/\partial t = -\partial_\alpha\overset{\circ}{c}_{3\alpha\beta}$. The quantities $\overset{\circ}{\underline{c}}_1,\overset{\circ}{\underline{c}}_2,\overset{\circ}{\underline{c}}_3$ are undetermined fluxes. Their specifications are called constitutive laws. The constitutive laws have to be invariant with respect to a symmetry of the macroscopic system considered. Moreover, we shall assume that $R(f)$ does not contain the derivatives with respect to $\underline{r}$ of the order $r \ge 3$ and $R(f)$ depends linearly on the second

derivatives. If the symmetry of the macroscopic system is O(3), then $\overset{\circ}{c}_1 = nu_\alpha$, $\overset{\circ}{c}_{2\alpha} = c_1\dot{u}_\alpha + \mathcal{P}_{\alpha\gamma}u_\gamma + \mathcal{Q}_\alpha$, $\overset{\circ}{c}_{3\alpha\gamma} = c_{3\alpha}u_\gamma + \mathcal{P}_{\alpha\gamma}$; $\mathcal{P}_{\alpha\gamma} = \hat{P}\delta_{\alpha\beta} - \eta_0\partial_\gamma u^*_\gamma\delta_{\alpha\beta} - \frac{1}{2}\eta_1(\partial_\alpha u^*_\beta + \partial_\beta u^*_\alpha)$, $\mathcal{Q}_\alpha = -\lambda\partial_\alpha e^*$, where e^*, $\hat{P}$, u^*, λ, η_0, η_1 are arbitrary, sufficiently regular functions of f, except the condition that all these functions are invariant with respect to $f \mapsto Jf$. The phenomenological quantities introduced by the Navier-Stokes-Fourier fluid mechanics are thus $\tilde{q}_{NSF} = (P, e^*, u^*, \eta_0, \eta_1, \lambda)$. The above constitutive laws imply that the dissipative equilibrium states $\tilde{\mathcal{E}}_+ = \{f \in H | e^* = \text{const.}, \underline{u}^* = \text{const}\}$. It can be shown now that if s is chosen as

$$e^* = \frac{\delta S}{\delta e}, \quad n^*_\alpha = \frac{\delta S}{\delta u_\alpha}, \quad -\frac{1}{n^2}\hat{P}e^* = \frac{\delta S}{\delta n}, \quad S = \int_D dxs$$

$$\lambda > 0, \quad \eta_0 > 0, \quad \eta_1 > 0 \quad \text{for all} \quad f \in H, \tag{1}$$

then all the properties of DDS are satisfied (with the exception of (DDS4) that is satisfied only locally as a consequence of (DDS7)). The phenomenological quantities q_{NSF} for which $\frac{\partial f}{\partial t} = R(f)$ possesses the properties of DDS are thus $\tilde{q}_{NSF}$ together with the compatibility relations (1). In the context of the non-equilibrium thermodynamics [5], (1) is postulated. The postulate is known as the assumption of the local equilibrium.

4. Applications

1. *Existence and Uniqueness Theory.* The structure of DDS does not seem to offer enough information to prove the existence and uniqueness of the solutions of the Cauchy problem for the $\frac{\partial f}{\partial t} = R(f)$. In particular, an information about the topological structure of H is missing. If we restrict however our attention only to a neighborhood of a regular equilibrium state and to the time evolution governed by the linearized equation, then (DDS7) together with the Hille-Yoshida-Phillips theorem [6] [Let $P: H \to H$ be a densely defined and closed linear operator defined in a Hilbert space H equipped with an inner product $(.,.)$. Moreover, let P and the adjoint of P be dissipative. Then there exists a semi-flow $\phi: \mathbb{R}_+ \times H \to H$ (ϕ is continuous) such that $\left.\frac{d}{dt}\right|_{t=0} \phi(t,x) = Px$ for all x in the domain of P] imply the existence of semi-flow. It is interesting to note that the non equilibrium thermodynamics potential W enters into the structure of H since $A = D^2_f W|_{F_0}$ and the inner product in (DDS7) is $\langle ., A. \rangle$. It has been shown by Friedrichs and Lax [7] for one special case of DDS called the system of conservation laws (f is defined as in fluid mechanics, $c_i = f$, $R^+ \equiv 0$, $\sigma \equiv 0$, $R^-(f)$ contains only the first order derivatives) that if s is convex and depends only on f, not on the derivatives of f, then the Cauchy initial value problem is well posed.

2. *Spectrum of P.* The Onsager-Casimir symmetry invites us to study P in the setting of indefinite inner product spaces [8]. A useful information about spectral properties of P has been obtained in this way [9].

3. *Compatibility Theory*. Every dynamical system possessing the structure of DDS is compatible with thermodynamics in the sense of (DDS6). The setting of (DDS6), without the dynamical context provided by other postulates of DDS, represents the "thermodynamic formalism" introduced by an der Waals [10].

The compatibility of the Boltzmann theory with the Navier-Stokes-Fourier theory has been discussed by Hilbert [11], Enskog & Chapman [2] and many others afterwards. We shall explain the idea of Enskog and Chapman on a model where H is a finite dimensional manifold and R is a differentiable vector field on H. We have thus two families $DT_1 = (H_1, R_1, Q_1)$ and $DT_2 = (H_2, R_2, Q_2)$ of dynamical systems possessing the structure of DDS, $\dim(H_1) = N_1$, $\dim(H_2) = N_2$. Observing the orbit spaces corresponding to DT_1 and DT_2, we would like to recognize the orbit space corresponding to DT_2 (or some of its qualitative features) inside the orbit space corresponding to DT_1. The identification of the recognized pattern in the orbit space of DT_1 with the orbit space of DT_2 would result in the mapping $h: R_1^{(2)} \to R_2^{(1)}$, $r: H_1^{(2)} \to H_2^{(1)}$, $v: Q_1^{(2)} \to Q_2^{(1)}$, where $H_1^{(2)} \subset H_1$ is a subset of H_1 on which the pattern is recognized, $R_1^{(2)}$ is the vector field on $H_1^{(2)}$ obtained as the restriction of R_1 to $H_1^{(2)}$, $Q_1^{(2)}$ is a subset of Q_1 for which the pattern can be recognized, $H_2^{(1)} \subset H_2$, $Q_2^{(1)} \subset Q_2$, $R_2^{(1)} = R_2\big|_{H_2^{(1)}}$.

The first, most obvious, qualitative feature of the phase portrait recognized in all dynamical systems possessing the structure of DDS is the set E (the thermodynamic equilibrium states). We shall require that the thermodynamics implied by DT_1 and the thermodynamics implied by DT_2 (i.e. the functional dependence of one of the thermodynamic fields $(p, T, \mu_1, \ldots, \mu_M)$ on the others) will be identical. This, together with physical considerations regarding the relations of the elements of H_1 and H_2 to measurements, usually implies the map h. We shall assume hereafter that h is known, well defined differentiable mapping. (In the case of KT and FM one obtains in this way $n(\underline{r},t) = \int d^3\underline{v}\, f(\underline{r},\underline{v},t)$, $\underline{u}(\underline{r},t) = \int d^3\underline{v}\, \underline{v} f(\underline{r},\underline{v},t) \cdot (n(\underline{r},t))^{-1}$, $e(\underline{r},t) = \frac{1}{2}\int d^3\underline{v} (\underline{v}-\underline{u}(r,t))^2 f(\underline{r},\underline{v},t)(n(\underline{r},t))^{-1}$.)

As the second step, we try to find an N_2-dimensional submanifold $H_1^{(12)}$ of H_1 containing E_1 such that

(1_G) $h \circ i = \hat{1}_2$, where i is the imbedding $H_2 \hookrightarrow H_1$, the image of i is $H_1^{(12)}$, 1_2 is the identity operator in H_2;

(2_G) the vectors of the vector field R_1 at the points of $H_1^{(12)}$ are sticking out of the tangent space to $H_1^{(12)}$ as less as possible.

The vector field $R_1^{(1)}$ is then constructed by requiring that the following diagram is commutative:

$$\begin{array}{ccccc} TH_2^{(1)} & \xrightarrow{Ti} & TH_1^{(12)} & \xrightarrow{Ti_n} & TH_1^{(2)} \\ \pi \big\downarrow\big\uparrow R_1^{(1)} & & \pi \big\downarrow\big\uparrow R_1^{(12)} & & \pi \big\downarrow\big\uparrow R_1^{(2)} \\ H_2^{(1)} & \xrightarrow[i]{} & H_1^{(12)} & \xrightarrow[i_n]{} & H_1^{(2)} \end{array}$$

π are the natural projection, i_n is the natural imbedding of $H_1^{(12)}$ in $H_1^{(2)}$, (1_G) implies $Th = (Ti)^{-1} \circ (Ti_n)^{-1}$. Three problems arise.

Problem 1 How to prove that $H_1^{(12)}$ exists?
How to find $H_1^{(12)}$?
In which sense the time evolution in $H_1^{(12)}$ characterizes the most important features of the time evolution in H_1?

Problem 2 Does $(H_2^{(1)}, R_1^{(1)}, Q_1^{(2)})$ possess the structure of DDS so that $R_1^{(1)}$ can be identified with $R_2^{(1)}$ and the maps r,v obtained?

Problem 3 Physical arguments indicate that the map v should be continuous on a dense subset of $Q_1^{(2)}$ (a kind of structural stability property).

Enskog and Chapman assumed that R^+ is the dominant part of R. They also observed that $\tilde{E}_+$ — the set of the dissipative equilibrium states — is isomorphic to H_2 and $h \circ i_0 = 1_2$, where $i_0 : H_2 \to \tilde{E}_+$. The submanifold $\tilde{E}_+$ is thus suggested as the first guess for $H_1^{(12)}$. The submanifold $\tilde{E}_+$ is then deformed so that the norm of the vector $e_0 = ((Ti_n)^{-1}R_1 - R_1)(x)$ is as small as possible for all $x \in \tilde{E}_+$. An iteration schema for deformations arises naturally. The first improvement on $\tilde{E}_+$ is $H_{11}^{(12)} = \{f \in H \mid R^+(f) = e_0\}$. Replacing now $\tilde{E}_+$ by $H_{11}^{(12)}$ the vector e_1 and the second iteration can be constructed,etc.

A deeper insight into the above three problems can be obtained by considering the whole problem of the compatibility only locally, in a neighborhood of a regular equilibrium state $X_0 \in H_1$. Let $Y_0 = h(X_0)$, H_1 resp. H_2 are the tangent spaces to H_1 resp. H_2 at X_0 resp. Y_0, (J_1, A_1, P_1) is the involution J acting on H_1, $A_1 = D_x^2 W_1 \big|_{X_0}$ (W_1 is the non equilibrium thermodynamic potential in DT_1) P_1 is the linearization of R_1 at X_0; similarly (J_2, A_2, P_2). The map h implies $y_i = \langle \xi^{(i)}, x\rangle$, where $(y_1, \ldots, y_{N_2}) \in \mathrm{H}_2$, $x \in \mathrm{H}_1$, $\xi^{(i)} \in \mathrm{H}_1$, $i=1,\ldots,N_2$ are linearly independent. The N_2-dimensional space spanned by $\xi^{(i)}$ will be denoted H_1'. The problem is now to find $\mathrm{H}_1^{(12)}$ such that

(1_L) None of the vectors $\xi^{(i)}$, $i=1,\ldots,N_2$ is orthogonal, with respect to $\langle .,. \rangle$

to $H_1^{(12)}$, $\dim H_1^{(12)} = N_2$. The mapping $p: H_1^{(12)} \to H_1'$ is one-to-one and $Jp = pJ$.

(2_L) $H_1^{(12)} \subset H_1$ is $J_1A_1P_1$-invariant, i.e. $x_1 \in H_1^{(12)}$ implies that $J_1x_1 \in H_1^{(12)}$, $P_1x_1 \in H_1^{(12)}$ and $\langle x, A_1x\rangle = \langle x_1, A_1x_1\rangle + \langle z, A_1z\rangle$, where $z \in H_1 \setminus H_1^{(12)}$. Moreover, $H_1^{(12)}$ is asymptotic with respect to the time evolution generated by P_1, i.e. the eigenspaces corresponding to the eigenvalues of P_1 that are closest to the origin are contained in $H_1^{(12)}$.

The conditions (1_L) and (2_L) are the local versions of the conditions (1_G) and (2_G) for the submanifold $H_1^{(12)}$. In this local setting, it is not difficult now to consider the above three problems. We shall prove here only that if $H_1^{(12)}$ exists then $P_1^{(1)}$ constructed from the above diagram possesses the properties of the local DDS (thus the postulate (DDS7)). More details to other aspects of the above problems, namely to the Problem 1 and the Problem 3 will be published elsewhere [12].

By $(J_1^{(12)}, A_1^{(12)}, P_1^{(12)})$ we denote (J_1, A_1, P_1) restricted to $H_1^{(12)}$ and represented in a basis $\eta^{(i)}$, $i=1,\dots,N_2$ of $H_1^{(12)}$. The matrices $J_1^{(12)}$, $A_1^{(12)}$, $P_1^{(12)}$ induce $(\tilde{J}', \tilde{A}', \tilde{P}')$ in H_1'; $\tilde{J}' = pJ_1^{(12)}p^{-1}$, $\tilde{A}' = (p^\dagger)^{-1}A_1^{(12)}p^{-1}$, $\tilde{P}' = pP_1^{(12)}p^{-1}$. The matrices $\tilde{J}', \tilde{A}', \tilde{P}'$ are represented in the basis $\zeta^{(i)} = p\eta^{(i)}$. The linear operator C such that $C\zeta^{(i)} = \xi^{(i)}$ is well defined, moreover $CJ_1 = J_1C$. The operators $C\tilde{J}'C^{-1} = J_1^{(12)}$, $(C^\dagger)^{-1}\tilde{A}'C^{-1}$, $C\tilde{P}'C^{-1}$ are the operators $\tilde{J}', \tilde{A}', \tilde{P}'$ represented in the basis $\xi^{(i)}$, $i=1,\dots,N_2$. Thus $J' = (C^\dagger)^{-1}J_1^{(12)}C^\dagger = J_1^{(12)}$, $A' = C(A')^{-1}C^\dagger$, $P' = (C^\dagger)^{-1}(\tilde{P}')^\dagger C^\dagger$ determine the time evolution of y_i, $i=1,\dots,N_2$. We now prove that (J', A', P') satisfy (DDS7). We first note that $(J_1^{(12)}, A_1^{(12)}, P_1^{(12)})$ satisfy (DDS7) since (2_L). We show that $\tilde{P}'$ possesses the Onsager-Casimir symmetry. $\tilde{A}'\tilde{P}' = (p^\dagger)^{-1}A_1^{(12)}P_1^{(12)}p^{-1}$, thus $(\tilde{A}'\tilde{P}')^\dagger = (p^\dagger)^{-1}(A_1^{(12)}P_1^{(12)})^\dagger p^{-1} = (p^\dagger)^{-1}J_1^{(12)}A_1^{(12)}P_1^{(12)}J_1^{(12)}p^{-1} = J_1^{(12)}(p^\dagger)^{-1}A_1^{(12)}P_1^{(12)}p^{-1}J_1^{(12)} = J_1^{(12)}\tilde{A}'\tilde{P}'J_1^{(12)}$. Consequently $A'P' = C(\tilde{A}')^{-1}(P')^\dagger C^\dagger$, thus $(A'P')^\dagger = C\tilde{P}'(\tilde{A}')^{-1}C$. Since $(\tilde{A}'\tilde{P}')^\dagger = J_1^{(12)}\tilde{A}'\tilde{P}'J_1^{(12)}$ implies $\tilde{P}'(\tilde{A}')^{-1} = (\tilde{A}')^{-1}J_1^{(12)}(\tilde{P}')^\dagger J_1^{(12)}$, we have $(A'P')^\dagger = J_1^{(12)}A'P'J_1^{(12)} = J'A'P'J'$.

4. *Extensions of the well established theories.* The setting of DDS is applied to construct extensions of the well established theories. We shall propose the following strategy:

(A) A well established theory (H, R, Q) is chosen.

(B) By following some physical considerations an extension $(H_{ext}, R_{ext}, Q_{ext})$ of (H, R, Q) is proposed.

(C) By requiring that the extended theory possesses the structure of DDS a compatibility condition among $q \in \tilde{Q}_{ext}$, thus a subset $Q_{ext} \subset \tilde{Q}_{ext}$, as well as

c_{iext} and s_{ext} are obtained.

(D) It has to be stressed that only candidates for well established dynamical theories of macroscopic systems are obtained in this way. Another consideration, mainly then the comparison with experiments, must follow.

Three different extensions of the Navier-Stokes-Fourier fluid dynamics (the fourteen state variable fluid dynamics, two point fluid dynamics [13], the higher order fluid dynamics [14]) and two different extensions of the Boltzmann theory [15] (the Enskog-like equations, the two-point kinetic theory) have been developed. Most of these results have not been published yet. We shall illustrate the discussion of an extension on the example of the Enskog-like extension of the Boltzmann theory, (A more general study of the Enskog-like extension will be published elsewhere [16]) and on the two-point extension of the Boltzmann theory.

Following Enskog [2], $R_{ext}(f) = R_B(f)+R_E(f)$, where $E_1(f) = \Sigma_j \int d^3\underline{v}\int d^2\underline{\varkappa}\cdot (v_\gamma - v_\gamma)\varkappa_\gamma \varkappa_\alpha \chi(i,j,n)[f(\underline{r},i,\underline{v}')\partial_\alpha f(\underline{r},j,\underline{v}') + f(\underline{r},j,\underline{v})\partial_\alpha f(\underline{r},j,\underline{v})] + \frac{1}{2}\Sigma_\ell \frac{\delta\chi}{\delta n(r,\ell)}\partial_\alpha n(\underline{r},\ell)\cdot$ $\cdot[f(\underline{r},i,\underline{v}')f(\underline{r},j,\underline{v}')+f(\underline{r},i,\underline{v})f(\underline{r},j,\underline{v})]$. We remind the reader that $n(\underline{r},i) = \int d^3\underline{v}f(\underline{r},i,\underline{v})$; χ is an arbitrary, sufficiently regular function of $i,j,n(\underline{r},\ell)$. The phenomenological quantities $\tilde{q}_{ext} = (q_B,\chi) \in \tilde{Q}_{ext}$. If $\chi \equiv 0$ then $R_{ext} \equiv R_B$ and $\tilde{Q}_{ext} \equiv Q_B$. We shall require that the extended theory possesses the structure of DDS. Note that the new term $R_E(f)$ belongs to $R^-_{ext}(f)$. Thus, R^+ and consequently also σ, $\tilde{E}_+$ remain the same as in the Boltzmann theory. It is convenient to start with (DDS6);

$$R^-(f)\Big|_{E_+} = \exp(N - \tfrac{1}{2}m_i b(\underline{r})v^2) \times$$

$$\times\begin{pmatrix} n(\underline{r},i)(\frac{1}{2}m_i v^2 - \frac{\partial N}{\partial b})v_\alpha\partial_\alpha b(\underline{r}) \\ -n(\underline{r},i)v_\alpha(\partial_\alpha \log n(\underline{r},i) + \frac{8\pi}{3}\sum_j(\chi(i,j,n)\partial_\alpha n(\underline{r},j) + \frac{4\pi}{3}\sum_\ell\frac{\delta\chi}{\delta n(\underline{r},\ell)}n(\underline{r},j)\partial_\alpha n(\underline{r},\ell))) \end{pmatrix} \tag{3}$$

We want to write (3) in such a way that (DDS6) will be satisfied. The second line can be written as

$$-n(\underline{r},i)v_\alpha\partial_\alpha(\log n(\underline{r},i)+ \frac{\delta\Xi}{\delta n(\underline{r},i)} + \sum_j n(\underline{r},j)\frac{\delta^2\Xi}{\delta n(\underline{r},i)\delta n(\underline{r},j)})$$

$$= -n(\underline{r},i)v_\alpha\partial_\alpha \frac{\delta(n(\underline{r},i)(\log n(\underline{r},i)-1)+ \sum_j n(\underline{r},j)\frac{\delta\Xi}{\delta n(\underline{r},j)})}{\delta n(\underline{r},i)} \tag{4}$$

provided

$$\frac{4\pi}{3}\chi(i,j,n) = \frac{\delta^2\Xi}{\delta n(\underline{r},i)\delta n(\underline{r},j)}\ , \tag{5}$$

where Ξ is an arbitrary function of $n(\underline{r},\ell)$. It is clear from (4) that if c_1, c_2 remain the same as in the case of the Boltzmann theory and $s_{ext} = f(\log f - 1) + \Sigma_j f(\underline{r},j,\underline{v}) \frac{\delta \Xi}{\delta n(\underline{r},j)}$, (indeed, if $\chi = 0$, s_{ext} reduces to s introduced in the Boltzmann theory) then (DDS6) is satisfied. A direct calculation shows that also (DDS7) is satisfied, therefore, also (DDS4) is satisfied in a small neighborhood of F_o. The compatibility relations restricting the choice of (i,j,n) are implied by (5):

$$\chi(i,j,n) = \chi(j,i,n)$$

and *(6)*

$$\frac{\delta \chi(i,j,n)}{\delta n(\underline{r},\ell)} = \frac{\delta \chi(i,\ell,n)}{\delta n(\underline{r},j)} = \frac{\delta \chi(j,\ell,n)}{\delta n(\underline{r},i)} .$$

The relations (6) define $Q_{ext} \subset \tilde{Q}_{ext}$. The thermodynamic interpretation of $\tau_1, \ldots, \tau_M, \tau$ remains the same as in the Boltzmann theory. Since s and therefore w is now different, the functional dependence of τ on $\tau_1, \ldots, \tau_M$ (i.e. the thermodynamic equation of state) is however different. Although the Enskog extension has been known for a long time, the compatibility conditions (6) and s_{ext} have not been known.

In the second example both f and q are extended; $f \equiv (h_1, h_2)$, where h_1 is the one particle distribution function introduced in the Boltzmann theory, $h_2 : \Omega_a \times \mathbb{R}^3 \times \mathbb{R}^3 \times \mathbb{R} \to \mathbb{R}_+$ is the long range two particle distribution function, $h_2(1,2) = h_2(2,1)$, $\Omega_a = \{(\underline{r}_1, \underline{r}_2) \in \Omega \times \Omega \mid |\underline{r}_1 - \underline{r}_2| > a$, a is the range of the short range repulsive potential between particles$\}$ Whenever it is possible, we use the abbreviation $(\underline{r}_1, \underline{v}_1) \equiv 1$, etc. The involution J is defined by $(h_1(\underline{r}_1,\underline{v}_1), h_2(\underline{r}_1,\underline{v}_1,\underline{r}_2,\underline{v}_2)) \overset{J}{\longmapsto} (h_1(\underline{r}_1,-\underline{v}_1), h_2(\underline{r}_1,-\underline{v}_1,\underline{r}_2,-\underline{v}_2))$. New coordinates (h,g) are introduced in H_{ext} by $h_1(1) = h(1)$, $h_2(1,2) = h_1(1)h_1(2)g(1,2)$. If $g(1,2) \equiv 1$ then the extended theory coincides with the Boltzmann theory. Physical considerations based for example on the BBGKY hierarchy [24] suggests the following equations governing the time evolution of f:

$$\frac{\partial h(1)}{\partial t} = -v_{1\alpha}\partial_{1\alpha}h + \frac{\partial h}{\partial v_{1\alpha}} \partial_{1\alpha} w_1 + B_1(h,g) \tag{7}$$

$$\frac{\partial g(1,2)}{\partial t} = -v_{i\alpha}\partial_{i\alpha}g + g(1,2)\frac{\partial}{\partial v_i}(\log h(i))\partial_{i\alpha}w_2 + g(1,2)\frac{\partial}{\partial v_{i\alpha}}(\log g(1,2))G_{i\alpha} + B_2(g,h).$$

The new phenomenological quantities $w_1, w_2, G_{i\alpha}$ introduced in (7) are arbitrary, sufficiently regular functions of (n,m), $n(\underline{r}_1) = \int d^3\underline{v}_1 h(\underline{r}_1,\underline{v}_1)$, $n(\underline{r}_1)n(\underline{r}_2)m(\underline{r}_1,\underline{r}_2) = \iint d^3\underline{v}_1 d^3\underline{v}_2 h(1)h(2)g(1,2)$. In (7) we have used the summation convention $v_{i\alpha}\partial_{i\alpha} = \Sigma_{i=1}^2 \Sigma_{\alpha=1}^3 v_{i\alpha}\partial_{i\alpha}$. If $B_1 \equiv B_2 \equiv 0$, then the right hand side of (7) forms R^-_{ext}. We have to find now B_1 and B_2 such that (i) B_1 and B_2 form R^+_{ext}, (ii) if $g \equiv 1$, then R^+_{ext} reduces to the Boltzmann collision operator R_B, (iii) R_{ext} will satisfy the postulates of DDS, (iv) one expects on the ground of physical considerations —

Gibbs postulates of equilibrium statistical mechanics — that

$$E_+ = \{h(1) = n(r_1)\exp(N - \tfrac{1}{2}bv_1^2),\ g(1,2) = m(r_1,r_2),\ N \text{ is determined by } \int d^3\underline{v}_1 h(1) = n(\underline{r}_1),\ b \text{ is an arbitrary function of } \underline{r}_1\}. \quad (8)$$

A large class of B_1 and B_2 satisfying (i)-(iv) above has been found [15]. We present here only one special solution that appears to be the simplest.

$$B_1(h,g) = \varepsilon_{11}\int d^3\underline{v}_3 \int_{(v_{3\gamma}-v_{1\gamma})\varkappa_\gamma>0} d^2\underline{\varkappa}\ (v_{3\gamma}-v_{1\gamma})\varkappa_\gamma[\mathring{h}(\underline{r}_1\underline{v}_1',\underline{r}_1\underline{v}_3') - \mathring{h}(\underline{r}_1,\underline{v}_1,\underline{r}_1,\underline{v}_3)]$$

$$B_2(g,h) \equiv 0; \quad (9)$$

$\mathring{h}(1,2) = \lambda\exp(h^*(1)+h^*(2))$, $h^*(1) = \frac{\delta S}{\delta h(1)}$, $S = \int d1\int d2\ s_{ext}$, s_{ext} depends on (h,g), if $g \equiv 1$ then s_{ext} has to reduce to $h(1\ gh-1)$, ε_{11}, λ depend on (n,m), $\varepsilon_{11} > 0$, $\lambda > 0$ for all $f \in H_{ext}$. For the sake of simplicity we consider here only a special form of k corresponding to the collisions of hard spheres. Similar calculations as those leading to the σ in the Boltzmann theory give

$$\int_D dx\sigma = \int d1\int d3 \int_{(v_{3\gamma}-v_{1\gamma})\varkappa_\gamma>0} d^2\underline{\varkappa}(v_{3\gamma}-v_{1\gamma})\varkappa_\gamma\varepsilon_{11}\lambda[h^*(\underline{r}_1,\underline{v}_1)+h^*(\underline{r}_1,\underline{v}_3)-h^*(\underline{r}_1,\underline{v}_1')-h^*(\underline{r}_1,\underline{v}_3')] \times$$
$$\times\ [\exp(h^*(\underline{r}_1,\underline{v}_1')+h^*(\underline{r}_1,\underline{v}_3'))-\exp(h^*(\underline{r}_1,\underline{v}_1)+h^*(\underline{r}_1,\underline{v}_3))]. \quad (10)$$

If $g \equiv 1$ then $s_{ext} = s$ and (10) is equivalent to $\int_D dx\sigma$ obtained in the Boltzmann theory. Since $(X-Y)(e^Y-e^X) \leq 0$, $X \in \mathbb{R}_+$, $Y \in \mathbb{R}_+$, and the equality holds if and only if $X=Y$, then $\int_D dx\sigma \leq 0$ and the equality holds if $f \in E_+^{(9)}$, where

$$E_+^{(9)} = \{f \in H | h^*(\underline{r}_1,\underline{v}_1)+h^*(\underline{r}_1,\underline{v}_3) = h^*(\underline{r}_1,\underline{v}_1')+h^*(\underline{r}_1,\underline{v}_3')\}. \quad (11)$$

A function s_{ext} has to be found so that all the postulates of DDS are satisfied and $E_+^{(9)}$ restricted to f satisfying $Jf = f$ is identical with E_+ defined in (8). If we restrict $R^-(f)$ to E_+ introduced in (8) we have

$$R^-(f)\Big|_{E_+} = \begin{pmatrix} \exp(N-\frac{1}{2}bv_1^2) & (n(\frac{1}{2}v_1^2 - \frac{\partial N}{\partial b})v_{1\alpha}\partial_{1\alpha}b) \\ \exp(N-\frac{1}{2}bv_1^2) & (-v_{1\alpha}\partial_{1\alpha}n - v_{1\alpha}bn\partial_{1\alpha}w_1) \\ & -v_{i\alpha}\partial_{i\alpha}m - v_{i\alpha}bm\partial_{i\alpha}w_2 \end{pmatrix}. \quad (12)$$

It can be directly verified that if $s_{ext} = h(1)(\log h(1)-1) + \frac{1}{2}h(1)h(2)[g(1,2)\log g(1,2) - (g(1,2)-1)]$, $c_{1ext} = c_1 = h(1)$, $c_{2ext} = \frac{1}{2}v^2h(1)+h(1)h(2)V$, $c_3 = \frac{1}{2}h(1)h(2)(g(1,2) -1)$,

$$w_2(\underline{r}_1,\underline{r}_2) = \frac{\delta V}{\delta m(\underline{r}_1,\underline{r}_2)}$$

$$w_1(\underline{r}_1) = \frac{1}{2}\,\frac{\delta\int_\Omega d^3\underline{r}_1'\int_\Omega d^3\underline{r}_2' n(\underline{r}_1')n(\underline{r}_2')V}{\delta n(\underline{r}_1)} + \frac{1}{\tau_2}\int_\Omega d^3\underline{r}_2 n(\underline{r}_2)[m(\underline{r}_1,\underline{r}_2)\log m(\underline{r}_1,\underline{r}_2)-m(\underline{r}_1,\underline{r}_2)-1)]$$
$$+\frac{\tau_3}{\tau_2}\int_\Omega d^3 r_2 n(\underline{r}_2)(m(\underline{r}_1,\underline{r}_2)-1), \qquad (13)$$

where V is an arbitrary sufficiently regular function of (n,m), then all the postulates of DDS are satisfied and (11) restricted to f such that Jf=f is identical with (8). The phenomenological quantities $q_{ext} = \{q_B, w_1, w_2,$ the compatibility conditions (13)$\}$, q_B are the phenomenological quantities introduced by the Boltzmann theory. The thermodynamical meanings of τ,τ_1,τ_2 are the same as in the Boltzmann theory. A new conserved quantity c_3 and thus also a new thermodynamic parameter τ_3 appeared in this extended theory.

5. Remarks

A one-component macroscopic system is regarded in classical mechanics as composed of N identical particles. A state of the system is described by $f_{CM} = (r,v) = (\underline{r}_1,\ldots,\underline{r}_N,\underline{v}_1,\ldots,\underline{v}_N)$, $\underline{r}_i \in \Omega$, $\underline{v}_i \in \mathbb{R}^3$, $i=1,2,\ldots,N$, $R_{CM} = \Omega^N \times \mathbb{R}^{3N}$. The time evolution of f is governed by

$$\begin{pmatrix}\dot r\\ \dot v\end{pmatrix} = \begin{pmatrix}0 & 1\\ -1 & 0\end{pmatrix}\begin{pmatrix}\frac{\partial h}{\partial r}\\ \frac{\partial h}{\partial v}\end{pmatrix}. \qquad (14)$$

The Hamiltonian function $h(r,v) = \frac{1}{2}\sum_{i=1}^{N} v_i^2 + V(\underline{r}_1,\ldots,\underline{r}_N)$; the potential energy V together with boundary conditions compose the phenomenological quantities entering the classical mechanics. The function V is assumed to be sufficiently regular. The involution J is defined by $(r,v) \overset{J}{\longmapsto} (r,-v)$. Thus in classical mechanics $R_{CM} = R^-_{CM}$, $R^+_{CM} \equiv 0$. If we denote by u_t the flow determined by (7), then $u_{-t}(J(r,v)) = u_t(r,v)$.

Before attempting to construct the orbit space determined by (14), it is convenient to look for deformations in H_{CM} so that the orbit space will be as simple as possible (e.g. the action-angle variables [17]). An ultimate achievement along this line is the rectification of the vector field on subsets of H [17]. In macroscopic systems $N \sim 10^{23}$. Following Maxwell, Boltzmann, Liouville and Koopman [2, 18] an attempt to solve (14) usually starts with the following reformulation of (14):

$$\begin{array}{ccc} H_{CM}\times R_+ & \xrightarrow{(u_t\times\hat{1})} & H_{CM}\times R_+ \\ \pi\downarrow\uparrow f & & \pi\downarrow\uparrow f_t \\ H_{CM} & \xrightarrow[u_t]{} & H_{CM} \end{array} \qquad (15)$$

where $\mathbb{R}_+$ is the positive real line, π a natural projection, f is the cross section of the trivial bundle constructed in the diagram, thus $f:\mathcal{H}_{CM} \to \mathbb{R}_+$, u_t is the flow determined by (14). The cross section f_t is obtained by requiring commutativity of the diagram (15). Thus $U_t f(r,v) = f_t(r,v) = f(u_{-t}(r,v))$. By differentiating this relation with respect to the time t, we obtain the Liouville equation

$$\frac{df}{dt} = \Big(\sum_{i=1}^{N} \Big(\frac{\partial V}{\partial \underline{r}_i}\frac{\partial}{\partial \underline{v}_i} - \underline{v}_i \frac{\partial}{\partial \underline{r}_i}\Big)\Big)f, \tag{16}$$

that is the reformulation of (14). The operator U_t is always linear, moreover, as a consequence of the Liouville theorem, U_t is unitary if we consider $f \in \mathcal{H}_L$, $\mathcal{H}_L$ is the Hilbert space equipped with the L_2 inner product denoted $\langle .,. \rangle$.. The diagram (15) can be seen as a universal linearization in expense of a substantial enlargement of the phase space. If we introduce $f(r,v) \overset{J}{\mapsto} f(r,-v)$, we obtain from (16) $\mathcal{R}_L = \mathcal{R}_L^-$, $\mathcal{R}_L^+ \equiv 0$. It is interesting to note that (16) itself can be regarded as a Hamiltonian system of the type introduced in the theory of the Korteweg de Vries equation [19]:

$$\frac{df(r,v)}{dt} = D\frac{\delta H_L}{\delta f(r,v)}, \tag{17}$$

where $H_L = S_L + \text{const}\int_{\Omega^N} dr \int dv\, f(\underline{r},v)$; $S = \frac{1}{2}\langle f,f\rangle$, $D = \Big(\sum_{i=1}^{N}\Big(\frac{\partial V}{\partial \underline{r}_i}\frac{\partial}{\partial \underline{v}_i} - \underline{v}_i\frac{\partial}{\partial \underline{r}_i}\Big)\Big)$.

The operator D plays the role of $\begin{pmatrix} 0 & 1 \\ -1 & 0 \end{pmatrix}$ in (14). With appropriate boundary conditions for f it can be proved easily that if we define $[F,G] = \Big\langle \frac{\delta F}{\delta f(r,v)}, D\frac{\delta G}{\delta g(r,v)}\Big\rangle$, $F,G:\mathcal{H}_L \to \mathbb{R}$, are differentiable, then $[F,G]$ possesses all the properties of the Poisson bracket. An immediate consequence of (17) is that H_L is a conserved quantity. We have thus obtained another proof that U_t is unitary. The phenomenological quantities (namely the potential energy $V(\underline{r}_1,\dots,\underline{r}_N)$) enters the Hamiltonian function h in (14) and $D = \begin{pmatrix} 0 & 1 \\ -1 & 0 \end{pmatrix}$ is independent of the phenomenological quantities. In (17) the situation is opposite. The Hamiltonian H_L is the same for all systems, the phenomenological quantity V enters the operator D. A timid comparison of (17) with DDS indicates that H_L is taking up some of the roles of the non equilibrium potential W. The Hamiltonian H_L seems to be therefore a nucleus of the non equilibrium thermodynamic potential. Two additional observations support this claim.

Let us consider $(f,g) \in \mathcal{H}_L \times \mathcal{H}_L = \tilde{\mathcal{H}}_L$ and

$$\frac{d}{dt}\begin{pmatrix} f \\ g \end{pmatrix} = \begin{pmatrix} 0 & D \\ D & 0 \end{pmatrix}\begin{pmatrix} \delta\tilde{H}_L/\delta f \\ \delta\tilde{H}_L/\delta g \end{pmatrix}, \tag{18}$$

where $\tilde{H}_L = \tilde{S}_L + \text{const}\int_{\Omega^N} dr\int dv\, f(r,v)$; $\tilde{S}_L = \int_{\Omega^N} dr\int dv(f(\log f - 1) + e^g)$. If (19) is restricted to the Lagrangian submanifold L of $\tilde{\mathcal{H}}_L$ defined by $L = \{(h,g) \in \tilde{\mathcal{H}}_L \mid g = \frac{\delta S_L}{\delta f}\}$, then (18) becomes identical with (17). The Hamiltonian function $\tilde{H}_L|_L$ has now the form clearly similar to W arising in the Boltzmann theory.

As the second observation we note that $R_L^+ \equiv 0$ and therefore $\tilde{E}_+ \neq R_L$. We shall however define ad hoc $E_+ = \{f \in H_L | f(r,v)=n(r)\exp(N-\frac{1}{2}b\Sigma_{i=1}^N v_i^2)$, N is defined by $\int dvf(r,v)=n(r)$, b is a constant$\}$. Now

$$R(f)\Big|_{E_+} = \sum_{i=1}^{N} (-n(r)v_{i\alpha}\partial_{i\alpha}(\log n+bV))\exp(N-\tfrac{1}{2}b\Sigma_{i=1}^N v_i^2),$$

thus the "thermodynamic equilibrium states" (we have to use quotation marks since E_+ has been chosen ad hoc) are $E = \{f \in H \,|\, \frac{\delta W}{\delta f} = 0\}$, where $W = \int_{\Omega^N} dr \int dv(f\cdot(\log f-1)+b(\frac{1}{2}\Sigma_{i=1}^N v_i^2 f+Vf)+\text{const.}f)$.

The reformulation (17) of (14) does not provide automatically an insight into the structure of the orbit space of (14). At least two different lines of thinking can now be followed.

The first method might be called the method of a limit. Usually the theory DT_2 that is to be derived is known (it is one of the realizations of DDS, e.g. the Boltzmann theory). A path in the phenomenological quantities Q_1 is found so that by following the path into a limit point (that is itself outside the domain of physically existing values of the phenomenological quantities) an exactly invariant submanifold $H_1^{(\infty)}$ of H_1 arises. The time evolution on $H_1^{(\infty)}$ is then identified with the time evolution introduced in DT_2 for one particular value $q \in Q_2$. An example of such a limit is discussed in [20]. The method of a limit has been also used in the problem KT → FM [21]. An advantage of the method of a limit is its relative simplicity; its disadvantages are the following: (i) There might be many different paths in Q_1 allowing to recognize DT_2, thus little information is obtained about the subset of Q_1 for which DT_2 is a "good" approximation. What a "good" approximation means at the points different from the limit point? (ii) No information about the map $v:Q_1 \to Q_2$ is obtained.

The second line of thinking is based on the argument of structural stability [12]. A subset of Q_1 is chosen. The qualitative features of the phase portrait of DT_1 that are as much as possible independent on $q \in Q_1$ and their dependence on $q \in Q_1$ is as smooth as possible are searched for. An example of the argument of the structural stability in the problem of the foundation of the Gibbs postulates referring to quantum mechanics → TH was discussed in [23]. The authors have shown that in a setting that is closely related, but not equivalent, to the quantum mechanics, the requirement of structural stability singles out the thermodynamic equilibrium state à la Gibbs among a large class of possible candidates for the thermodynamic equilibrium states. The application of the structural stability argument in the context KT → FM (see the Problem 3 in the section 4.3) is discussed in [12].

REFERENCES

[1] S.Carnot, Reflections on the motive power of fire (edited by E.Mendoza, Dover Publ.)(1960).

[2] S.G.Brush, Kinetic theory, Vol.1,2,3, Pergamon Press (1966).

[3] D.Ruelle, Statistical mechanics, Rigorous results, Benjamin (1969).

R.Bowen, Equilibrium states and the ergodic theory of Anosov diffeomorphisms, Lecture Notes in Math. 470, Springer (1975).

[4] L.Onsager, Phys.Rev.37 (1931) 305, 38 (1931) 2265.

H.G.Casimir, Rev.Mod.Phys.17 (1945) 343.

[5] S.R.de Groot and P.Mazur, Non-equilibrium thermodynamics, McGraw-Hill, New York (1969).

[6] K.Yoshida, Functional Analysis, Springer (1968).

[7] K.O.Friedrichs and P.D.Lax, Proc.Nat.Acad.Sci. USA 68 (1971) 1686.

[8] J.Bognar, Indefinite Inner Product Spaces, Springer-Verlag (1974).

[9] M.Grmela and I.Iscoe, Ann.Inst.Henri Poincare 28 (1978) 111.

[10] J.D. Van der Waals, Ph.D. dissertation, Leiden (1873).

N.G.Van Kampen, Phys.Rev. 135 (1964) A362.

[11] D.Hilbert, Math.Ann. 72 (1912) 562.

C.Cercignani, Theory and Application of the Boltzmann Equation, Elsevier, New York (1975).

[12] M.Grmela and R.Rosen, Preprint CRM-753 (1978).

[13] M.Grmela, Preprint CRM-788 (1978).

[14] M.Grmela, Helv.Phys.Acta 50 (1977) 393.

[15] M.Grmela, Preprint CRM-783 (1978).

[16] M.Grmela and L.S.Garcia-Colin (in preparation).

[17] V.Arnold, Méthodes mathématiques de mécanique classique, Edition Mir, Moscow (1976).

V.Arnold, Equations différentielles ordinaires, Edition Mir, Moscow (1974).

[18] B.O.Koopman, Proc.Nat.Acad.Sci.U.S. 17 (1931) 315.

[19] C.S.Gardner, J.Math.Phys.12 (1971) 1548.
P.D.Lax, Comm.Pure Appl.Math.28 (1975) 141.

[20] H.Grad in "Handbuch der Physik" vol XII, Springer-Verlag, Berlin (1958).
O.E.Lanford, in Lecture Notes in Physics vol.38 (1975),ed.by J.Moser,Springer-Verlag.

[21] M.A.Pinsky, SIAM-AMS Proceedings Vol.10 (1976) 119.

[22] A.Andronov and L.Pontryagin, Dohl.Akad.Nauk SSSR 14 (1937) 247.

R.Thom, Stabilité structurelle et morphogénèse, W.A.Benjamin (1972).

[23] R.Haag, D.Kastler and E.B.Trych-Pohlmeyer, Commun.Math.Phys.38 (1974) 173.

[24] J.Yvon, Les corrélations et l'entropie en mécanique statistique classique, Dunod, Paris (1966).

ON THE PREVALENCE OF APERIODICITY IN SIMPLE SYSTEMS

by

Edward N. Lorenz

0. Introduction

As the lone meteorologist at a seminar of mathematicians, I feel that a few words regarding my presence may be in order. Let me begin with some remarks about the mathematics of meteorology.

One of the most familiar problems of interest to meteorologists is weather forecasting. Mathematically this is an initial-value problem. The atmosphere and its surroundings are governed by a set of physical laws which in principle can be expressed as a system of integro-differential equations. At the turn of the century, the forecast problem was identified by Bjerknes [1] as the problem of solving these equations, using initial conditions obtained from observations of current weather. Detailed numerical procedures for solving these equations were formulated during World War I by Richardson [8], but the practical solution of even rather crude approximations had to await the advent of computers.

Another problem of interest is climate. This is a problem in dynamical systems. The climate is often identified with the set of all long-term statistical properties of the atmospheric equations. It is commonly assumed that one can devise a finite system of ordinary differential equations whose typical solutions nearly duplicate those of the more realistic system. In the phase space of such a system, weather forecasting, particularly at short range, is a local problem, while climate is global.

The atmosphere is a forced dissipative system; the forcing is thermal, while the dissipation is thermal and mechanical. Any system of equations whose general solution can hope to approximate the atmosphere must likewise contain forcing and dissipation. The various orbits in phase space are therefore not confined to separate energy surfaces, as they would be in a conservative system.

During my early exposure to theoretical meteorology, I had wondered whether there might instead be a single surface which a few special orbits would occupy, and which the remaining orbits would approach. I had even hoped to discover some smooth function which would vanish on this surface, and would assume positive values on one side and negative values on the other. Needless to say I was unsuccessful, and, in the light of more recent results, the search for such a function seems rather naive. I presently turned to other matters.

My return to dynamical systems was prompted by an interest in weather forecasting rather than climate. By the middle 1950's "numerical weather prediction", i.e., forecasting by numerically integrating such approximations to the atmospheric equations as could feasibly be handled, was very much in vogue, despite the rather mediocre results which it was then yielding. A smaller but determined group favored statistical prediction, and especially prediction by linear regression, using large numbers of predictors. Apparently because of a misinterpretation of a paper by Wiener [12], the idea became established that the statistical method could duplicate the performance of the dynamical method, despite the essential nonlinearity of the dynamic equations. I was skeptical, and decided to test the idea by applying the statistical method to a set of artificial data, generated by solving a system of equations numerically. Here the dynamical method would consist of solving the equations all over again, and would obviously give perfect results. I doubted very much that the statistical method would do likewise.

The first task was to find a suitable system of equations to solve. In principle any nonlinear system might do, but a system with some resemblance to the atmospheric equations offered the possibility of some useful by-products. The system would have to be simple enough to be able to simulate a rather long stretch of weather with a reasonable amount of computation. Moreover, the general solution would have to be aperiodic, since the statistical prediction of a periodic series would be a trivial matter, once the periodicity had been detected. It was not obvious that these conditions could be met.

At about this time our group was fortunate enough to secure its own digital computer, which still sits across the hall from my office. The computer was slow by today's standards, but we were competing with no one for its use. Moreover, its very slowness enabled us to watch the output being printed as it was produced, and we could stop the computation and introduce new numbers whenever the present output appeared uninteresting.

We first chose a system which had been used for numerical weather prediction. The system represented the three-dimensional structure of the atmosphere by two horizontal surfaces, and we proceeded to expand the horizontal field of each atmospheric variable in a series of orthogonal functions. We then reduced the system to manageable size by discarding all terms of the series except those representing the largest few horizontal scales, and programmed the resulting system for the little computer.

For a while our search produced nothing but steady or periodic solutions, but at last we found a system of twelve equations whose solutions were unmistakably aperiodic. It was now a simple matter to put the statistical forecasting method to test, and we found, incidentally, that it failed to reproduce the numerically generated weather data.

During our computations we decided to examine one of the solutions in greater detail, and we chose some intermediate conditions which had been typed out by the computer and typed them in as new initial conditions. Upon returning to the computer an hour later, after it had simulated about two months of "weather", we found that it completely disagreed with the earlier solution. At first we suspected machine trouble, which was not unusual, but we soon realized that the two solutions did not originate from identical conditions. The computations had been carried internally to about six decimal places, but the typed output contained only three, so that the new initial conditions consisted of old conditions plus small perturbations. These perturbations were amplifying quasi-exponentially, doubling in about four simulated days, so that after two months the solutions were going their separate ways.

It soon became evident that the instability of the system was the cause of its lack of periodicity. The variables all had limited ranges, so that near repetitions of some previous conditions were inevitable. Had the system been stable, the difference between the original occurrence and its near repetition would not have subsequently amplified, and essentially periodic behavior would have resulted.

I immediately concluded that, if the real atmospheric equations behaved like the model, long-range forecasting of specific weather conditions would be impossible.

The observed aperiodicity of the atmosphere, once the normal diurnal and annual variations are removed, suggests that the atmosphere is indeed an unstable system. The inevitable small errors in observing the current weather should therefore amplify and eventually dominate.

Still, I felt that we could better appreciate the problems involved by studying a simpler example. The ideal system would contain only three variables, whence we could even construct models of orbits in phase space, or of the surface, if any, which these orbits would approach. However, our attempts to strip down the twelve-variable system while retaining the aperiodicity proved fruitless.

The break came when I was visiting Dr. Barry Saltzman, now at Yale University. In the course of our talks he showed me some work on thermal convection, in which he used a system of seven ordinary differential equations [5]. Most of his numerical solutions soon acquired periodic behavior, but one solution refused to settle down. Moreover, in this solution four of the variables appeared to approach zero.

Presumably the equations governing the remaining three variables, with the terms containing the four variables eliminated, would also possess aperiodic solutions. Upon my return I put the three equations on our computer, and confirmed the aperiodicity which Saltzman had noted. We were finally in business.

1. A Physical System with a Strange Attractor.

In a changed notation, the three equations with aperiodic solutions are

$$dX/dt = -\sigma X + \sigma Y \quad , \tag{1.1}$$

$$dY/dt = -XZ + rX - Y \quad , \tag{1.2}$$

$$dZ/dt = XY - bZ \quad . \tag{1.3}$$

Although originally derived from a model of fluid convection, (1.1)-(1.3) are more easily formulated as the governing equations for a laboratory water wheel, constructed by Professor Willem Malkus of M.I.T. to demonstrate that such equations are physically realizable. The wheel is free to rotate about a horizontal or tilted axis. Its circumference is divided into leaky compartments. Water may be introduced from above, whereupon the wheel can become top-heavy and begin to rotate. Different compartments will then move into position to receive the water. Depending upon the values of the constants of the apparatus, the wheel may be observed to remain at rest, rotate continually in one direction or the other, or reverse its direction at regular or irregular intervals.

The equations are written for a wheel of radius a with a horizontal axis, and with its mass confined to the rim. Its angular velocity $\Omega(t)$ may be altered by the action of gravity g on the nonuniformly distributed mass and by frictional damping proportional to Ω. The mass $\rho(t,\theta)$ per unit arc of circumference may be altered by a mass source increasing linearly with height, a mass sink proportional to ρ, and, at a fixed location in space, by rotation of the wheel. Here t is time and θ is arc of circumference, measured counterclockwise. The wheel then obeys the equations

$$d(a^2\bar{\rho}\,\Omega)/dt = -\,g\,a\,\overline{\rho\cos\theta} - k\,a^2\bar{\rho}\,\Omega \quad , \qquad (1.4)$$

$$\partial\rho/\partial t + \Omega\,\partial\rho/\partial\theta = A + 2B\sin\theta - h\rho \quad , \qquad (1.5)$$

representing the balances of angular momentum and mass, where $\overline{}$ denotes an average with respect to θ, and A, B, k, and h are positive constants. From (1.5) it follows that $\bar{\rho}$ approaches A/h exponentially; assuming that $\bar{\rho}$ has reached A/h, (1.4) and (1.5) yield the three ordinary differential equations

$$d\Omega/dt = -k\,\Omega - (gh/aA)\,\overline{\rho\cos\theta} \quad , \qquad (1.6)$$

$$d\,\overline{\rho\cos\theta}\,/\,dt = -\,\Omega\,\overline{\rho\sin\theta} - h\,\overline{\rho\cos\theta} \quad , \qquad (1.7)$$

$$d\,\overline{\rho\sin\theta}\,/\,dt = \Omega\,\overline{\rho\cos\theta} - h\,\overline{\rho\sin\theta} + B \quad . \qquad (1.8)$$

With a suitable linear change of variables, (1.6)-(1.8) reduce to (1.1)-(1.3), with $b=1$.

In the convective model the motion takes place between a warmer lower surface and a cooler upper surface, and is assumed to occur in the form of long rolls with fixed parallel horizontal axes and quasi-elliptical cross sections. The water wheel is therefore like a "slice" of a convective roll. The variables X, Y, Z measure the rate of convective overturning and the horizontal and vertical temperature variations. The damping results from internal viscosity and conductivity, and σ denotes the Prandtl number, while r is proportional to the Rayleigh number. Because the horizontal and vertical temperature structures differ, Y and Z need not damp at the same rate, whence b need not equal unity. The equations may afford a fair representation of real convection when r is near unity, but they become unrealistic when r is large, since real convective rolls would then break up into smaller eddies.

Although we have discussed (1.1)-(1.3) in detail elsewhere [4], we shall repeat some of the results needed for the later discussion. First, it follows that

$$\tfrac{1}{2}\, d\,[X^2 + Y^2 + (Z-\sigma-r)^2]\,/\,dt = -[\sigma X^2 + Y^2 + b(Z-\tfrac{1}{2}\sigma - \tfrac{1}{2}r)^2] + b(\tfrac{1}{2}\sigma + \tfrac{1}{2}r)^2 \quad . \tag{1.9}$$

The ellipsoid E in (X,Y,Z) - phase-space defined by equating the right side of (1.9) to zero passes through the center of the sphere S_o whose equation is $X^2 + Y^2 + (Z-\sigma-r)^2 = c^2$, and hence lies wholly in the region R_o enclosed by S_o , provided that c exceeds the maximum diameter of E . It follows from (1.9) that every point exterior to E , and hence every point exterior to S_o , has a component of motion toward the center of S_o , so that every orbit ultimately becomes trapped in R_o.

Next, if S is a surface enclosing a region R of volume V ,

$$dV/dt = -(\sigma + b + 1)V \quad . \tag{1.10}$$

Hence, following the passage of time intervals Δt, $2\Delta t$, ..., S is carried into surfaces S_1, S_2, ... enclosing regions R_1, R_2, ... of volumes V_1, V_2, ... , where $V_n \to 0$ exponentially. If $S = S_o$, $R_o \supset R_1 \supset R_2 \supset \ldots$, whereupon every orbit is ultimately trapped in a set $R_\infty = R_o \cap R_1 \cap R_2 \cap \ldots$ of zero volume. This set could be a point, a curve, a surface, or a complex of points, curves, or surfaces.

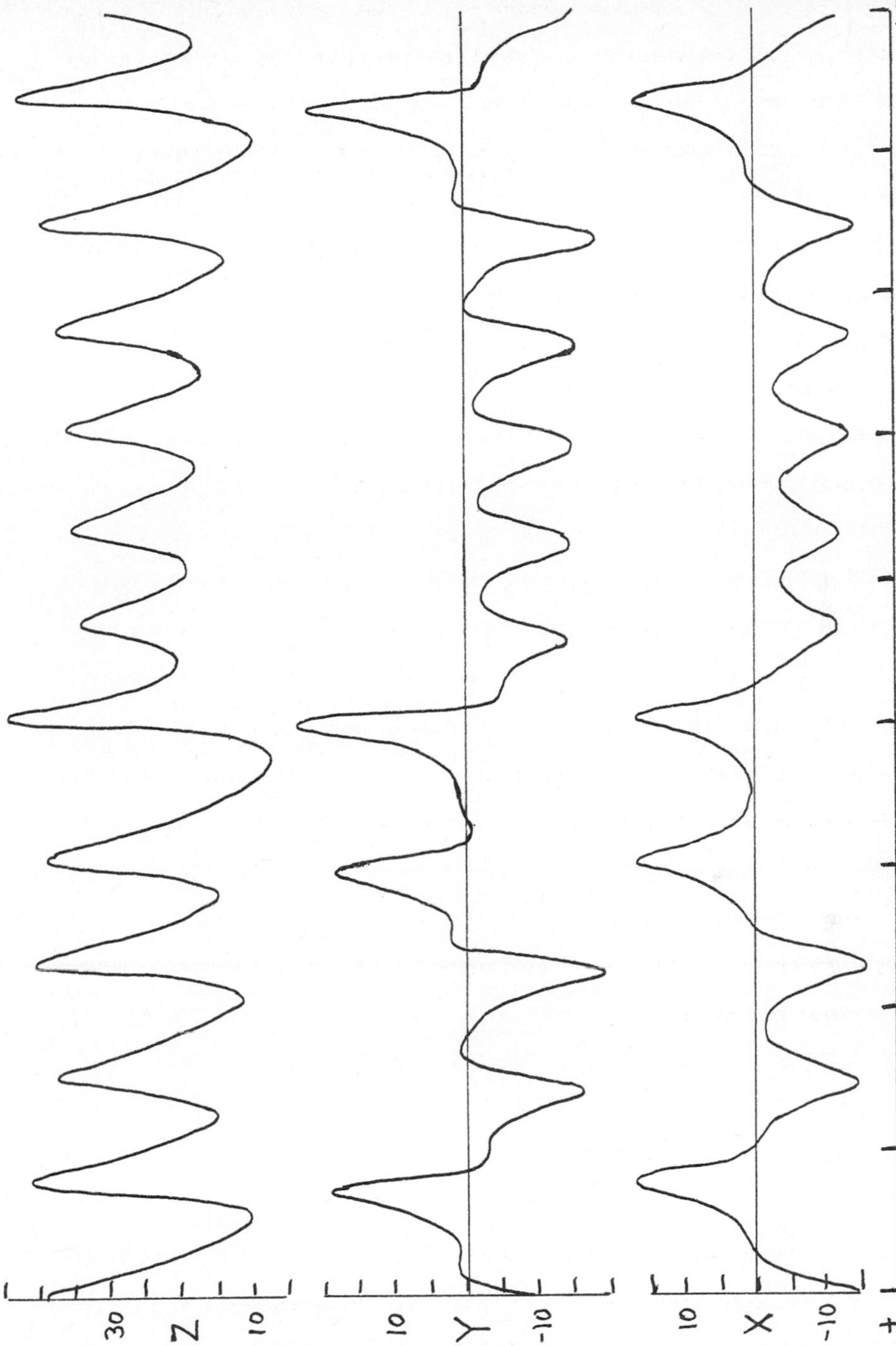

Fig. 1.1. Variation of X, Y, Z for particular solution of Eqs. (1.1)-(1.3).

The attractor set is R_∞ , or a portion of R_∞.

Eqs. (1.1)-(1.3) possess the obvious steady solution $X = Y = Z = 0$; this becomes unstable when $r > 1$. In this event there are two additional steady solutions $X = Y = \pm (br-b)^{\frac{1}{2}}$, $Z = r-1$; these become unstable when r passes its critical value

$$r_c = \sigma(\sigma + b + 3) (\sigma - b - 1)^{-1} \quad . \tag{1.11}$$

This can occur only if $\sigma > b + 1$. We shall consider only solutions where $r \geqq r_c$; these are most readily found by numerical integration.

In the first example we shall use Saltzman's values $b = 8/3$ and $\sigma = 10$, whence $r_c = 470/19 = 24.74$; as in [4] we shall use the slightly supercritical value $r = 28$. Here we note another lucky break; Saltzman used $\sigma = 10$ as a crude approximation to the Prandtl number (about 6) for water. Had he chosen to study air, he would probably have let $\sigma = 1$, and the aperiodicity would not have been discovered.

For advancing in time we use the alternating 4-cycle scheme [6], equivalent to a fourth-order Runge-Kutta scheme, with a time increment $\delta t = 0.005$. Our initial point $X = Y = 6.0$, $Z = 13.5$ lies on the parabola passing through the fixed points.

Fig. 1.1 shows the variations of X, Y, and Z from $t = 9$ to $t = 18$; the behavior seems to be typical. Evidently Z is always positive, and possesses a succession of unambiguously defined maxima and minima, spaced at fairly regular but not exactly equal intervals. In absolute value X and Y behave somewhat like Z, but they change sign rather irregularly.

Fig. 1.2 shows the projection of the orbit on the Y-Z plane, from $t = 9$ to $t = 14$. The three unstable fixed points are at 0, C, and C'. The curve spirals outward rather regularly from C' or C until it reaches a critical distance, whereupon it crosses the Z-axis and merges with the spiral about C or C'.

Following a brief initial interval the orbit should be virtually confined to the attractor set. Fig. 1.3 shows the topography of the attractor, as seen from

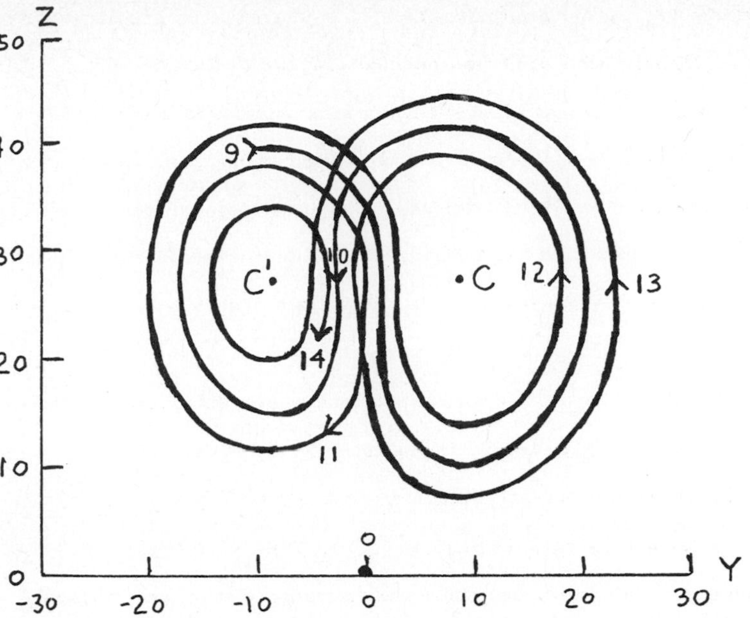

Fig. 1.2. Projection of segment of solution of Eqs. (1.1)-(1.3) on Y-Z plane. Numbers 9-14 indicate values of t. Unstable fixed points are at 0, C, and C'.

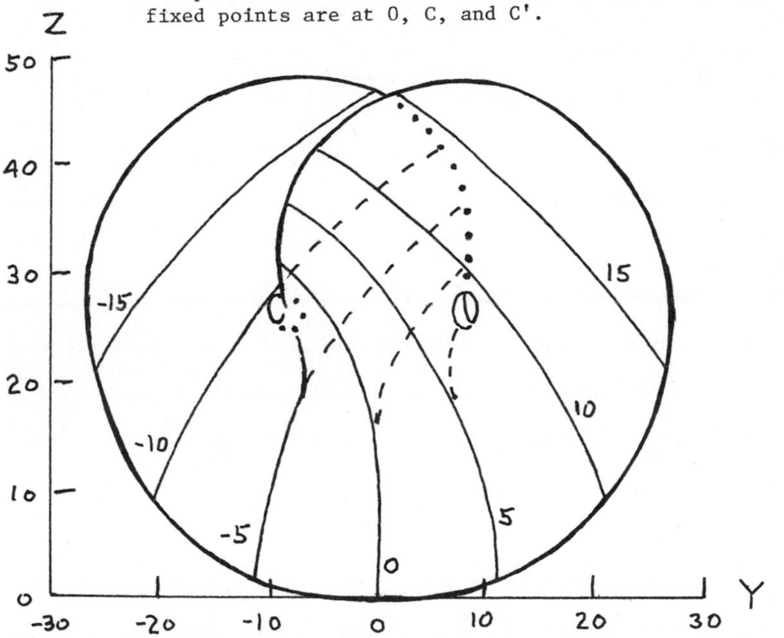

Fig. 1.3. Topography of the attractor for Eqs. (1.1)-(1.3). Solid lines are contours of X; dashed lines are contours of lower value of X where two values occur. Heavy curve is natural boundary of attractor.

the positive X-axis; the curves are contours of X. Where there are two values of X, the higher one occurs if the orbit is just completing a circuit about C. As one follows an orbit, the two sheets of the attractor appear to merge; however, this would require pairs of orbits to merge, which is impossible. Hence what appears to be a single sheet must be composed of two sheets, extremely close together, so that what looks like two merging sheets must contain four sheets. Continuing with this reasoning, we find that these four sheets must be eight sheets, then sixteen, etc, and we conclude that there is actually an infinite complex of sheets. The closure of these sheets forms the attractor set; a curve normal to the sheets would intersect it in a Cantor set. Attractors of this sort have become known as strange attractors [9].

The regularity of the spirals about C and C' in Fig. 1.2 implies that the value Z_n of Z at its nth maximum determines with fair precision the value Z_{n+1} at the following maximum, as well as indicating whether Y will change sign before the next maximum occurs. Fig. 1.4 is constructed as a scatter diagram of successive maxima of Z, but in fact reveals no scatter. It appears to define a difference equation

$$Z_{n+1} = F(Z_n) \tag{1.12}$$

whose analytic form cannot however easily be determined. We shall base our subsequent conclusions rather heavily on the appearance of Fig. 1.4, and on the assumption that it is for practical purposes a curve.

Maxima of Z are intersections of the orbit with the conic $bZ = XY$. The curve in Fig. 1.4 is therefore a form of Poincaré map; we shall call it a Poincaré curve. The conic intersects a surface of constant $Z > 0$ in a hyperbola. Since the attractor has zero volume, it intersects the hyperbola in a set of measure zero, which must be a Cantor set. The orbits emanating from this Cantor set reintersect the conic in a set whose Z-coordinates form another Cantor set. It follows that a vertical line in Fig. 1.3 intersects the Poincaré

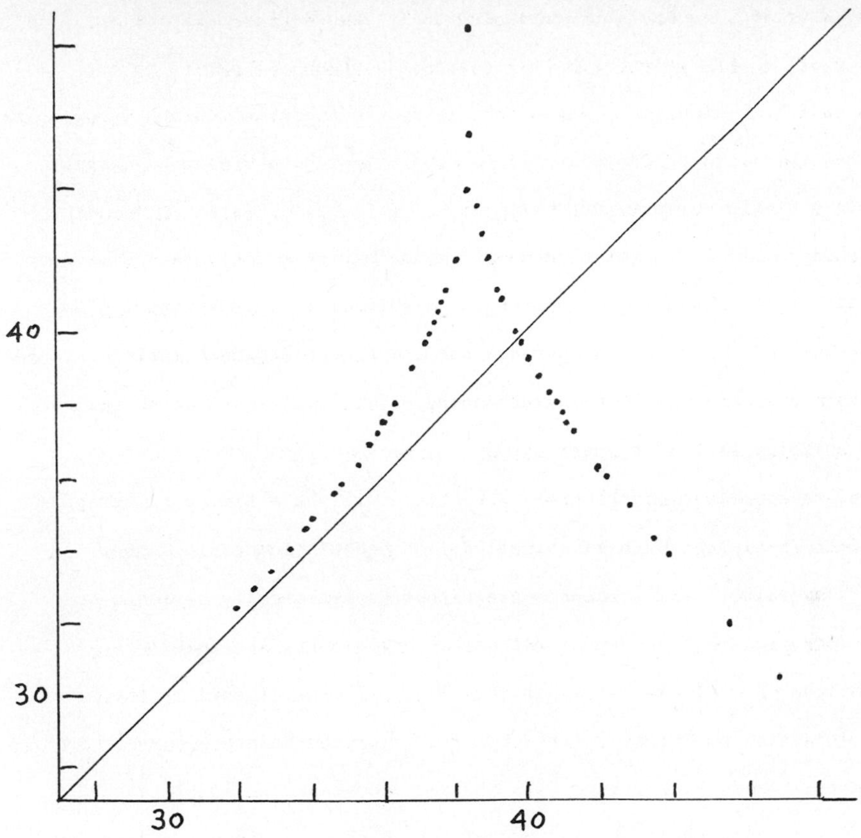

Fig. 1.4. "Scatter diagram" for successive maxima of Z for Eqs. (1.1)-(1.3), revealing lack of scatter.

curve in a Cantor set, so that the curve is really composed of a Cantor set of curves. However, the extreme horizontal distance between two curves on the same side of $Z_n = 38.5$ is about 10^{-4} times the distance between two curves on opposite sides, so that the Poincaré curve is closely approximated by a pair of merging curves, i.e., a single curve with a cusp.

A sequence $Z_1, Z_2, \ldots$ of successive maxima may be exactly periodic, i.e., $Z_m = Z_n$ with $m \neq n$. It may be asymptotically periodic, i.e., asymptotic to a periodic sequence $Z_1', Z_2', \ldots$, in which case the latter sequence must be stable. Finally it may be aperiodic. The corresponding orbits will likewise be periodic (exactly or asymptotically) or aperiodic.

Assuming reasonable smoothness of F, there will be a finite number of exactly periodic sequences of a given period, and hence a countable number altogether. Thus almost all sequences are either asymptotically periodic or aperiodic. If no periodic sequences are stable, almost all sequences are aperiodic.

If a maximum Z_n is altered slightly, Z_{n+1} will be altered by the same amount, multiplied by the slope $\lambda_n = d\,Z_{n+1}/d\,Z_n$ of the Poincaré curve. An exactly periodic sequence of period N is therefore unstable or stable according to whether or not the product Λ_N of the slopes at the N points exceeds unity in absolute value. Since it appears from Fig. 1.4 that the slope exceeds unity everywhere, there are no stable periodic sequences, and the general solution of (1.1)-(1.3) is aperiodic.

2. Aperiodicity in a Quadratic Difference Equation

In the previous section we used a difference equation, defined graphically rather than analytically, to draw conclusions regarding a system of differential equations. We now turn to an analytically defined difference equation.

The single first order equation

$$x_{n+1} = a(x_n - x_n^2) \tag{2.1}$$

to which any quadratic difference equation may be reduced by a linear change of variables, has been extensively studied as the "simplest" nonlinear equation; a comprehensive discussion is given by Guckenheimer [2]. The solutions of (2.1), like those of (1.12), may be exactly or asymptotically periodic, or aperiodic.

We shall replace (2.1) by

$$X_{n+1} = \tfrac{1}{2} X_n^2 - A \qquad , \tag{2.2}$$

where $X_n = a(1-2x_n)$ and $A = \frac{1}{2} a^2 - a$. The variable X_n is then the slope of the graph of X_{n+1} against X_n. It is evident that if $-\frac{1}{2} \leqq A \leqq 4$ and $|X_0| \leqq a$, $|X_n| \leqq a$ for all n; if also $A > 0$, $|X_n| \leqq A$ for large enough n. If

$$\Lambda_N = \prod_{n=1}^{N} X_n \qquad , \tag{2.3}$$

a periodic solution with $X_N = X_0$ is unstable or stable according to whether or not $|\Lambda_N| < 1$.

Our principal concern is with the probability P that if A and X_0 are chosen randomly from (0, 4) and (0, A), the sequence $X_0, X_1, X_2, \ldots$ will be aperiodic; specifically we are interested in whether $P = 0$ or $P > 0$. In an earlier study [5] we conjectured that $P > 0$. We are as yet unable to prove the conjecture, so we shall simply present supporting evidence, which will at times lack the rigor needed for a proof.

Our interest in this question stems from the existence of relations between difference and differential equations, as illustrated by Eqs. (1.1)-(1.3) and (1.12). We believe that the answer for a large class of difference equations is the same as the answer for (2.2), and that many systems of differential equations, including some representing physical systems, give rise to difference equations of this sort. In a sense, then, we are asking whether aperiodicity is an exceptional or a normal phenomenon.

The point where $X_n=0$ is called a <u>singularity</u>. For any A , we shall call the solution with $X_0=0$ the <u>singular</u> solution. A useful theorem [3, 5, 11] tells us that if a stable periodic solution exists, the singular solution approaches it asymptotically. A corollary is that there is at most one stable periodic solution.

We shall call a value of A periodic if a stable periodic solution exists, and aperiodic otherwise. If the (Lebesque) measure of the set of aperiodic values of A in (0, 4) exceeds zero, $P > 0$.

It is easily shown that a stable solution of period 1 (steady) exists if $-\frac{1}{2} < A < \frac{3}{2}$; this bifurcates to a period 2 which is stable if $\frac{3}{2} < A < \frac{5}{2}$, and thence to period 4, 8, ... , the sequence of intervals terminating at $A = 2.802$. Within (2.802, 4) there are some aperiodic values of A.

Numerical solutions of (2.2) suggest that if A' and A'' are distinct aperiodic values of A , the corresponding singular solutions eventually acquire opposite signs. It follows that for some intermediate value A_c , the singular solution is exactly periodic, and stable, since $\Lambda_N = 0$. Such a value of A will be called central.

By continuity there is an interval enclosing A_c where $|\Lambda_N| < 1$, whence a continuum of periodic values of A separates A' from A'' . The set of aperiodic values is therefore nowhere dense.

For example, $X_3=0$ when $A = A_c = 3.510$ and $X_0=0$, and the interval where period 3 is stable extends from $A = A_a = 3.5$ (exactly) to $A = 3.538$.
For slightly higher values of A , period 6, then period 12, etc. are stable, and for still higher values up to $A = A_e = 3.581$, the singular solution is semiperiodic of period 3, i.e., there are three nonoverlapping intervals such that $X_0, X_3, X_6, \ldots$ occupy one, $X_1, X_4, \ldots$ occupy another, and $X_2, X_5, \ldots$ occupy the other. Within the semiperiodic range there are some periodic values of A , the periods being multiples of 3. For the aperiodic but semiperiodic values, a variance spectrum would contain lines superposed on a continuun. We shall call the interval from A_a to A_e a semiperiodic band.

A similar semiperiodic band encloses each other central value of A . The band for period 1 is the entire interval $(-\frac{1}{2}, 4)$, since a completely aperiodic solution may be considered semiperiodic of period 1. Moreover, every semiperiodic band (except for period 2) is virtually a small copy of $(-\frac{1}{2}, 4)$, containing within it the same structure. Thus there are bands within bands within bands, etc. A band which lies within no other band except $(-\frac{1}{2}, 4)$ will be called <u>prime</u>; other bands will be called <u>composite</u>. The period of a composite band is obviously a composite number; the converse does not hold.

Because of the similarity of the bands, the measure of the aperiodic values of A is positive if and only if the measure of the values of A exceeding 3/2 and not contained in prime bands is positive. We might then attempt to answer our question by summing the lengths of the prime bands. Table 1 presents these for periods ≤ 7; the band for period 2 ends at 3.0874; and it is evident that the remaining bands do not fill much of the space in (3.0874, 4). However, for any large period there exist a few prime bands, located very close to prime bands of much lower period, which, although narrow, are exceptionally wide for their period. For example, of the 26,817,356,775 bands of period 41, whose average width is certainly $< 3.4 \times 10^{-11}$, one, with $A_c = 3.49788$, has a width of 1.73×10^{-7}. We have not been able to show that these exceptional bands, taken together, do not fill the space which the "normal" bands leave nearly empty.

Our conjecture that $P > 0$ was originally prompted by the observation that when a value of A in (3, 4) was chosen at random, the resulting singular solution was usually aperiodic. We must therefore note that with the usual computer precision most solutions become incorrect before 100 iterations. The inevitable round-off errors introduced in the early iterations amplify by a factor whose average may approach 2.0 per iteration, until the noise drowns the signal. Indeed, May [7] regards the computer solutions as simulations, and suggests that there may be periodicities considerably higher than 100 which the simulations fail to reveal.

To test this possibility we have repeated some of the computations with a special multiple-precision program, using as many as 500 decimal places, and carrying upper and lower bounds to the true value of X . These bounds remain close together for 1000 and sometimes 3000 iterations. For no tested values of

Table 1. Limiting values A_a, A_e, central values A_c, and widths $A_e - A_a$ of prime semiperiodic bands of period $\leqq 7$, for Eq. (2.2).

N	A_a	A_c	A_e	$A_e - A_a$
2	1.50000	2.00000	3.08738	1.58738
7	3.14943	3.14978	3.15255	0.00312
5	3.24879	3.25083	3.26672	0.01793
7	3.34791	3.34813	3.34991	0.00200
3	3.50000	3.50976	3.58066	0.08066
7	3.66458	3.66463	3.66502	0.00044
5	3.72117	3.72156	3.72466	0.00349
7	3.76958	3.76961	3.76978	0.00020
6	3.81450	3.81456	3.81503	0.00053
7	3.85428	3.85430	3.85441	0.00013
4	3.88110	3.88160	3.88552	0.00442
7	3.90740	3.90741	3.90747	0.00007
6	3.93353	3.93355	3.93369	0.00016
7	3.95436	3.95436	3.95438	0.00002
5	3.97082	3.97085	3.97108	0.00026
7	3.98363	3.98363	3.98364	0.00001
6	3.99275	3.99275	3.99277	0.00002
7	3.99819	3.99819	3.99819	0.00000

A where we had not found a periodicity less than about 30 did we discover any higher periodicities. If the interval (3, 4) is filled by semiperiodic bands, the periods must be high indeed.

What we did generally observe in these solutions was that the product Λ_N continued to increase quasi-exponentially with N . The periodic bands seem to consist of those rare values of A where, after many iterations, we suddenly encounter a value of X so close to zero that it cancels the remaining factors in Λ_N. Encountering a value which partially cancels the product, and then another value which completes the cancelation, is also possible but seems less likely.

Accordingly, for our final bit of evidence supporting our conjecture we have constructed a statistical model of the difference equation (2.2). We take $A > 0$ and choose $X_1, X_2, \ldots$ randomly and independently from the interval $(-A, A)$. Letting Λ_N again be given by (2.3), we seek the probability $P(A)$ that $|\Lambda_N| > 1$ for all N. This model cannot prove or disprove our conjecture, since successive values of X generated by (2.2) are not independent, and the distribution of these values of X in $(-A, A)$ is not uniform. The model can be regarded as highly indicative.

We find that $P(A) = 0$ if $A \leq e$, but $P(A) = 1 - A'/A > 0$ if $A > e$, where $A' < e$ is a number such that $(\log A')/A' = (\log A)/A$. For example, if $A = 4$, $b = 2$ and $P = 1/2$; if $A = 3.375$, $A' = 2.25$ and $P \approx 1/3$. To establish this result we let P_N be the probability that $|\Lambda_N| < 1$ for $n \leq N$, and note that $1 - P_1 = 1/A$, while by direct integration $P_N - P_{N+1}$ equals $1/A$ times a function of $(\log A)/A$. Hence $1 - P(A)$ and $1 - P(A')$ differ only by the factor A'/A. Since the (geometric) mean of X is <1 when $A < e$, the result follows.

The implication is that for (2.2), in the vicinity of $A = 3.375$ about one third of the values of A should be aperiodic, while near $A = 4$ about one half should be aperiodic. The more general implication is that aperiodicity is a normal phenomenon. It is remarkable that aperiodic values for (2.2) first appear at $A = 2.802$, which is so close to e. Actually the numerical solutions suggest that nearly all values of A near 4 are aperiodic; the discrepancy may occur because the model has assumed a uniform distribution of X, while in reality the larger values tend to occur more frequently.

3. Some Attractors are Stranger than Others

It is apparent that there is a wide variety of systems of equations with aperiodic general solutions. There should therefore be a wide variety of strange attractors.

Let us consider (1.1)-(1.3) for other values of the parameters. For the values of b and σ previously used, but for $r = r_c$, the Poincaré curve would look about like Fig. 1.4, but with unit slope at the fixed point. Since it would still be concave upward everywhere, no stable periodic solutions would be introduced.

For any b, r_c is large if σ is near $b+1$ or if σ is large, and there is a value

$$\sigma_m = b+1 + [2(b+1)(b+2)]^{\frac{1}{2}} \tag{3.1}$$

for which r_c assumes a minimum value r_{cm}. To keep our study manageable, we shall vary b, letting $\sigma = \sigma_m$ and $r = r_{cm}$ in all cases. It is of interest that when $b = 8/3$, the values $\sigma = 10$ and $r = 28$ used previously are not far from $\sigma_m = 9.52$ and $r_{cm} = 24.72$.

We shall again base our conclusions mainly on the Poincaré maps. For purposes of comparison we shall divide X, Y, Z by their values at C, so that the fixed points become (0,0,0), (1,1,1), and (-1,-1,1).

In our numerical integrations we have chosen initial conditions on the parabola $Y = X$, $Z = X^2$ passing through the fixed points. We assume that after passing one or two maxima of Z the orbit is close to the attractor, and we study the remainder of the solution. By suitably adjusting the initial point along the parabola, we can force the orbit to visit the rarely visited portions of the attractor.

Fig. 3.1 shows the Poincaré curves for $b = 2$, 1, and 1/2. The curve for $b = 2$ is much like Fig. 1.4, and there is no possibility of a stable periodic solution. For $b = 1$ and 1/2, singularities have appeared, at $Z = 1.75$ and $Z = 1.60$, and the possibility of periodic solutions with $|\Lambda_N| < 1$ arises. Actually these do not occur, because solutions with points close to the singularity also contain points close to the cusp.

However, for $b = 1/2$ the solution with a maximum of Z exactly at 1.60 has the eighth subsequent maximum close to 1.60. If the curve were a single curve rather than a Cantor set of curves, we could be sure that by changing b slightly we would obtain a solution of period 8 with $\Lambda_8 = 0$. Lacking this

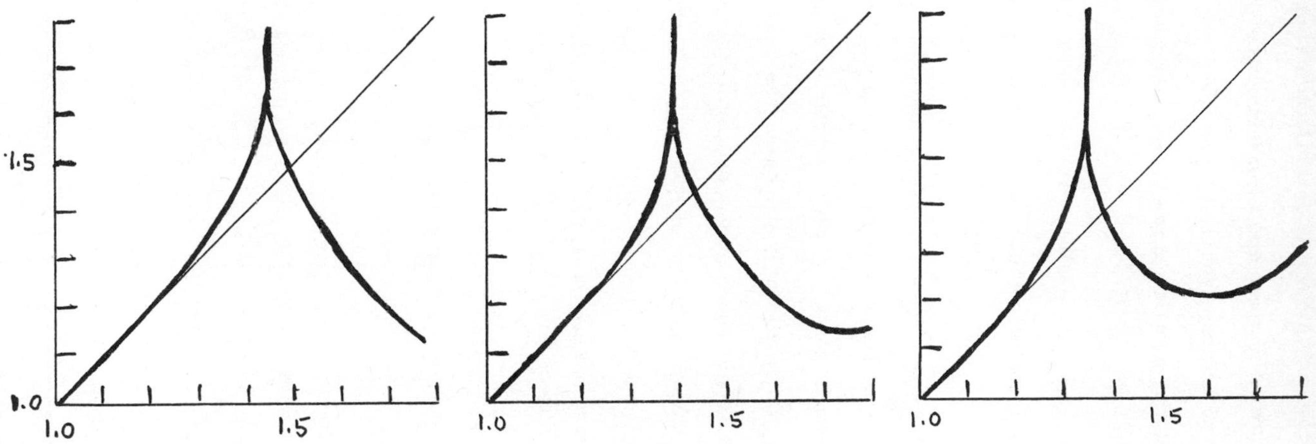

Fig. 3.1. Poincaré curves for Eqs. (1.1)-(1.3) for b = 2 (left), b = 1 (center), and b = 1/2 (right), with $\sigma = \sigma_m$ and $r = r_{cm}$.

assurance, we find for b = 1/2, by adjusting the initial conditions along the parabola, a solution where the third maximum of Z exactly hits the singularity. We then adjust b and repeat the process, until we have found the desired solution of period 8. Finally we extend the solution through several more periods to be sure that it is indeed stable.

With some search we located a stable period 8 at b = 0.498007. Varying b by intervals of 0.25×10^{-6}, we found no period 8 at b = 0.49800625, and a stable period 8 only from 0.49800650 to 0.49800750. The solution becomes a stable period 16 from 0.49800775 to 0.49800825, and is semiperiodic at 0.49800850 and 0.49800875, with apparently a stable period 24 at the latter value. At 0.49800900 the stable periodicity is gone. We have, in fact, found an extremely narrow semiperiodic band of b which evidently possesses the same structure as the semiperiodic bands of A noted in the previous section.

Lowering b to 1/4 and then 1/8, we encounter in Fig. 3.2 some possibly unexpected additional cusps. Points near or to the right of these cusps can be reached only after a nearly direct hit on the first cusp, and represent extremely rare events.

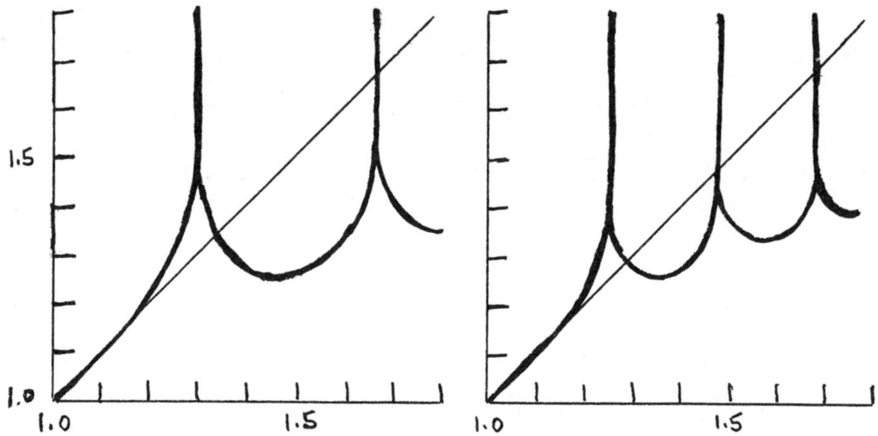

Fig. 3.2. Poincaré curves for Eqs. (1.1)-(1.3) for b = 1/4 (left) and b = 1/8 (right), with $\sigma = \sigma_m$ and $r = r_{cm}$.

For b = 1/4, points near the singularity at Z = 0.145 are preceded by points near the first cusp, but not so near as when b = 1/2. The semiperiodic bands near b = 1/4 should therefore be somewhat wider. We find, in fact, a stable period 3 from 0.2596 to 0.2609, becoming period 6 from 0.2609 to 0.2612. Another stable period 3 occurs from 0.2685 to 0.2697, becoming period 6 from 0.2697 to 0.2704. One periodic solution has one point slightly to the left of the cusp; the other has one slightly to the right.

Near b = 1/8 stable periodic solutions are abundant. In fact, at b = 0.115 the fixed point acquires a slope of -1, and for b < 0.115 period 1 is stable. Period 2 is stable from 0.120 to 0.135, period 4 is stable at 0.140, and at 0.145 the solution is semiperiodic with period 2. Periodicity disappears by 0.150. At 0.165 and 0.170 a second period 2 appears, becoming period 4 at 0.175 and 0.180. The solution is semiperiodic at 0.185, and aperiodic at 0.190.

We find, then, that the range of b from 0.1 to 1.0 is teeming with semiperiodic bands. It seems probable that, as with the difference equation (2.2), any two aperiodic values of b are separated by a semiperiodic band, in most cases very narrow. Below 0.25 most values of b are periodic, and above 0.5 most are aperiodic. In any event, aperiodicity is not an exceptional phenomenon, even below b = 1.1. Above b = 1.1 the singularity disappears, and all values of b are aperiodic.

Since period 2 is stable at b = 1/8, one may wonder how Fig 3.2 can show a curve instead of just two points. When the general solution is aperiodic, say at b = 0.15, an arbitrary orbit rapidly approaches the attractor set. At b = 1/8 such an orbit approaches a set which is essentially an analytic extension of the attractor set from higher values of b . Only somewhat later does it become trapped by the stable periodic orbit, which is the true attractor. Fig. 3.2 describes its behavior in the meantime.

A point on the Poincaré map corresponds to a segment of an orbit between two maxima of Z . For b = 2, as with Figs. 1.3 and 1.4, points to the right or left of the cusp correspond to segments which do or do not cross from one wing of the attractor (the region around C or C') to the other. The cusp corresponds to an orbit which makes a direct hit on the origin and terminates. Orbits emanating from the origin form a natural boundary for the attractor, and leave holes surrounding C and C'.

For b = 1 or b = 1/2, where a singularity occurs, the orbits emanating from the origin still form natural boundaries, but the holes at C and C' , as viewed from the X-axis, are bounded by another orbit which corresponds to the singularity. Lines parallel to the X-axis are tangent to the surface at the edge of a hole, and away from the edge there are two values of X for a given Y and Z , on orbits corresponding to points to the left and right of the singularity. As one continues around C or C' , these two orbits appear to merge, i.e., the curved surface becomes folded, while they also appear to join up with an orbit from the other wing.

At b = 1/4, points to the right of the second cusp correspond to orbits, including those emanating from the origin, which cross the X-Y and Y-Z planes and then cross back again in descending from a maximum of Z ; seen from the Z-axis they spiral downward. The second cusp itself corresponds to another direct hit on the origin; hence the two cusps must have equal heights. A topographic map of the attractor would, in some locations, have to show four values of X , which would be on orbits proceeding from a large maximum and a small maximum of Z in one wing of the attractor, and two distinct intermediate maxima in the other wing. At b = 0.4, where the second cusp first appears, the trajectory emanating from the origin would run back into it; above and below b = 0.4, the attractors are thus topographically distinct.

Strange attractors appear to be characteristic of forced dissipative systems with aperiodic general solutions, such as systems describing turbulent flow. Presumably the attractor can become stranger as the number of variables increases. What we have shown is that we need not go beyond three equations, nor even change the form of the equations, to find more complicated attractors than the one which we originally presented.

Acknowledgement. I wish to thank Dr. James Curry for many stimulating conversations, and for critically reading the manuscript. The work was supported by the Climate Dynamics Program of the National Science Foundation under Grant NSF-g 77 10093 ATM.

REFERENCES

1. V. Bjerknes, Das Problem der Wettervorhersage, betrachtet vom Standpunkt der Mechanik und der Physik, Meteorol. Zeit. (1904), 107.

2. J. Guckenheimer, On bifurcation of maps of the internal, Invent. Math., 39 (1977), 165-178.

3. G. Julia, Mémoire sur l'itération des fonctions rationelle, J. Math. Pures. Appl., 4 (1918), 47-245.

4. E.N. Lorenz, Deterministic nonperiodic flow, J. Atmos. Sci., 20 (1963), 130-141.

5. E.N. Lorenz, The problem of deducing the climate from the governing equations, Tellus, 16 (1964), 1-11.

6. E.N. Lorenz, An N-cycle time-differencing scheme for stepwise numerical integration. Mon. Weather Rev., 99, 644-648.

7. R.M. May, Simple mathematical models with very complicated dynamics, Nature, 261 (1976), 459-467.

8. L.F. Richardson, Weather prediction by numerical process. Cambridge Univ. Press (1922).

9. D. Ruelle and F. Takens, On the nature of turbulence, Commun. Math. Physics, 20 (1971), 167-192.

10. B. Saltzman, Finite amplitude free convection as an initial value problem -I, J. Atmos. Sci., 19 (1962), 329-341.

11. D. Singer, Stable orbits and bifurcation of maps of the interval, SIAM J. Appl. Math., 35 (1978), 260-267.

12. N. Wiener, Nonlinear prediction and dynamics. Proc. 3rd Berkeley Sympos. Math. Statistics and Probability, Vol. 3 (1956), 247-252.

MASSACHUSETTS INSTITUTE OF TECHNOLOGY

On the Geometry of the Liapunov-Schmidt Procedure[†]

Jerrold E. Marsden*

1. Introduction

One of the most commonly used procedures in bifurcation theory is the Liapunov-Schmidt procedure (which we review in §2). However, in differential topology this procedure is also regularly used, but of course, under a different name, namely "transversality." The aim of this short note is to make this link explicit.

There are two reasons for geometrizing the Liapunov-Schmidt procedure. First of all, it is useful in some applications the author, A. Fischer and V. Moncrief have made to relativity (work in progress). Secondly, the dynamic analogue of the procedure, namely center manifold theory, already has a geometric flavor (i.e. it makes intrinsic sense on manifolds), so it is natural to bring the classical Liapunov-Schmidt procedure in line with it.

2. Review of the Liapunov-Schmidt Procedure

Let X and Y be Banach spaces and $f: X \times R^p \to Y$ a C^k map, $k \geq 1$. Let $D_x f(x,\lambda)$ be the (Fréchet) derivative of f with respect to x, a continuous linear map of X to Y. Let $f(x_0,\lambda_0) = 0$ and let

$$X_1 = \ker D_x f(x_0,\lambda_0).$$

[†]The lectures presented by the author are not reproduced here since that material is available in J. Marsden, Qualitative Methods in Bifurcation Theory, Bull. Am. Math. Soc. 84 (1978), 1125-1148, R. Abraham and J. Marsden, *Foundations of Mechanics* , Second Edition, Addison Wesley (1978), and in J. Marsden and M. McCracken, The Hopf Bifurcation and its Applications, Springer Applied Math Sciences #19 (1976).

*Partially supported by the National Science Foundation.

Assume X_1 is finire dimensional with a complement X_2 so that $X = X_1 \oplus X_2$. Also, assume

$$Y_1 = \text{Range } D_x f(x_0,\lambda_0)$$

is closed and has a finite-dimensional complement Y_2. In other words, $D_x f(x_0,y_0)$ is a Fredholm operator. Write $Y = Y_1 \oplus Y_2$ and let $P: Y \to Y_1$ be the projection. By the implicit function theorem, the equation

$$Pf(x_1 + x_2,\lambda) = 0$$

has a unique solution $x_2 = u(x_1,\lambda)$ near x_0,λ_0, where $x = x_1 + x_2 \in X = X_1 \oplus X_2$. Thus, the equation $f(x,\lambda) = 0$ is equivalent to the *bifurcation equation*

$$(I - P)f(x_1 + u(x_1,\lambda),\lambda) = 0,$$

a system of dim Y_2 equations in dim X_1 unknowns. This reduction of $f(x,\lambda) = 0$ to the bifurcation equation is the *Liapunov-Schmidt procedure.*

For purposes of this procedure alone, the assumption that X_1 and Y_2 are finite dimensional is, of course, irrelevant. This is made with the theory of Fredholm operators waiting in the wings. Similarly, the parameter space R^p may be replaced by a Banach space Z. In fact, the parameter is just "along for the ride."

3. A Topological Procedure

Let X and Z be Banach manifolds, Y a Banach space and $f: X \times Z \to Y$ be a C^1 map. (More generally, one can replace $X \times Z$ by a fiber bundle over Z). We are interested in solving the equation

$f(x,\lambda) = 0$ for $(x,\lambda) \in X \times Z$. Let (x_0,λ_0) be a known solution and let

$$X_1 = \ker D_x f(x_0,\lambda_0)$$

and assume X_1 splits; i.e. $T_{x_0}X = X_1 \oplus X_2$ for a closed subspace $X_2 \subset T_{x_0}X$. Let

$$Y_1 = \text{Range } D_x f(x_0,\lambda_0)$$

and assume Y_1 is closed and splits; i.e. $Y = Y_1 \oplus Y_2$ for a closed subspace Y_2. (This, of course, involves a choice of Y_2, as it did above.) Let $P: Y \to Y_1$ be the projection.

The map f is, for fixed λ_0, transversal to the subspace Y_2 at (x_0,λ_0). Therefore, in a neighborhood of (x_0,λ_0),

$$S_P = \{(x,\lambda) \mid Pf(x,\lambda) = 0\}$$

is a smooth submanifold of $X \times Z$ tangent to $X_1 \times T_{\lambda_0}Z$ at (x_0,λ_0). [In the notation of §2, $S_P = \{(x_1 + u(x_1,\lambda),\lambda)\}$].

Let f_P denote the restriction of f to S_P. Clearly $f(x,\lambda) = 0$ iff $(I - P)f_P(x,\lambda) = 0$ iff $f_P(x,\lambda) = 0$. The later condition is the geometric version of the bifurcation equation. It has proven to be useful, at least to the author.

4. More General

We can allow Y to be an arbitrary Banach manifold to clarify the choices involved. (I don't know any examples where this is necessary or useful.) Now we fix $y_0 \in Y$ and attempt to solve $f(x,\lambda) = y_0$ near a known solution (x_0,y_0). Let X_1, X_2 be as above and let $T_{y_0}Y = Y_1 \oplus Y_2$ where $Y_1 = \text{Range } D_x f(x_0,\lambda_0)$.

Now choose a submanifold $M \subset Y$ tangent to Y_2 at y_0. The procedure only depends on this choice; it is analogous to the choice in §2 of a *linear* complement to Y_1.

Again, f is transversal to M (with λ a parameter), so

$$S_M = \{(x,\lambda) \mid f(x,\lambda) \in M\}$$

is a submanifold of $X \times Z$. Let f_M be the restriction of f to S_M, regarded as a map of S_M to M. Then the obvious assertion that

$$f(x,\lambda) = y_0 \in Y \quad \text{iff} \quad f_M(x,\lambda) = y_0 \in M$$

is the abstract Liapunov-Schmidt procedure.

5. A Sample Calculation

In the usual Liapunov-Schmidt theory of §2, to analyze zeros of the map

$$g(x_1,\lambda) = (I - P)f(x_1 + u(x_1,\lambda),\lambda)$$

we need to compute its derivatives at (x_{10},λ_0) (x_{10} is the first component of x_0). It is usually assumed that $D_\lambda f(x_0,\lambda_0) = 0$. Thus (x_{10},λ_0) is a critical point for g. By implicit differentiation one finds that

$$D^2 g(x_{10},\lambda_0) = (I - P)D^2 f(x_0,\lambda_0)$$

(The right hand side appropriately restricted.)

In the context of §3, it is clear that if $D_\lambda f(x_0,\lambda_0) = 0$ then (x_0,λ_0) is a critical point (in the sense of zero derivative) of f_P. Moreover, it is now *obvious* from the fact that Hessians are well-defined at critical points that $D^2f_P(x_0,\lambda_0) = (I - P)D^2f(x_0,\lambda_0)$ restricted to $T_{(x_0,z_0)}S_P \times T_{(x_0,z_0)}S_P = (X_1 \times T_{\lambda_0}Z) \times (X_1 \times T_{\lambda_0}Z)$; i.e. we recover the same conclusion as above. The procedure can, of course, be repeated to obtain a formula for D^kf_P if the k-1 jet of f_P vanishes, as is well-known in bifurcation theory. Note that once the structure of the zeros of g is found, this still has to be lifted via the graph of u to obtain the zeros of f. In the geometric setting this is not necessary; the zeros of f_P *are* the zeros of f.

While these results are essentially nothing more than new language for well-known material, the geometric setting seems to clarify and even simplify what is going on.

6. Potential Operators and Vector Fields

If $X = Y$ are Hilbert spaces and the original map f in Section 2 is the gradient of a potential function ϕ in the variable x, then the reduced function $g(x_1,\lambda) = (1-P)f(x_1 + u(x_1,\lambda),\lambda))$ is also a gradient, with a modified potential $\bar{\Phi}$ (depending on u); see, for instance. P. Rabinowitz, J. Funct. An. 25 (1977) 412-424. We wish to explain the geometric meaning of this. In such cases, the function f may be thought of as a vector field on a manifold X, depending on the parameters λ.

For vector fields the procedure in §3 can be refined somewhat. At a zero (x_0,λ_0) of a (parametrized) vector field $f: X \times Z \to TX$, the derivative $D_xf(x_0,\lambda_0)$ makes intrinsic sense as a linear map of $T_{x_0}X$ to itself. One can now seek an invariant manifold S_{λ_0} for f tangent to the subspace $X_1 = \ker D_xf(x_0,\lambda_0)$. To do this for nearby λ as well, it is convenient to suspend f to the vector field $\bar{f}: X \times Z \to T(X \times Z)$ by setting $\bar{f}(x,\lambda) = (f(x,\lambda),0)$ and find an invariant manifold $\bar{S}$ associated to $\ker D\bar{f}(x_0,\lambda_0)$; one can then take $S_\lambda = \bar{S} \cap X \times \{\lambda\}$, an invariant manifold for each λ. One may regard S_λ as implicitly defined in the same way as the function $u(x_1,\lambda)$ in the Liapunov-Schmidt procedure is implicitly defined.

Zeros of f near (x_0,λ_0) necessarily lie on $\bar{S}$, so the problem reduces to finding zeros of $f|\bar{S}$, the analogue of the bifurcation equation. For finding fixed points , this is a geometric formulation of the Liapunov-Schmidt procedure. The fact that we are dealing with vector fields entails that the choice of Y_2 (or M in Section 4) is now automatically made; both S_P and M are now replaced by $\bar{S}$.

In order to capture dynamic bifurcations as well as static ones, it is necessary to enlarge $\bar{S}$ to the full center manifold, as is explained in, for example, J. Marsden and M. McCracken, *The Hopf Bifurcation and Its Applications*, Springer Appl. Math. Sciences #19 (1976). (For operators with real eigenvalues, such as potential operators, $\bar{S}$ equals the center manifold.)

The fact that the reduction of a potential operator by the Liapunov-Schmidt procedure results in a potential operator is now clear. In fact, if one uses the space $\bar{S}$, a modification of ϕ is *not necessary*; one needs only to restrict it to $\bar{S}$. This is because of the following obvious fact: the restriction of a gradient vector field to an invariant submanifold is a gradient vector field whose potential is the restriction of the original one; i.e. $(\nabla\phi)|S = \nabla(\phi|S)$ if $\nabla\phi$ is tangent to S.

INTEGRABLE SYSTEMS AND ALGEBRAIC CURVES

H. P. McKean *

Courant Institute of Mathematical Sciences

New York

The voice of him that crieth in the wilderness, Prepare ye the way of the Lord, make straight in the desert a highway for our God. Every valley shall be exalted and every mountain and hill shall be made low: and the crooked shall be made straight and the rough places plain. Isaiah 40: 3-4.

PREFACE

The recent surprising discoveries linking special many-body problems and shallow water waves to algebraic geometry have their mathematical roots in the theory of integrable systems created by Hamilton and Jacobi , and in the ultimately dominant geometrical ideas injected into complex function theory by Riemann. The purpose of these lectures is to describe this connection. It is a pleasure to thank the Canadian Mathematical Congress for its invitation to do so at its June, 1978 meeting at Calgary. I also wish to thank P. van Moerbeke, J. Moser, and E. Trubowitz with whom most of the subjects treated below have been discussed, much to my profit. The present account differs somewhat from the actual lectures with a view to affording a more comprehensive picture of a slightly narrowed field. The principal changes are 1) the omission of the sine-Gordon equation; 2) the inclusion of the Toda lattice and the Boussinesq equation, as I had planned but had not sufficient time to present; 3) the addition of an outline of the group-theoretical picture of the Calogero and Toda lattices.

Calgary and So. Landaff, July 1978.

---

* The research for this paper was carried out at the Courant Institute of Mathematical Sciences, Grant NSF-MCS-76-07039.

CONTENTS

ACKNOWLEDGEMENT. It is a pleasure to thank Constance Engle for her customary first-class typing.

1.

1. LATTICES AND SHALLOW WATER WAVES

1.1 The Simple Harmonic Lattice.

Let n particles of equal mass m = 1 be placed in a circle and connected, each to its nearest neighbors, by springs obeying Hooke's law with common spring constant K = 1, say, and let q_i, p_i $(i=1,\ldots,n)$ be their positions and momenta. The total energy is $H = p^2/2 + U$ with $p^2 = p_1^2 + \cdots + p_n^2$ and $2U = (q_1-q_2)^2 + (q_2-q_3)^2 + \cdots + (q_n-q_1)^2$, so*

$$\begin{pmatrix} \dot{q} \\ \dot{p} \end{pmatrix} = \begin{pmatrix} 0 & 1 \\ -1 & 0 \end{pmatrix} \begin{pmatrix} \partial H/\partial q \\ \partial H/\partial p \end{pmatrix},$$

i.e.,

$$\dot{q}_i = p_i, \quad \dot{p}_i = q_{i-1} - 2q_i + q_{i+1} \qquad (i=1,\ldots,n)$$

with $q_0 = q_n$ and $q_{n+1} = q_1$. The quadratic form U may be brought to principal axes by the substitution $q_i - q_{i+1} = q_i'$ $(i=1,\ldots,n)$. The equations of motion decouple:

$$(q_i')^{\bullet} = p_i', \quad (p_i')^{\bullet} = -q_i' \qquad (i=1,\ldots,n),$$

from which it is plain that the energies

$$H_i = \frac{1}{2}[(p_i')^2 + (q_i')^2] \qquad (i=1,\ldots,n)$$

of the individual modes are constants of the motion and that each mode executes a simple harmonic motion:

$$q_i'(t) = \sqrt{2H_i}\,\sin t, \quad p_i'(t) = \sqrt{2H_i}\,\cos t$$

up to a phase; especially, the motion is periodic on the n-dimensional torus determined by fixing the individual $H_i = h_i$ $(i=1,\ldots,n)$. This is the simplest imaginable illustration of an integrable Hamiltonian system, which is to say that it admits

---

* $\bullet$ signifies differentiation with respect to time.

$$n = \frac{1}{2} \times \text{the number of degrees of freedom}$$

independent integrals H_j $(j=1,\ldots,n)$ and that the individual motions $\dot{q} = \partial H_j/\partial p$, $\dot{p} = -\partial H_j/\partial q$ so produced commute; see, e.g., ARNOLD [1974] for general information on this subject. The purpose of the present lectures is to describe a number of more complicated examples of this state of affairs arising in diverse places, e.g., many-body problems, shallow water theory, etc. CHU-SCOTT-McLAUGHLIN [1973] provide an excellent review of the subject as of that date. The subject is full of surprising connects with, e.g., isospectral classes of Jacobi matrices or Hill's operators, complex function theory, and algebraic geometry, but more of this below.

1.2 The Fermi-Pasta-Ulam Lattice

Now it is a part of the conventional wisdom that adding terms of degree ≥ 3 in q_i $(i=1,\ldots,n)$ to the Hamiltonian H of §1.1 will produce a sharing and an ultimate equipartition of energy between the modes. Let $H = p^2/2 + U$ with

$$U = u(q_2 - q_1) + u(q_3 - q_2) + \cdots + u(q_1 - q_n) ,$$

so that

$$\dot{q}_i = p_i , \quad \dot{p}_i = f(q_{i+1} - q_i) - f(q_i - q_{i-1}) \quad (i=1,\ldots,n)$$

with $f = -u'$. FERMI-PASTA-ULAM [1965] studied numerically the case $f = ax + bx^2$ with $n = 64$ and $q_0 = q_n = 0$, discovering, to their astonishment, not an ultimate equipartition of energy, but a quasi-periodic behavior like that of an integrable system. ZAKHAROV [1973] conjectured that while the lattice is not itself an integrable system, it is so close to one that the discrepenacy cannot be detected in practical computer time. Be that as it may, it is now known that a number of such many-body problems are integrable, the most striking being the lattice of TODA [1967] $[f(x) = a(1 - e^{-bx})]$

and the many-body problem of CALOGERO [1971] and MOSER [1975] $[f(x) = x^{-3}]$.

1.3 Approximation by Partial Differential Equations

The simplest example is provided by the lattice of §1.1: if $q_i(nt)$ is well approximated for $n \uparrow \infty$, moderate $t > 0$, and $i/n \to x$ by the values of a smooth function $q(t,x)$ of period 1, then the equations of motion $n^2\ddot{q}_i = q_{i-1} - 2q_i + q_{i+1}$ $(i=1,\ldots,n)$ go over into the wave equation $\partial^2 q/\partial t^2 = \partial^2 q/\partial x^2$, which is itself an inte-integrable system, only now in an infinite number of degrees of freedom. The total energy is*

$$H = \frac{1}{2} \int_0^1 [p^2 + (q')^2]\, dx ,$$

the equations of motion are

$$\begin{pmatrix} \dot{q} \\ \dot{p} \end{pmatrix} = \begin{pmatrix} 0 & 1 \\ -1 & 0 \end{pmatrix} \begin{pmatrix} \partial H/\partial q \\ \partial H/\partial p \end{pmatrix} = \begin{pmatrix} 0 & 1 \\ -1 & 0 \end{pmatrix} \begin{pmatrix} -q'' \\ p \end{pmatrix} = \begin{pmatrix} p \\ q'' \end{pmatrix},$$

and counting the number of degrees of freedom as $2\times\infty$, there is a self-evident ∞ of integrals of the motion:** $H_0 = \frac{1}{2}|\hat{p}(0)|^2$ and, for $j \geq 1$,

$$H_j^- = \frac{1}{2}|\text{real part of } \hat{p}(j)|^2 + 2\pi^2 j^2 |\text{real part of } \hat{q}(j)|^2$$

$$H_j^+ = \frac{1}{2}|\text{imag part of } \hat{p}(j)|^2 + 2\pi^2 j^2 |\text{imag part of } \hat{q}(j)|^2 ;$$

indeed, the substitution $qp \to \hat{q}\hat{p}$ reduces the motion to that of infinitely many uncoupled (complex) harmonic oscillators:

$$\hat{q}(j)^{\bullet} = \hat{p}(j) , \quad \hat{p}(j)^{\bullet} = -4\pi^2 j^2 \hat{q}(j) \qquad (j \in Z) .$$

---

* ' signifies differentiation with respect to x.
$\partial H/\partial q$ is the gradient in function space; see §1.4 below.

** $\hat{p}(j) = \int_0^1 p(x) \exp(-2\pi\sqrt{-1}\, jx)\, dx$, *etc*.

A similar approximation is believed to take place for the nonlinear lattice

$$\ddot{q}_i = f(q_{i+1} - q_i) - f(q_i - q_{i-1}) \qquad (i=1,\ldots,n).$$

Let $f(0) = 0$, $f'(0) = a > 0$, $f''(0) = b \neq 0$, and $q_i(nt) = q(t,i/n)$ as before. Then, up to terms in n^{-3},

$$\begin{aligned}\frac{\partial^2 q}{\partial t^2} = n^2\ddot{q}_i &= an^2[q_{i-1} - 2q_i + q_{i+1}] \times [1 + b(q_{i+1} - q_{i-1})] \\ &= a\left[\frac{\partial^2 q}{\partial x^2} + \frac{1}{12n^2}\frac{\partial^4 q}{\partial x^4}\right] \times \left[1 + b\left(\frac{2}{n}\frac{\partial q}{\partial x} + \frac{1}{3n^3}\frac{\partial^3 q}{\partial x^3}\right)\right] \\ &= a\left[\frac{\partial^2 q}{\partial x^2} + \frac{2b}{n}\frac{\partial q}{\partial x}\frac{\partial^2 q}{\partial x^2} + \frac{1}{12n^2}\frac{\partial^4 q}{\partial x^4}\right],\end{aligned}$$

and there are 3 levels of approximation:

1) $$\frac{\partial^2 q}{\partial t^2} = a\frac{\partial^2 q}{\partial x^2} + O(n^{-1}),$$

<u>i.e.</u>, the wave equation;

2) $$\frac{\partial^2 q}{\partial t^2} = a\left[1 + \frac{2b}{n}\frac{\partial q}{\partial x}\right]\frac{\partial^2 q}{\partial x^2} + O(n^{-2}),$$

which is subject to shocks and so rejected as not reflecting the actualities of the lattice;

3) $$\frac{\partial^2 q}{\partial t^2} = a\frac{\partial}{\partial x}\left[\frac{\partial q}{\partial x} + \frac{b}{n}\left(\frac{\partial q}{\partial x}\right)^2 + \frac{1}{12n^2}\frac{\partial^3 q}{\partial x^3}\right] + O(n^{-3}),$$

which is accepted provisionally, subject to further investigation.

3) is equivalent to

4) $$\frac{\partial^2 q}{\partial t^2} = \frac{\partial^2}{\partial x^2}\left[q + q^2 + \frac{\partial^2 q}{\partial x^2}\right]$$

by scaling, an extra $\partial/\partial x$, and the substitution $\partial q/\partial x \to q$.

4) is the equation favored by BOUSSINESQ [1872] for the description of long waves in shallow water. It permits waves moving in both directions; for uni-directional waves $q(t, x \pm t)$ of slowly changing

shape,* it reduces to

$$5)\qquad \pm\frac{\partial q}{\partial t} = q\,\frac{\partial q}{\partial x} + \frac{1}{2}\,\frac{\partial^3 q}{\partial x^3}\ ,$$

which is the equation introduced by KORTEWEG-DE VRIES [1895] to describe the observations of SCOTT-RUSSELL [1844] upon solitary waves in a canal; see, also RAYLEIGH [1876], and, for a critical discussion of shallow water wave theory, KRUSKAL [1975] and WHITHAM [1974]. 5) is known to be integrable. ZAKHAROV [1973] conjectured that 4) is, too, and that this explains the observations of FERMI-PASTA-ULAM [1965] cited above. The plan of the present lectures is to deal chiefly with 5) as an integrable system. The Toda lattice is similar and will be described more briefly. 4) is harder, its study being only in its infancy, but I will come back to it.

1.4 Aside on Calculus in Function Space

A smooth numerical function $h(q)\colon R^n \to R^1$ has a gradient $\partial h/\partial q = [\partial h/\partial q_i\colon i=1,\dots,n]$ with values in R^n and a Hessian $\partial^2 h/\partial q^2 = [\partial^2 h/\partial q_i \partial q_j\colon 1 \le i,j \le n]$ which is a symmetric map of R^n into itself, and

$$h(q + \varepsilon p) = h(q) + \varepsilon\,\frac{\partial h}{\partial q}\cdot p + \frac{\varepsilon^2}{2}\,p\cdot\frac{\partial^2 h}{\partial q^2}\,p + o(\varepsilon^2)$$

p being any direction in R^n. The same considerations apply in function spaces such as** $C^\infty_\downarrow$. I adopt an unconventional notation which parallels that for R^n. The technical assumptions are kept deliberately vague. Let $H\colon C^\infty_\downarrow \to R^1$ be a numerical function. The gradient of H at $q \in C^\infty_\downarrow$ is a hopefully pleasant function $\partial H/\partial q(x)$ such that

---

* $\partial^2 q/\partial t^2$ negligible is the construction to be put upon this phrase.

** $C^\infty_\downarrow$ is the class of infinitely differentiable functions from R^1 to itself vanishing rapidly at $\pm\infty$.

$$H[q+\varepsilon p] = H[q] + \varepsilon \int_{-\infty}^{\infty} \frac{\partial H}{\partial q(x)} p(x)\, dx + o(\varepsilon)$$

for every function $p \in C_{\downarrow}^{\infty}$; in particular, if X is a nice vector field on $C_{\downarrow}^{\infty}$, then the action of the flow $\partial q/\partial t = Xq$ upon H is expressed by the chain rule:

$$H^{\bullet} = XH = \int \frac{\partial H}{\partial q} Xq\, dx = \int \frac{\partial H}{\partial q} \dot{q}\, dx .$$

The Hessian of H is defined similarly: it is a hopefully pleasant symmetric operator $\partial^2 H/\partial q^2$ with kernel $\partial^2 H/\partial q(\xi)\partial q(\eta)$ such that

$$H[q+\varepsilon p] = H[q] + \varepsilon \int \frac{\partial H}{\partial q(\xi)} p(\xi)\, d\xi + \frac{\varepsilon^2}{2} \iint p(\xi) \frac{\partial^2 H}{\partial q(\xi)\partial q(\eta)} p(\eta)\, d\xi\, d\eta + o(\varepsilon^2) ;$$

for example, if

$$H_2 = \int \left[\tfrac{1}{2} q^3 + \tfrac{1}{4} (q')^2 \right] dx ,$$

then $\partial H_2/\partial q = (3/2)q^2 - (1/2)q''$ and* $\partial^2 H/\partial q^2 =$ (multiplication by $3q) - (1/2)\, D^2$.

---

* $Df = f'$.

2. KORTEWEG-DEVRIES SOLITARY WAVES AND INTEGRALS OF THE MOTION

2.1 The Solitary Wave

The next task is to report upon a remarkable feature of the Korteweg-de Vries equation on the line, namely the existence of interacting solitary waves which preserve their form through collisions. The solitary wave was first described by SCOTT-RUSSELL [1844]:

> "I was observing the motion of a boat which was rapidly drawn along a narrow channel by a pair of horses, when the boat suddenly stopped — not so the mass of water in the channel which it had put in motion; it accumulated round the prow of the vessel in a state of violent agitation, then suddenly leaving it behind, rolled forward with greater velocity, assuming the form of a large solitary elevation, a rounded, smooth and well-defined heap of water, which continued its course along the channel apparently without change of form or diminution of speed. I followed it on horseback, and overtook it still rolling on at a rate of some eight or nine miles an hour, preserving its original figure some thirty feet long and a foot to a foot and a half in height. Its height gradually diminished, and after a chase of one or two miles I lost it in the windings of the channel. Such, in the month of August 1834, was my first chance interview with that singular and beautiful phenomenon... "

The mathematical description of the solitary wave is easy. The equation is taken in the form $\partial q/\partial t = 3qq' - (1/2)q'''$ with $q \in C_{\downarrow}^{\infty}$, say, and a traveling wave $q(t,x) = q(x - ct)$ progressing to the right is sought. The equation reduces to $cq' + 3qq' - (1/2)q''' = 0$, which is easily integrated to obtain

$$q(x - ct) = - c \, \mathrm{ch}^{-2}[\sqrt{(c/2)} \, (x - ct)] \, .$$

The important feature <u>speed</u> = <u>amplitude</u> is emphasized: little waves travel slowly, large waves quickly.

2.2 Solitons: Numerical Experiments

KRUSKAL-ZABUSKY [1965] made an extensive numerical study of the equation with special attention to trains of solitary waves. The results are striking. Let the initial disturbance be the approximate sum of two solitary waves with speeds $a < b$, widely separated so as not to interact at first, the big wave being placed to the left of the small wave. The left-hand wave moves more quickly and catches up to the small wave. A complicated collision takes place, and then comes the surprise: after a sufficient length of time, the big wave emerges in approximately its original shape and moves rapidly away to the right, leaving the little wave in <u>its</u> original shape in its wake, the only change being the introduction of a <u>phase shift</u> to the separation $(b - a)t + d(\pm)$ $(t \to \pm\infty)$ between them; see Fig. 1 depicting $-q$ at a) $t = 0$, b) moderate t, and c) $t \uparrow \infty$.

FIG. 1

a)

b)

c)

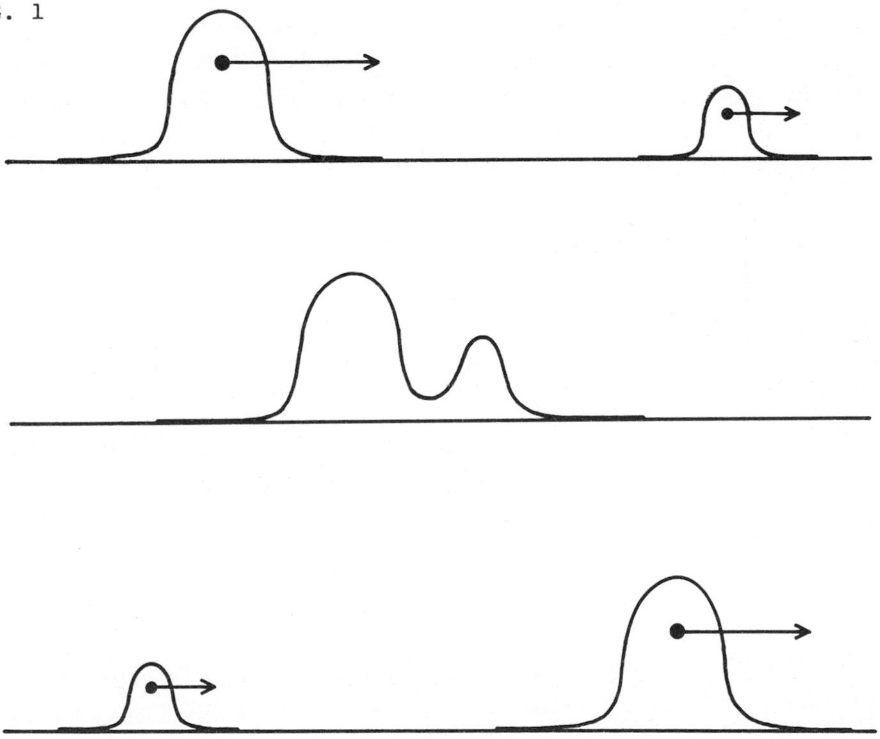

The moral is that solitary waves behave like particles which may collide but cannot lose their individuality. KRUSKAL-ZABUSKY [1965] coined the word soliton to describe this new mathematical entity. I cite one more numerical experiment with initial disturbance in the form of a triangle; see Fig. 2 for $-q$ at a) $t = 0$, b) moderate t, and c) $t \uparrow \infty$: Fig. 2b depicts the formation of solitons on the back of the triangle and a small oscillating wake; Fig. 2c shows the

FIG. 2

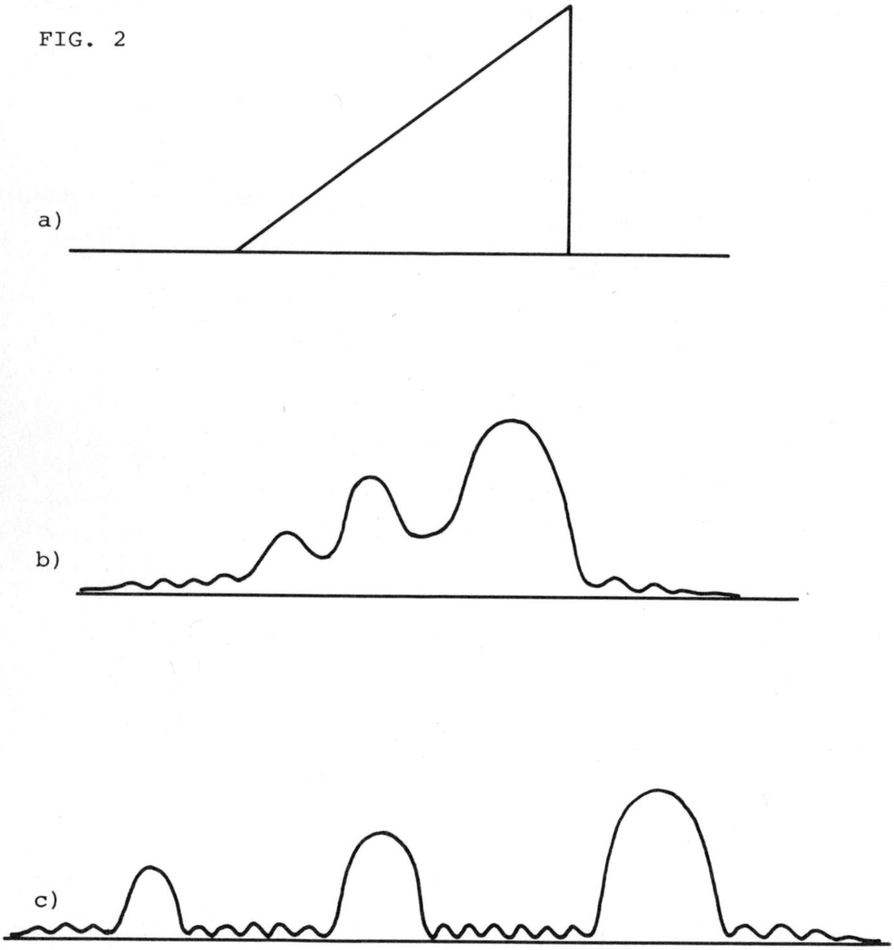

separating out of a train of solitary waves located in order of amplitude (= speed) and the wakes being damped. The best pictures of these phenomena may be found in TAPPERT [197]; see, also, FORNBERG-WHITHAM [1978].

2.3 Many Solitions: Hirota's Formula

GARDINER-GREENE-KRUSKAL-MIURA [1973] expressed the interaction of g solitons in closed form; see, also, TANAKA [1972/3]. The present version is due to HIROTA [1971]; see §5.2 for verification and further discussion. Let $0 < k_1 < k_2 < \dots < k_g$ be fixed and define

$$\vartheta(\mathfrak{x}) = \sum e^{\mathfrak{x}\cdot n} \prod_{i<j} \left(\frac{k_i - k_j}{k_i + k_j}\right)^{2n_i n_j} ,$$

in which the sum is over $n = (n_1, \dots, n_g)$ with $n_i = 0$ or 1 $(i=1,\dots,g)$. Let $k = (k_1,\dots,k_g)$ and $k^3 = (k_1^3,\dots,k_g^3)$. Then the interaction of g solitons with speeds $c_i = 2k_i^2$ $(i=1,\dots,g)$ may be expressed as*

$$q(t,x) = -2[\ell g\, \vartheta\, (\mathfrak{x} - 2kx + 4k^3 t)]'' ;$$

for example, if $g = 1$, then $\vartheta(\mathfrak{x}) = 1 + e^{x_1}$ and $q = -2(\ell g\, \vartheta)''$ is

$$-2\,[\ell g(1 + e^{x_1 - 2k_1 x + 4k_1^3 t})]'' = -2k_1^2\, \mathrm{ch}^{-2}[k_1(x - 2k_1^2 t)] ,$$

up to a phase shift, in agreement with the formula for the solitary wave of speed $c = 2k_1^2$. The reader is invited to look at the case $g = 2$; especially, to verify that $q = -2\,(\ell g\, \vartheta)''$ separates out into the sum of two solitary waves and to compute the phase shift [answer: $d(+) - d(-) = \sqrt{2}\,(k_1^{-1} + k_2^{-1})\, \ell g\, (k_2 + k_1)(k_2 - k_1)^{-1}$].

2.4 Integrals of the Motion

The persistence of the individuality of solitons requires, or at least suggests, the existence of a huge number of integrals of the motion, so that the individual solitary wave can, so to say, <u>remember its shape</u>. This was confirmed by GARDINER-KRUSKAL-MIURA [1968]. The conventional wisdom asserts that integrals reflect the existence of symmetries, but the present symmetries, if such they be, are hidden, and it is an open problem to interpret the integrals exhibited below

---

* ' signifies differentiation with respect to x.

in such a style. The equation is $\partial q/\partial t = 3qq' - (1/2)q''' = X_2 q$, and and it is easy to check that $H_0 = \int q$, $H_1 = (1/2)\int q^2$, and $H_2 = \int [(1/2)q^3 + (1/4)(q')^2]$ are conserved; for example,

$$\begin{aligned} H_2^\bullet &= \int [\tfrac{3}{2} q^2 X_2 q + \tfrac{1}{2} q'(X_2 q)'] \, dx \\ &= -\tfrac{3}{4}\int q^2 q''' - \tfrac{3}{2}\int qq'q'' + \tfrac{1}{4}\int q''q''' \\ &= 0 . \end{aligned}$$

To obtain more integrals of the motion in a systematic way, I pull out of the air the operator $Q = -D^2 + q$ and the elementary solution $p(t,x,y) = [\exp(-tQ)]_{xy}$ of the heat equation $\partial p/\partial t = -Qp$ for fixed $q \in C_\downarrow^\infty$. Now

$$p(t,x,y) = p^o(t,x,y) - \int_0^t ds \int p^o(t-s,x,y')\, q(y')\, p(s,y',y)\, dy' ,$$

p^o being the elementary solution for $q = 0$: $(4\pi t)^{-1/2}\exp[-(x-y)^2/4t]$, and it is easy to verify by means of this identity that $p(t,x,y)$ enjoys a development $(4\pi t)^{-1/2}[1 + a_1 t + a_2 t^2 + \ldots]$ for $t \downarrow 0$, a_1, a_2, etc. being universal polynomials in q, q', q'', etc., without constant term; in particular, the quantity $H = \int [p(t,x,y) - (4\pi t)^{-1/2}]$ enjoys a development $(4\pi t)^{-1/2}[b_1 t + b_2 t^2 + \ldots]$ for $t \downarrow 0$. H is preserved by translation of q [$\dot{q} = q'$]; it is to be proved that it is also preserved by the Korteweg-de Vries motion [$\dot{q} = 3qq' - \frac{1}{2} q'''$]. The proof and its further development is divided into a number of steps; it is adapted from McKEAN-MOERBEKE [1975].

STEP 1.* $\partial H/\partial q(x) = -t\, p(t,x,x)$.

PROOF. Let $\dot{q}$ be an infinitessimal change in q and $\dot{p}$ the corresponding change in p. Then $\partial p^\bullet/\partial t + Q\dot{p} = -\dot{q}\, p$, so

---

* $\partial H/\partial q$ is explained in §1.4.

$$\dot{p}(t,x,y) = \int_0^t ds \int p(t-s,x,y') \, [-\dot{q}(y') \, p(s,y',y)] \, dy' ,$$

and, by integration with regard to x,

$$H^{\bullet} = \int \dot{p}(t,x,x) \, dx = -\int_0^t ds \int dx \int p(t-s,x,y) \, \dot{q}(y) \, p(s,y,x) \, dy$$

$$= -\int_0^t ds \int p(t,y,y) \, \dot{q}(y) \, dy$$

$$= - t \int p(t,y,y) \, \dot{q}(y) \, dy .$$

STEP 2. $\bar{p} = p(t,x,x)$ <u>satisfies</u> $-2\partial^2\bar{p}/\partial t\partial x = K\bar{p}$ <u>with</u> $K = qD + Dq - (1/2)D^3$.

PROOF. Let subscripts [0,1,2] denote partial differentiation. Then, on the diagonal $x = y$, $p_0 = p_{11} - qp = p_{22} - qp$ implies $p_{01} = p_{111} - q'p - qp_1 = p_{221} - qp_1$ and $p_{02} = p_{222} - q'p - qp_2 = p_{112} - qp_2$, whence

$$\begin{aligned} 4\partial^2\bar{p}/\partial t\partial x = 4(p_{01} + p_{02}) &= p_{111} - q'p - qp_1 + 3(p_{221} - qp_1) \\ &\quad + p_{222} - q'p - qp_2 + 3(p_{112} - qp_2) \\ &= p_{111} + 3p_{112} + 3p_{122} + p_{222} \\ &\quad -2q'p - 4q(p_1 + p_2) \\ &= -2K\bar{p} . \end{aligned}$$

STEP 3. H <u>is preserved by the Korteweg-de Vries flow.</u>

PROOF. Let q move under the stated flow $\dot{q} = 3qq' - (1/2)q''' = Kq$ with K as in step 2. Then $H^{\bullet} = \int (\partial H/\partial q) \, \dot{q}$ by the chain rule in function space, and by the skew-symmetry of K,

$$H^{\bullet} = \int \frac{\partial H}{\partial q} Kq = - \int K \frac{\partial H}{\partial q} q = - \int K(-t\bar{p})q$$

$$= - 2t \int \frac{\partial \bar{p}'}{\partial t} q = - 2t \frac{\partial}{\partial t} \int \bar{p}'q = 2t \int \bar{p}q' .$$

The final integral vanishes by the self-evident invariance of H under translation: $\int (\partial H/\partial q)q' = - t \int \bar{p}q' = 0$.

STEP 4. The development $H = (4\pi t)^{-1/2}[b_1 t + b_2 t^2 + \ldots]$ now makes plain that b_1, b_2, etc. are integrals of the motion. To compute them, use steps 1 and 2 to verify that $c_j = \partial b_j/\partial q$ satisfies $Kc_j = - (2j-1)c'_{j+1}$ $[c_1 = -1]$. Now let* $H_j = (-1)^j(2j-1)(2j-3) \cdots 3\cdot 1\cdot b_{j+1}$ and deduce the recipe of A. Lenard:**

$$X_j q = (\partial H_j/\partial q)' = K(\partial H_{j-1}/\partial q) \qquad (j=1,2,3,\ldots).$$

The fact that $\partial H_j/\partial q$ is a universal polynomial in q,q',q'', etc. without constant term permits the unambiguous computation of these gradients, inductively, and it is an elementary lemma that $\partial H_j/\partial q$ has an unambiguous primitive H_j $(j=0,1,2,\ldots)$ vanishing for $q = 0$. The first few integrals are listed in the table.***

j	H[q]	Xq
0	$\int q$	0
1	$\int \frac{1}{2} q^2$	q'
2	$\int [\frac{1}{2} q^3 + \frac{1}{4} (q')^2]$	$3qq' - \frac{1}{2} q'''$
3	$\int [\frac{5}{8} q^4 + \frac{5}{4} q(q')^2 + \frac{1}{8} (q'')^2]$	$\frac{15}{2} q^2q' - 5q'q'' - \frac{5}{2} qq''' + \frac{1}{4} q'''''$

* $((2j-1)\cdots 3\cdot 1 = 1$ if $j = 0$.

** See GARDNER-GREENE-KRUSKAL-MIURA [1974].

*** See LAX [1975].

The quantities displayed have an isobaric character if degrees are ascribed as follows: deg q = 2, deg ' = 1; for instance, $3qq' - \frac{1}{2} q'''$ is of degree 5. The developments of $\bar{p} = p(t,x,x)$ and H are recorded for future use:*

$$\bar{p} = (4\pi t)^{-1/2} \sum_{j=0}^{\infty} (-1)^j \frac{\partial H_j}{\partial q} t^j \Big/ (2j-1) \cdots 3\cdot 1$$

$$H = \int [\bar{p} - (4\pi t)^{-1/2}] = (4\pi t)^{-1/2} \sum_{j=1}^{\infty} (-1)^{j-1} H_{j-1} t^j \Big/ (2j-3) \cdots 3\cdot 1.$$

* $(2j-1)\cdots 3\cdot 1 = 1$ if $j = 0$.

3. HAMILTONIAN MECHANICS AND THE COMMUTATOR FORMALISM

3.1 Classical Mechanics; see ARNOLD [1974] for general information on this topic. The chief ingredient of classical Hamiltonian mechanics is the symplectic form $\omega = \sum_{i=1}^{n} dq_i \wedge dp_i$ giving rise to the Poisson bracket $[A,B]$ between the mechanical quantities $A,B \in C^\infty(R^{2n})$:

$$[A,B] = \sum_{i=1}^{n} \left[\frac{\partial A}{\partial q_i}\frac{\partial B}{\partial p_i} - \frac{\partial A}{\partial p_i}\frac{\partial B}{\partial q_i}\right] .$$

This satisfies the Jacobi identity

$$[[A,B],C] + [[B,C],A] + [[C,A],B] = 0 ,$$

and for a Hamiltonian system $\dot{q} = \partial H/\partial p$, $\dot{p} = -\partial H/\partial q$ enables the expression of the rate of change of any mechanical quantity F as $\dot{F} = [F,H]$. It follows that if two Hamiltonians H_1 and H_2 are involutive, meaning that $[H_1,H_2] = 0$, then the corresponding vector fields

$$X = \sum_{i=1}^{n} \left[\frac{\partial H}{\partial p_i}\frac{\partial}{\partial q_i} - \frac{\partial H}{\partial q_i}\frac{\partial}{\partial p_i}\right]$$

commute, as to the actual flows they produce in R^{2n}; in fact, for the fields, Jacobi's identity implies

$$\begin{aligned}[X_1,X_2]F &= (X_1X_2 - X_2X_1)F = X_1[F,H_2] - X_2[F,H_1] \\ &= [[F,H_2],H_1] - [[F,H_1],H_2] \\ &= [[F,H_2],H_1] + [[H_2,H_1],F] + [[H_1,F],H_2] \\ &= 0 ,\end{aligned}$$

and this lifts to the flows since, for small ab ,

$$e^{aX_1}e^{bX_2}e^{-aX_1}e^{-bX_2} = 1 - ab[X_1,X_2] + o(ab) ,$$

so that if X_1 and X_2 commute, then for large ab, the left-hand commutator may be estimated by dividing the rectangle $a \times b$ into little cells $a_i \times b_j$ $(i,j=1,\ldots,n)$ and appraising

$$\prod[1 + o(a_i b_j)] = 1 + \sum o(a_i b_j) = 1 + o(ab) = 1 + o(1) .$$

3.2 Integrable Systems

Let the system $\dot{q} = \partial H/\partial p$, $\dot{p} = -\partial H/\partial q$ have 2n degrees of freedom and n independent integrals of the motion $H_1,\ldots,H_n$ in involution. The adjective independent means that grad H_i $(i=1,\ldots,n)$ are independent of each point qp of R^{2n}. Then $H_i = h_i$ $(i=1,\ldots,n)$ defines an n-dimensional submanifold $M \subset R^{2n}$ and the commuting vector fields $X_i: F \in C^\infty(M) \to [F,H_i]$ $(i=1,\ldots,n)$ span its tangent space at each point. Pick a point $o \in M$ as origin. Then $\exp(t_1X_1+\cdots+t_nX_n) \equiv e(t\cdot X)$ acting upon o fills out a whole connected piece M_1 of M as $t = (t_1,\ldots,t_n)$ runs over R^n, and M may be identified via the map $t \to \exp(t\cdot X)o$ of $R^n \to M_1$ as the quotient $J = R^n/L$ of R^n by the lattice $L \subset R^n$ of so-called periods $t \in R^n$ such that $\exp(t\cdot X)$ acts as the identity on M_1. This proves part of a theorem of ARNOLD [1974]: each component of M may be identified as the product of an a-dimensional torus and a copy of R^b of total dimension $a + b = n$ in such a way that the flows $\exp(tX_i)$ $(i=1,\ldots,n)$ reduce to straight line motions at constant speed on J. This is what is meant by an integrable system. Clearly, it is a very exceptional thing in that the n integrals of motion satisfy $n(n-1)/2$ conditions $[H_i,H_j] = 0$ $(1 \leq i < j \leq n)$, which is a highly over-determined situation. The famous examples of integrable systems include JACOBI's proof of the addition theorem for hyperelliptic integrals and his discussion of the geodesic flow on an ellipsoid [1884], KOVALEVSKAYA's discussion of the top [1889]*, and NEUMANN's treatment of constrained harmonic oscillators [1859]; see §10, below. POINCARÉ [1892] killed the subject in his time by proving the non-integrability of the 3-body problem, and it is only recently that interest revives.

3.3 Korteweg-De Vries: Hamiltonian Formalism

The present article interprets the integrals of §2.4 in the mechanical language of §3.2. This was done by GARDINER [1971] and FADDEEV-ZAKHAROV [1971]. The integral $H_2 = \int [(1/2)q^3 + (1/4)(q')^2]$ is related to the KORTEWEG-de VRIES flow by the rule

$$\dot{q} = X_2 q \equiv D(\partial H_2/\partial q) = 3qq' - (1/2)q''' .$$

This is reminiscent of the classical prescription

$$\begin{pmatrix} \dot{q} \\ \dot{p} \end{pmatrix} = \begin{pmatrix} 0 & 1 \\ -1 & 0 \end{pmatrix} \begin{pmatrix} \partial H/\partial q \\ \partial H/\partial p \end{pmatrix} = \begin{pmatrix} \partial H/\partial p \\ -\partial H/\partial q \end{pmatrix} ,$$

except that 1) the phase space $C^\infty_\downarrow$ is now infinite-dimensional, 2) there is no p any more, only a q, and 3) the skew-symmetric matrix $\begin{pmatrix} 0 & 1 \\ -1 & 0 \end{pmatrix}$ is replaced by the skew-symmetric operator D = '. This suggests the introduction of the nonclassical symplectic form

$$\omega[\partial/\partial q(x), \partial/\partial q(y)] = -1/2 \quad (y < x) \quad = +1/2 \quad (y > x)$$

if you like that language;** in any case, the Poisson bracket required to make

$$\dot{q} = X_2 q = D\frac{\partial H_2}{\partial q} = [q, H_2]$$

is

$$[A,B] = \int \frac{\partial A}{\partial q} D \frac{\partial B}{\partial q} dx$$

The bracket satisfies Jacobi's identity, as is easily verified using†

* See GOLUBEV [1960].

** See DIKII-GELFAND [1975] for a more conventional format.

† $\partial^2 H/\partial q^2$ is the Hessian operator of §1.4.

$$[[A,B],C] = \int \left(\frac{\partial C}{\partial q}\right)' \left[\frac{\partial^2 A}{\partial q^2}\left(\frac{\partial B}{\partial q}\right)' - \frac{\partial^2 B}{\partial q^2}\left(\frac{\partial A}{\partial q}\right)'\right] dx ,$$

and it is a source of satisfaction that <u>the integrals</u> H_j $(j=0,1,2,\ldots)$ <u>of</u> §2.4 <u>are involutive</u>.

PROOF. Using the rule $X_{j+1}q = K\,\partial H_j/\partial q$ of §2.4 and the skew-symmetry of $K = qD + Dq - (1/2)D^3$,

$$\begin{aligned}
[H_i,H_j] = \int \frac{\partial H_i}{\partial q} D \frac{\partial H_j}{\partial q} &= \int \frac{\partial H_i}{\partial q} X_j q \\
&= \int \frac{\partial H_i}{\partial q} K \frac{\partial H_{j-1}}{\partial q} \\
&= -\int \frac{\partial H_{j-1}}{\partial q} K \frac{\partial H_i}{\partial q} \\
&= -\int \frac{\partial H_{j-1}}{\partial q} X_{i+1} q \\
&= -\int \frac{\partial H_{j-1}}{\partial q} D \frac{\partial H_{i+1}}{\partial q} \\
&= \int \frac{\partial H_{i+1}}{\partial q} D \frac{\partial H_{j-1}}{\partial q} \\
&= [H_{i+1},H_{j-1}] ,
\end{aligned}$$

and, by repetition,

$$[H_i,H_j] = [H_{i+j},H_0] = \int \frac{\partial H_{i+j}}{\partial q} D \frac{\partial H_0}{\partial q} = 0$$

in view of $\partial H_0/\partial q = 1$. The upshot is that the flows $\partial q/\partial t = X_j q$ $(j=1,2,\ldots)$ commute, the question of <u>existence</u> being left aside: it will be disposed of in §5 by means of explicit formulae. The question of <u>integrability</u> is also postponed: the mere existence of infinitely many integrals of the motion is not decisive since there are infinitely many degrees of freedom and precisely $\infty/2$ integrals are required, not one more or less, whatever that may mean,

3.4 Korteweg-De Vries: Commutator Formalism

The somewhat mysterious intervention of the operator $Q = -D^2 + q$ in §2.4 is clarified by the discovery of LAX [1968] that the KORTEWEG-de VRIES flow is equivalent to $Q^\bullet = [Q, K_2]$ with* $K_2 = 2D^3 - (3/2)(qD + Dq)$. The proof is elementary. The skew-symmetry of K_2 reveals the deep fact that Korteweg-de Vries can be viewed as an isospectral flow of Q: $-K_2$ is the infinitessimal transformation reflecting the continual orthogonal similarity of Q during the motion. This looks accidental, but there is something deeper to it. Let $J = \sqrt{-Q}$ be the formal square root $J = D + c_0 + c_1 D^{-1} + c_2 D^{-2} + \dots$ determined by $J^2 = (D + c_0 + \cdots)^2 = -Q$: up to terms in D^{-2},
$J^2 = D^2 + 2c_0 D + (c_0' + c_0^2 + 2c_1) + (c_1' + 2c_2 + 2c_0 c_1)D^{-1}$,
so $c_0 = 0$, $c_1 = -q/2$, $c_2 = q'/4$. The local part J^+ of J is simply D, and

$$Q^\bullet = [Q, J^+] = [Q, D] = -q'$$

is just the flow of translation run backwards. Now comes the cute thing: up to terms in D^{-1},

$$\begin{aligned} J^3 = J(-Q) &= (D - (q/2)D^{-1} + (q'/4)D^{-2})(D^2 - q) \\ &= D^3 - Dq - (q/2)D + q'/4 \\ &= (1/2)K_2 , \end{aligned}$$

i.e., the local part $(J^3)^+$ is $(1/2)K_2$, so

$$Q^\bullet = [Q, 2(J^3)^+] = X_2 q$$

is just the Korteweg-de Vries flow. GELFAND-DIKII [1975 & 1976]** have elevated these remarks into a general principle; up to a constant multiple, $Q^\bullet = [Q, (J^{2j-1})^+]$ is equivalent to the flow $\partial q/\partial t = X_j q$ for $j = 1,2,3,\dots$, up to an inessential scaling.

---

*K_1 is similar to, but not the same as, the operator $K = qD + Dq - \frac{1}{2}D^3$ of §2.4.

**See, also, ADLER [1978 & 1979] and SYMES [1978] for a more accessible explanation.

PROOF. Let $n = 2j - 1$ be odd. Then J^n is the sum of a differential operator $(J^n)^+ = D^n + a_1 D^{n-1} + \ldots + a_n$ and an operator $(J^n)^- = b_1 D^{-1} + b_2 D^{-2} + \ldots$ of degree ≤ -1. Now $[Q, J^n] = [-J^2, J^n] = 0$, so $[Q,(J^n)^+] = -[Q,(J^n)^-]$ is at the same time a differential operator and of degree ≤ 0, <u>i.e.</u>, it is an operator of multiplication. Now let G be the Green function $(Q - \lambda)^{-1}_{xy}$ on the diagonal $x = y$. Then* $K\bar{G} = 2\lambda\bar{G}'$, so $[Q,(\bar{G}D - (1/2)\bar{G}')(Q - \lambda)^{-1}] = 2G'$ by an easy calculation, and* $[Q,K_j] = X_j q$ $(j=1,2,3,\ldots)$ with a differential operator K_j of degree $2j - 1 = n$ and top coefficient 2^{j-1}. The upshot is that $L = 2^{-j+1}K_j$ and $L = (J^n)^+$ have the same top term and share the property that $[Q,L] = M$ is an operator of multiplication. This permits their identification: if $L = D^n + c_1 D^{n-1} + \ldots + c_n$, then the fact that $[Q,L] = M$ is of degree ≤ 0 allows the identification of c_i $(i = 1,2,\ldots,n)$ in that order without specific knowledge of M, as is readily verified.

$K_j = 2^{j-1}(J^n)^+$, like K_2, is skew-symmetric: in fact, $(J^*)^2 = -Q$, so $J^* = -J$ by the uniqueness of J and the skew-symmetry of D and this skew-symmetry is inherited by J^n and $(J^n)^+$; in particular, $\partial q/\partial t = X_j q$ $(j=1,2,3,\ldots)$ <u>can be viewed as an isospectral flow of</u> Q. The dimension $d = 1$ is favorable to isospectral flows: in dimensions $d \ge 2$, a spectral rigidity sets in; see, <u>e.g.</u>, GUILLEMIN-KAZHDAN [1979], McKEAN [1972], and SINGER [1980] for compact manifolds, and MUMFORD [1979] for lattices; compare §11.6.

* $\bar{G} = \int_0^\infty e^{\lambda t}\,\bar{p}\,dt$ $(\lambda < 0)$; now see step 2 of §2.4 and the development of $\bar{p}$ at the end of that article.

4. SCATTERING

The material presented here is prepared with a view to the explicit solution of the Korteweg-de Vries equation $\partial q/\partial t = X_2 q$; it may be found in FADDEEV [1974]; see, also, DEIFT-TRUBOWITZ [1979] for corrections and new ideas.

4.1 Forward Scattering

Let $Q = -D^2 + q$ with $q \in C_{\downarrow}^{\infty}$, as before, and let $k \neq 0$ be real. $Qf = k^2 f$ has two solutions f_- and f_+ with the behavior indicated in the table. The numbers $s_{ij}(k)$ $(i,j=1,2)$ are not specified beforehand; they are uniquely determined by the prescription. The scattering matrix $s = (s_{ij})$ is unitary and satisfies the reality condition $s_{ij}^*(k) = s_{ij}(-k)$ $(i,j=1,2)$.

	$x \downarrow -\infty$	$x \uparrow +\infty$
f_+	$e^{\sqrt{-1}kx} + s_{12}(k)e^{-\sqrt{-1}kx}$	$s_{11}(k)\, e^{\sqrt{-1}\,kx}$
f_-	$s_{22}(k)\, e^{-\sqrt{-1}\,kx}$	$e^{-\sqrt{-1}kx} + s_{21}\, e^{\sqrt{-1}\,kx}$

The spectrum of Q acting upon $L^2(R^1)$ is always continuous in $0 \leq k^2 < \infty$ and may include $g < \infty$ additional eigenvalues or bound states $-k_g^2 < \ldots < -k_1^2 < 0$ to the left. The number of such bound states is overestimated by $\int \sqrt{-\min(q,0)}$; see BARGMANN [1949]. The transmission coefficient $s_{11} = s_{22} = 1 + O(k^{-1})$ is an outer Hardy function in the open upper half-plane apart from poles arising from the bound states:

$$s_{11}(k) = \exp\left[\frac{1}{2\pi\sqrt{-1}} \int \frac{\ell g|s_{11}(k')|^2}{k'-k}\, dk'\right] \prod_{i=1}^{g} \frac{k+\sqrt{-1}\,k_i}{k-\sqrt{-1}\,k_i} :$$

The unitarity and reality of s implies $1 = |s_{11}|^2 + |s_{12}|^2$ and

$s_{12}=-s_{21}s_{11}/s^*_{11}$, so the <u>reflection coefficient</u> s_{21} and the bound states $-k^2_i$ $(i=1,\dots,g)$ determine the whole of the scattering matrix. s_{21} is infinitely differentiable and vanishes rapidly at $k = \pm\infty$, and every such function satisfying a) $s_{21}(k) = s^*_{21}(-k)$, b) $|s_{21}| \le 1$, and c) $\int \lg(1-|s_{21}|^2)(1+k^2)^{-1}\,dk > -\infty$ arises in this way.

Now the function $e_\pm = \exp(\mp\sqrt{-1}\,kx)\, f_\pm/s_{11}$ is Hardy in the upper half plane; in fact, $e_\pm = 1+O(k^{-1})$ and 1)* $e_\pm - 1 \in H^{2+}$. The entries of the table for $x \uparrow +\infty$ indicate that $f_- = (f_+/s_{11})^* + s_{21}(f_+/s_{11})$, so, in the absence of bound states, $e^*_+ + s_{21}\exp(2\sqrt{-1}\,kx)e_+ = f_-\exp(\sqrt{-1}\,kx)$ $= 1+O(k^{-1})$, and 2) $e^*_+ + s_{21}\exp(2\sqrt{-1}\,kx)e_+ - 1 \in H^{2+}$, too. DEIFT-TRUBOWITZ [1979] isolated 2) and emphasized its importance: it lies at the heart of scattering theory. 1) and 2) lead easily to the fact that <u>in the absence of bound states, the map</u> $Q \to s_{21}$ <u>is</u> 1:1. LEVINSON [1944] proved the analogue for the half-line; see, also, LEVINSON [1953]. The present proof is from DEIFT-TRUBOWITZ [1979].

PROOF. Let e be the difference of the functions e_+ for operators Q_1 and Q_2 having the <u>same</u> reflection coefficient s_{21}. Then $e \in H^{2+}$ by 1) and $e^* + s_{21}\exp(2\sqrt{-1}\,kx)e \in H^{2+}$ by 2). Now $\int h_1h_2 = 0$ for functions of class H^{2+}, so $\int e[e^* + s_{21}\exp(2\sqrt{-1}\,kx)e]\,dk = 0$, whence $0 \ge \int |e|^2(1-|s_{21}|)\,dk$, and that is impossible unless $e = 0$ $[Q_1 = Q_2]$ since, by unitarity, $|s_{21}|^2 = 1-|s_{11}|^2$ and $|s_{11}|^2 > 0$ almost everywhere in view of $\int \lg|s_{11}|^2(1+k^2)^{-1}\,dk > -\infty$.

DEIFT-TRUBOWITZ [1979] confirmed the importance of 1) and 2) in the following striking statement: <u>if</u> $e - 1: x \in R^1 \to H^{2+}$ <u>is twice differentiable in that space, if</u> $e - 1 \sim c(x)/2\sqrt{-1}\,k$ <u>near</u> $k = \infty$, <u>and</u>

---

* H^{2+} is the space of boundary values of functions h analytic in the open upper half-plane with $\|h\|^2 = \sup_{b>0} \int |h(a+\sqrt{-1}\,b)|^2\,da < \infty$. H^{2+} is identical to the class of transforms $h(k) = \int_0^\infty \exp(\sqrt{-1}\,kx)\,\hat{h}(x)\,dx$ with $\int_0^\infty |\hat{h}|^2\,dx < \infty$; see DYM-McKEAN [1972] for such matters.

if $e^* + r(k)\exp(2\sqrt{-1}\,kx)e - 1 \in H^{2+}$ *with* $|r(k)| < 1$ *a.e., then, necessarily,* $Qf = k^2 f$ *with* $f = \exp(\sqrt{-1}\,kx)e$ *and* $q = c'$.

PROOF. The function $h = e'' + 2\sqrt{-1}\,k\,e' - q\,e \in H^{2+}$ satisfies

$$h^* + r\exp(2\sqrt{-1}\,kx)h = \left(D^2 - 2\sqrt{-1}\,kD - q\right)\left[e^* + r\exp(2\sqrt{-1}\,kx)e\right] .$$

The left-hand side belongs to $L^2(R^1)$, while the right-hand side is Hardy in the upper half-plane, so the common value is of class H^{2+}. Now h vanishes by the preceding proof. $Qf = k^2 f$ is plain from that.

4.2 Some Gradients

$\partial s(k)/\partial q(x)$ is computed for future use and also to clarify the meaning of s_{21}. Let $f^\bullet(y) = \partial f_-(y)/\partial q(x)$ and $s^\bullet(f) = \partial s(k)/\partial q(x)$ for fixed k and x. Then $Qf^\bullet(y) = k^2 f^\bullet(y)$ $(y \neq x)$, so, by the table of §4.1,

$$f^\bullet(y) = s^\bullet_{22} f_-(y)/s_{22} \quad (y < x) \qquad = s^\bullet_{21} f_+(y)/s_{11} \quad (y > x) ,$$

and

$$\frac{s^\bullet_{21}}{s_{11}} f_+(x) - \frac{s^\bullet_{22}}{s_{22}} f_-(x) = f^\bullet(x+) - f^\bullet(x-) = 0$$

$$\frac{s^\bullet_{21}}{s_{11}} f'_+(x) - \frac{s^\bullet_{22}}{s_{22}} f'_-(x) = (f^\bullet)^+(x) - (f^\bullet)^-(x) = f_-(x) ,$$

which may be solved for

$$s^\bullet_{11} = \frac{\partial s_{11}}{\partial q(x)} = \frac{f_-(x) f_+(x)}{2\sqrt{-1}\,k} , \qquad s^\bullet_{21} = \frac{\partial s_{21}}{\partial q(x)} = \frac{f_-^2(x)}{2\sqrt{-1}\,k}$$

by use of $s_{11} = s_{22}$ and $f'_+ f_- - f_+ f'_- = s_{11} 2\sqrt{-1}\,k$; the latter is read off the table. The first formula leads to

$$\partial\, \ell g\, s_{11}(k)/\partial q(x) = (Q - k^2)^{-1}_{xx} = \int_0^\infty e^{k^2 t}\, p(t,x,x)\, dt$$

on the positive imaginary axis $[k^2 < 0]$. The second formula indicates that s_{21} is some kind of non-linear FOURIER transform: $\partial s_{21}/\partial q = (2\sqrt{-1}\,k)^{-1} \exp(-2\sqrt{-1}\,kx)$ for $q = 0$, so the chain rule suggests that for small q and moderate $k \neq 0$,

$$s_{21}(k) \simeq (2\sqrt{-1}\,k)^{-1} \int \exp^{-2\sqrt{-1}kx} q(x)\,dx .$$

4.3 Backward Scattering

1) $e_+ - 1 \in H^{2+}$ and 2) $e_+^* + s_{21}\exp(2\sqrt{-1}\,kx)e_+ - 1 \in H^{2+}$ are now employed to compute the inverse map $s_{21} \to q$ in the absence of bound states. 1) states that $f_+/s_{11} - \exp(\sqrt{-1}\,kx) \in \exp(\sqrt{-1}kx)H^{2+}$, so

$$f_+/s_{11} = e^{\sqrt{-1}kx} + \int_x^\infty K(x,y)\, e^{\sqrt{-1}ky}\,dy$$

with real $K(x,y) = 0$ $(y < x)$ and $\int_x^\infty K^2(x,y)\,dy < \infty$. 2) states that $(f_+/s_{11})^* + s_{21}(f_+/s_{11}) - \exp(-\sqrt{-1}\,kx) \in \exp(-\sqrt{-1}\,kx)\,H^{2+}$, so

$$\int K(x,y)e^{-\sqrt{-1}ky}\,dy + s_{21}\int K(x,y)e^{-\sqrt{-1}ky} + s_{21}e^{\sqrt{-1}kx} \in e^{-\sqrt{-1}kx}\,H^{2+}.$$

s_{21} is now expressed as $\int L(y)\exp(-\sqrt{-1}\,ky)\,dy$ with real L subject to $\int L^2(y)\,dy < \infty$, whence

$$\int [K(x,y) + \int K(x,y')L(y'+y)\,dy' + L(x+y)]e^{-\sqrt{-1}ky}\,dy \in e^{-\sqrt{-1}kx}\,H^{2+},$$

i.e., $K + KL + L = 0$ $(y > x)$ with a self-evident notation. This is the Gelfand-Levitan-Marcenko equation; see GELFAND-LEVITAN [1951] and AGRANOVICH-MARCENKO [1963]. It is used to determine K from L. Then q is determined from K as follows: $f_+ = s_{11}(I+K)\exp(\sqrt{-1}kx)$ satisfies $Qf = k^2 f$, and it is easy to see that this requires $q(x) = -2D\,K(x,x)$. Now let Q have ground states $-k_g^2 < \ldots < -k_1^2 < 0$. The spectral data s_{21} must be augmented by the numbers k_i $(i=1,\ldots,g)$

and the norming constants c_i $(i=1,\ldots,g)$ determined by $c_i^2\int|(f_+/s_{11})(x,\sqrt{-1}k_i)|^2dx=1$. The map from q to the augmented spectral data a) $s_{21}\in C_\downarrow^\infty$ with $s_{21}(k)=s_{21}^*(-k)$, $|s_{21}|\le 1$, and $\int \ell g(1-|s_{21}|^2)(1+k^2)^{-1}\,dk>-\infty$, b) $0<k_1<\ldots<k_g$, and c) $0<c_i<\infty$ $(i=1,\ldots,g)$ is now 1:1 and onto, and the only change in the backward scattering recipe is that the sum $\sum c_i^2\exp(-k_ix)$ must be added to $L(x)=(2\pi)^{-1}\int s_{21}(k)\exp(-\sqrt{-1}\,kx)\,dk$. DEIFT-TRUBOWITZ [1979] may be consulted for details.

4.4 Dyson's Formula

The recipe has been put into a more attractive form by DYSON [1976].* $K=-L^+-(KL)^+$, so

$$K=-L^++(L^+L)^+-((L^+L)^+L)^++\cdots,$$

and with the temporary notation M for the restricted kernel $L(\xi+\eta)$: $\xi,\eta\ge x$,

$$\begin{aligned}K(x,x)&=-L(2x)+\int_x^\infty L(x+y_1)L(y_1+x)\,dy_1-\int_x^\infty\int_x^\infty L(x+y_1)L(y_1+y_2)L(y_2+x)\,dy_1\,dy_2+\ldots\\&=D\,\mathrm{sp}(M-\tfrac12M^2+\tfrac13M^3-\cdots)\\&=D\,\mathrm{sp}\,\ell g(I+M)\\&=D\,\ell g\det(I+M),\end{aligned}$$

sp being the trace and det the Fredholm determinant. The upshot is Dyson's formula:

$$q(x)=-2D^2\,\ell g\det[I+L(\xi,\eta):\xi,\eta\ge x].$$

---

* J^+ signifies the upper triangular part of the kernel J.

5. INTEGRABILITY OF KORTEWEG-DE VRIES

5.1 Application of Scattering

GARDINER-GREENE-KRUSKAL-MIURA [1967] made the remarkable discovery that under the Korteweg-de Vries flow $\partial q/\partial t = X_2 q = 3qq' - (1/2)q'''$, the spectral data of q move in a very simple way: $s_{11}(k)$, or, what is equivalent, $|s_{21}(k)|$, and the numbers k_i $(i=1,\ldots,g)$ are fixed, while the phase of $s_{21}(k)$ is advanced by $4\sqrt{-1}\,k^3 t$ and $\ell g\, c_i^{\,2}$ by $4k_i^{\,3}t$ $(i=1,\ldots,g)$. This effects a complete solution of the problem via the recipe of backward scattering;* in my opinion, it represents one of the decisive mathematical events of the present century. The stated motion of s_{21} will now be verified. The commutator rule** $Q^{\bullet} = [Q, K_2]$ with $K_2 = 2D^3 - (3/2)(qD + Dq)$ states that Q moves in an isospectral way, $-K_2$ being the corresponding infinitessimal orthogonal transformation; in particular, the eigenfunction f_-^{o} for $Q^o = Q$ at time $t = 0$ evolves according to $f^{\bullet} = -K_2 f$. Now

$$f \simeq a\, e^{-\sqrt{-1}\,kx} + b\, e^{\sqrt{-1}\,kx} \qquad (x \uparrow \infty)$$

with $a = 1$ and $b = s_{21}^{o}$, initially, and $K_2 \simeq 2D^3$ near $x = \infty$, so

$$\dot{a}\, e^{-\sqrt{-1}kx} + \dot{b}\, e^{\sqrt{-1}kx} \simeq f^{\bullet} = -K_2 f \simeq -2D^3[a\, e^{-\sqrt{-1}kx} + b\, e^{\sqrt{-1}\,kx}]$$

$$= -2\sqrt{-1}\,k^3\, a\, e^{-\sqrt{-1}kx} + 2\sqrt{-1}\,k^3\, b\, e^{-\sqrt{-1}kx},$$

whence $a^{\bullet} = -2\sqrt{-1}\,k^3 a$, $b^{\bullet} = 2\sqrt{-1}\,k^3 b$, and consequently,

$$f \simeq e^{-2\sqrt{-1}k^3 t}\, e^{-\sqrt{-1}kx} + e^{2\sqrt{-1}k^3 t}\, s_{21}^{o}\, e^{\sqrt{-1}kx} \qquad (x \uparrow \infty);$$

similarly,

$$f \simeq e^{-2\sqrt{-1}k^3 t}\, s_{11}^{o} e^{-\sqrt{-1}kx} \qquad (x \downarrow -\infty).$$

---

* See TANAKA [1973] and MURRAY [1978 & 1979] for detailed investigation of the solution using this recipe.

** See §3.4.

The upshot is that Q has a solution of the form

$$f_- \simeq s_{11}^o e^{-\sqrt{-1}kx} \quad (x \downarrow -\infty) \qquad \simeq e^{-\sqrt{-1}kx} + e^{4\sqrt{-1}k^3 t} s_{21}^o e^{\sqrt{-1}kx} \quad (x \uparrow \infty)$$

<u>i.e.</u>, $s_{11} = s_{11}^o$ and $s_{21} = \exp(4\sqrt{-1}\, k^3 t) s_{21}^o$, as stated. The computation of norming constants is similar. The same procedure may be used to solve the higher flows $\partial q/\partial t = X_j q$ $(j=3,4,5,\dots)$: up to a constant multiplier,* $Q^\bullet = [Q, K_j]$ with $K_j = [(-Q)^{n/2}]^+$ and $n = 2j-1$, and this is equivalent to $s_{21} = \exp(\sqrt{-1}\, k^n t) s_{21}^o$ and $c_i = \exp(k_i^n t) c_i^o$ $(i=1,\dots,g)$.

<u>Amplification</u>. I want to emphasize what has been achieved here: $x^\bullet = \sqrt{1-x^2}$ looks complicated but is reduced to $y^\bullet = 1$ by the substitution $x = \sin y$. Tnis is perfectly elementary, but notice that $y = \sin^{-1} x$ is really a very complicated transcendental substitution, not at all self-evident to the novice. The substitution $q \to s_{21}$ is much less self-evident, though it is just such a trick, only now in infinitely many dimensions, reducing 1) $q^\bullet = X_2 q = 3qq - (1/2)q'''$ to 2) $s_{21}^\bullet = 4\sqrt{-1}\, k^3 s_{21}$. Notice that 2) is equivalent to 3) $\dot{q} = -4q'''$ which is just 1) with the nonlinearity crossed out, up to a factor of 8. It may be hoped that this is only the first of many applications of the nonlinear transform $q \to s_{21}$.

5.2 Reflectionless Potentials

The purpose of this article is to carry out the recipe of §5.1 in the reflectionless case: $s_{21} \equiv 0$. The existence of such Q is due to BARGMANN [1949]. L(x) is now the sum $\sum c_i^2 \exp(-k_i x)$, and it is easy to solve** $K + (KL)^+ + L^+ = 0$ for

$$K(x,x) = D\, \ell g \det \left[I + \frac{c_i c_j}{k_i + k_j} e^{-(k_i+k_j)x} : 1 \le i,j \le n \right];$$

---

* See §3.4.

** KAY-MOSES [1956]; see, also, GARDINER-GREENE-KRUSKAL-MIURA [1974] and TANAKA [1972, 1973].

for the proof, just put $K(x,y) = \sum f_i(x) \exp(-k_i y)$ and grind it out. It follows that if $c_i = \sqrt{2k_i} \exp(x_i/2)$ $(i=1,\ldots,g)$ and if

$$\vartheta(\mathfrak{x}) = \det \left[I + \frac{2\sqrt{k_i k_j}}{k_i + k_j} e^{(x_i+x_j)/2} : 1 \leq i,j \leq n \right]$$

with $\mathfrak{x} = (x_1,\ldots,x_g)$, then the solution of $\partial q/\partial t = X_2 q$ may be expressed as

$$q(t,x) = -2D^2 \ell g \, \vartheta(\mathfrak{x} - 2kx + 4k^3 t)$$

with $k = (k_1,\ldots,k_g)$ and $k^3 = (k_1^3,\ldots,k_g^3)$. This is precisely the formula of HIROTA [1971] stated in §2.3, as may be verified by developing the determinant in the style $\det(I + M) = 1 + \mathrm{sp}\, M + \ldots + \det M$. The identification of the reflectionless potentials as the many-soliton functions of §2.3 is an unexpected bonus.

5.3 Isospectral Classes

The specification of s_{11} and $-k_i^2$ $(i=1,\ldots,g)$ defines an isospectral class P of operators Q preserved by the flows $\partial q/\partial t = X_j q$ $(j=1,2,3,\ldots)$. P is the general leaf of a foliation* of $C_\downarrow^\infty$. P is of finite dimension g if and only if $s_{21} \equiv 0$ $[s_{11} \equiv 1]$ and there are g bound states; this is the many-soliton case. The function phase $s_{21}(k)$ and the numbers $\ell g\, c_i$ $(i=1,\ldots,g)$ play the role of coordinates on P, and inasmuch as they move in straight lines at constant speed under the commuting flows $\partial q/\partial t = X_j q$ $(j=1,2,3,\ldots)$, it is natural to declare, e.g., $\partial q/\partial t = X_2 q$ to be integrable. This is the statement of FADDEEV-ZAKHAROV [1971]. The actual situation is more complicated: the number of degrees of freedom is infinite and it is not clear how to count $\infty/2$ involutive integrals of the motion. The naive picture is that P is specified by fixing the

*The nomenclature is abused in that a local product structure is absent.

integrals $H_j = h_j$ $(j=0,1,2,\dots)$ so that its normal bundle is spanned by the gradients $\partial H/\partial q$ and its tangent bundle by the fields $X\colon q \to D\ \partial H/\partial q$. This is plainly incorrect if, e.g., $q = 0$ near $x = \pm\infty$: then $\partial H/\partial q$ and Xq vanish in the same way and cannot span the ambient space. The correct tangent space is probably built out of $Xq = D[\partial\ \ell g\ s_{11}(k)/\partial q] = D[(k^2 - Q)^{-1}_{xx}]$ for k^2 off the spectrum of Q, but this awaits investigation; see, however, §8.8 for a similar situation which is fully understood.

The map from spectral data to Q is most elegantly expressed by means of Dyson's formula:* for example, in the absence of bound states, if $\not{x}$ = phase s_{21} and $\vartheta(\not{x}) = \det[I+L$ restricted to $[0,\infty)]$, then $q(x) = -2D^2\ \ell g\ \vartheta(\not{x} + x\not{x}_1)$, $\not{x}_1 = 2\sqrt{-1}\ k = X_1[\text{phase } s_{21}(k)]$ being the $\not{x}$-direction corresponding to the infinitessimal translation $X_1\colon q \to q'$. The $\not{x}$-direction corresponding to KORTEWEG-DE VRIES is $\not{x}_2 = 4\sqrt{-1}\ k^3 = X_2[\text{phase } s_{21}(k)]$, so $\exp(tX_2)q = -2D^2\ \ell g\ \vartheta(\not{x}+x\not{x}_1+t\not{x}_2)$. The notation may seem peculiar; it is adopted for comparison with Hirota's formula [§5.2] and later formulas of the same type for Korteweg-de Vries on the circle [§8].

5.4 Integrals of the Motion

$s_{11}(k)$ and $-k_i^2$ $(i=1,\dots,g)$ are integrals of the motion, and it is interesting to relate these to the old integrals H_j $(j=1,2,3\dots)$. This was done by FADDEEV-ZAKHAROV [1971]; see, also, FADDEEV-BUSLAEV [1960]. The formula

$$\ell g\ s_{11}(k) = \frac{1}{2\pi\sqrt{-1}} \int \frac{\ell g |s_{11}(k')|^2}{k'-k}\, dk' + \sum_{i=1}^{g} \ell g \frac{k + \sqrt{-1}\ k_i}{k - \sqrt{-1}\ k_i}$$

was noted in §4.1; also, from §4.2,

$$(\partial/\partial q)\ \ell g\ s_{11}(k) = (k^2 - Q)^{-1}_{xx} = -\int_0^\infty e^{k^2 t}\, p(t,x,x)\, dt$$

---

* §4.4.

for, e.g., positive imaginary k. The expansion of §2.4 is now applied with the result that

$$(\partial/\partial q)\ \lg\ s_{11}(k) \sim -\sum_{j=0}^{\infty} (-1)^j\ \partial H_j/\partial q\ 2^{-j-1}\ k^{-2j-1} \qquad (k\to\sqrt{-1}\infty).$$

The gradients are easily removed to obtain a development of $\lg s_{11}$ itself, and comparison with the first formula for $\lg s_{11}$ as $k \to \sqrt{-1}\infty$ leads to the trace formulas of FADDEEV-ZAKHAROV [1971]. The following is typical:

$$H_2 = \frac{4}{\pi}\int \lg |s_{11}(k)|^2\ k^4\ dk + \frac{16}{5}(k_1^5 + \ldots + k_g^5)\ .$$

5.5 The Bäcklund Transformation

The Bäcklund transformation produces new solutions of $\partial q/\partial t = X_2 q$ from old; see ESTABROOK-WAHLQUIST [1973], FLASCHKA-McLAUGHLIN [1976], and, for historical information, LAMB [1976]. The substitution $A: p \to q = c + p' + p^2$ of MIURA [1968] carries a solution p of the modified Korteweg-de Vries equation $\partial p/\partial t = 3(c+p^2)p' - (1/2)p'''$ into a solution of $\partial q/\partial t = X_2 q$. The modified equation is also solved by $-p$, so the Bäcklund transformation $B: q \to A(-A^{-1}q)$ maps one solution of $\partial q/\partial t = X_2 q$ into another. The transformation is well-defined if $c = -k_\infty^2 < 0$ lies to the left of the spectrum of Q. Then $-D^2+Bq$ has an additional bound state at $-k_\infty^2$; in particular, the successive application of the Bäcklund transformation with diminishing $c = -k_i^2$ $(i=1,\ldots,g)$ produces out of the vacuum $[q = 0]$ the many-soliton functions of §2.3. MIURA [1976] may be consulted for additional information; see, also, DEIFT-TRUBOWITZ [1978] for a different approach.

6. KDV ON THE CIRCLE: ISOSPECTRAL CLASSES AND BORG'S THEOREM

The Korteweg-de Vries equation $\partial q/\partial t = X_2 q$ is to be solved in the class C_1^∞ of infinitely differentiable functions of period 1, say. This was done for finitely many lacunae (see below) by DUBROVIN-NOVIKOV [1974], ITS-MATVEEV [1975], McKEAN-van MOEBERKE [1975], and NOVIKOV [1974]; see, also, AKHIEZER [1961], BAKER [193], LAX [1975], STEIMANN [1957], and, for excellent reviews, DUBROVIN-MATVEEV-NOVIKOV [1976] and MATVEEV [1976]. The simplest solution of $\partial q/\partial t = X_2 q$ of period 1 is the <u>cnoidal wave</u>: The solitary wave equation $cq' + 3qq' - (1/2)q''' = 0$ is integrated twice to obtain
1) $a + bq + (c/2)q^2 + (1/2)q^3 - (1/4)(q')^2 = 0$. This reduces to
2) $(p')^2 = 4(p - e_1)(p - e_2)(p - e_3)$ with real $e_1 > e_2 > e_3$ for $q = 2p + d$ and suitable a, b, c, d. 2) is solved by the Weierstrassian function p with primitive periods $2\omega_1 = 1$ and $2\omega_2 \in \sqrt{-1}\, R^1$, evaluated at $x + \omega_2$ $(0 \le x < 1)$ so as to keep p real and smooth. The cnoidal wave is $q(t,x) = 2p(x - ct + \omega_2) + d$.

6.1 <u>Periodic and Anti-Periodic Spectrum;</u> see MAGNUS-WINKLER [1966].

$Q = -D^2 + q$ is called Hill's operator in honor of G. W. Hill who studied it in connection with the motion of the moon. Let $y_1(x,\lambda)$ $[y_2(x,\lambda)]$ be the solution of $Qy = \lambda y$ with $y_1(0) = 1$, $y_1'(0) = 0$ $[y_2(0) = 0,\ y_2'(0) = 1]$. The <u>periodic</u> and <u>anti-periodic</u> spectra of Q are determined by solving $Qf = \lambda f$ with periodic or anti-periodic f $[f(x + 1) = \pm f(x)]$. They comprise a simple periodic ground state λ_0 followed by alternately anti-periodic and periodic pairs $\lambda_1 \le \lambda_2 < \lambda_3 \le \lambda_4 < \dots \uparrow \infty$ of simple or double eigenvalues; the intervals $(-\infty,\lambda_0]$ and $[\lambda_{2i-1},\lambda_{2i}]$ $(i=1,2,\dots)$ are termed <u>lacunae</u> from the fact that the spectrum of Q acting in $L^2(R^1)$ is their closed complement. The infinite differentiability of q is reflected in the fact that $\lambda_{2i} - \lambda_{2i-1}$ vanishes rapidly as $i \uparrow \infty$;* in fact, there is a common development

$$\lambda_{2i-1}, \lambda_{2i} = i^2\pi^2 + c_0 + c_1 i^{-2} + c_2 i^{-4} + \cdot \qquad (i \uparrow \infty).$$

The spectrum is determined by the existence of eigenvalues $e = \pm 1$ of the monodromy matrix

$$M = \begin{bmatrix} y_1(1,\lambda) & y_2(1,\lambda) \\ y_1'(1,\lambda) & y_2'(1,\lambda) \end{bmatrix};$$

the general eigenvalue $e = \Delta \pm \sqrt{\Delta^2-1}$ of M, viewed as a 2-valued function of λ, is the so-called <u>multiplier</u> of Q. M has determinant 1, so the periodic [anti-periodic] spectrum comprises the roots of $\Delta(\lambda) = +1$ [-1], Δ being the <u>discriminant</u> (1/2) sp M $= (1/2)\,[y_1(1,\lambda) + y_2'(1,\lambda)]$; see Figure 3. Δ is an integral function of order 1/2 and type 1 with $\Delta(\lambda) \sim \cos\sqrt{\lambda}$ $(\lambda \downarrow -\infty)$, so it can be

FIG. 3

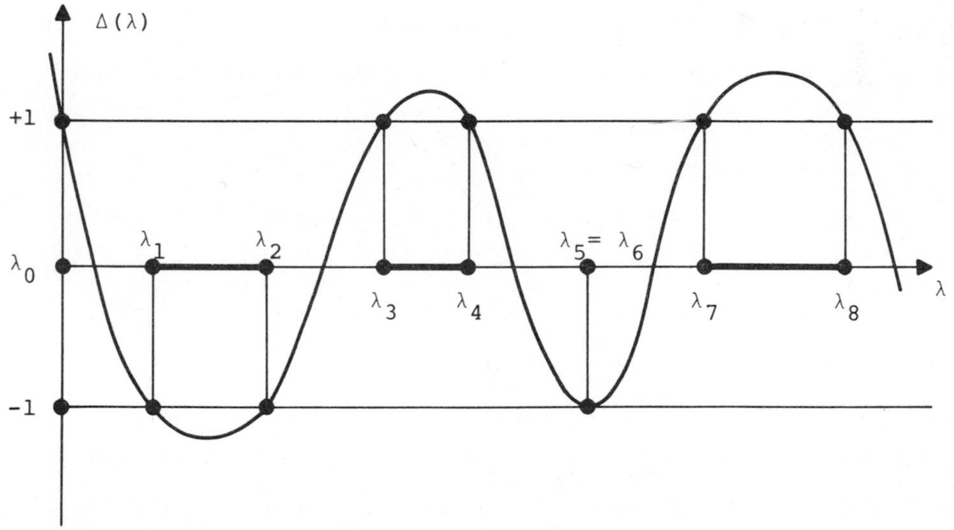

---

* HOCHSTADT [1965].

recovered from its roots, i.e., Δ and the periodic and anti-periodic spectra are equivalent pieces of information. The assumption is made that there are only $g < \infty$ non-trivial lacunae, i.e., that Q has only $2g + 1$ simple eigenvalues $\lambda_0 = \lambda_0^o < \lambda_1^o < \lambda_2^o < \ldots < \lambda_{2g-1}^o < \lambda_{2g}^o$, the rest of the spectrum being double, and the problem is posed: to determine the so-called isospectral class P of operators Q with fixed simple spectrum λ_i^o $(i=0,1,\ldots,2g)$. INCE [1940] provides an example; if p is the Weierstrassian function with primitive periods $2\omega_1 = 1$ and $2\omega_2 \in \sqrt{-1}\, R^1$ and if $q(x) = g(g+1)\, p(x+\omega_2)$ with $g = 0,1,2,3,\ldots$, then $\lambda_0 < \lambda_1 < \ldots < \lambda_{2g}$ is precisely the simple spectrum of Q; the spectrum is purely simple for every other choice of g. HOCHSTADT [1963] noticed that the simple spectrum determines the double: in fact, if $\Delta = \cos\theta$, then* $\Delta^{\bullet} = -\sin\theta\,\theta^{\bullet}$, so that, by cancellation of the double spectrum in $\Delta^{\bullet}$ and $\sqrt{\Delta^2-1}$,

$$\theta(\lambda) = \pm\sqrt{-1}\int_{\lambda_0}^{\lambda} \frac{\Delta^{\bullet}\, d\lambda'}{\sqrt{\Delta^2-1}} = \pm\frac{\sqrt{-1}}{2}\int_{\lambda_0}^{\lambda} \frac{\prod_{i=1}^{g}(\lambda'-\lambda_i^{\bullet})}{\sqrt{\prod_{i=0}^{2g}(\lambda'-\lambda_i^o)}}\, d\lambda' ,$$

$\lambda_i^{\bullet} \in [\lambda_{2i-1}^o, \lambda_{2i}^o]$ $(i=1,\ldots,g)$ being the nontrivial roots of $\Delta^{\bullet}(\lambda) = 0$. The proof is finished by noticing that these are specified by the simple spectrum: $\Delta^{\bullet}\, d\lambda / \sqrt{\Delta^2-1}$ is real in the lacunae, so

$$0 = \int_{\lambda_{2i-1}^o}^{\lambda_{2i}^o} \frac{\Delta^{\bullet} d\lambda}{\sqrt{\Delta^2-1}} \qquad (i = 1,\ldots,g) .$$

This represents g equations for the elementary symmetric functions of $\lambda_i^{\bullet}$ $(i=1,\ldots,g)$, with determinant

$$\det\left[\int_{\lambda_{2i-1}^o}^{\lambda_{2i}^o} \lambda^{j-1}\left(-\prod_{k=0}^{2g}(\lambda-\lambda_k^o)\right)^{-1/2} d\lambda :\ 1 \le i,j \le g\right] \neq 0 .$$

---

* $\bullet$ signifies differentiation with respect to λ.

The corollary that λ_i^o $(i=0,\ldots,2g)$ determines Δ is used below.

6.2 Auxiliary Spectrum

The auxiliary spectrum of Q is determined by solving $Qf = \mu f$ with $f(0) = f(1) = 0$; it comprises the roots of $y_2(1,\mu) = 0$, and it is easy to see that there is precisely one such root μ_i in each lacuna $[\lambda_{2i-1},\lambda_{2i}]$ $(i=1,2,\ldots)$. The majority are caught at double eigenvalues $\lambda_{2i-1} = \lambda_{2i}$; the rest are denoted by $\mu_i^o \in [\lambda_{2i-1}^o,\lambda_{2i}^o]$ $(i=1,\ldots,g)$. BORG [1945] proved that the auxiliary spectrum μ_i $(i=1,2,\ldots)$ and the norming constants $c_i^2 = [\int_0^1 y_2^{\,2}(x,\mu_i)\,dx]^{-1}$ $(i=1,2,\ldots)$ specify q. Now *

$$\int_0^1 y_2^2(x,\mu_i)\,dx = y_2^{\bullet}(1,\mu_i)\,y_2'(1,\mu_i)\ ,$$

and if the auxiliary spectrum is known, then so is $y_2^{\bullet}(1,\mu)$, $y_2(1,\mu) \sim (1/\sqrt{\mu})\sin\sqrt{\mu}$ being an integral function of order $1/2$; also, $y_2'y_1 - y_2y_1' = 1$ and $y_1 + y_2' = 2\Delta$, so that $y_2' = \Delta \pm \sqrt{\Delta^2-1}$ on the auxiliary spectrum, and if the simple spectrum is fixed, i.e., if Δ is known,** then the only new information contributed by the norming constant c_i^2 is the signature of the radical $\sqrt{\Delta^2(\mu_i)-1}$; the latter is ambiguous only if $\lambda_{2i-1} < \mu_i < \lambda_{2i}$. Now $\Delta^2(\lambda)-1$ splits into $-(\lambda-\lambda_0^o)\cdots(\lambda-\lambda_{2g}^o)\times$ a product of squares arising from the double spectrum. This suggests the introduction of the points $p_i = [\mu_i^o, y(\mu_i^o)]$ $(i=1,\ldots,g)$ with $y(\lambda) = \sqrt{-(\lambda-\lambda_0^o)\cdots(\lambda-\lambda_{2g}^o)}$ to specify the relevant information; it also suggests Figure 4 in which the lacuna $[\lambda_{2i-1}^o,\lambda_{2i}^o]$ is opened up into a circle, $y(\mu_i^o) > 0$ $[< 0]$ corresponding to the upper [lower] semicircle $(i=1,\ldots,g)$. The upshot of Borg's theorem is that $q \to p_i$ $(i=1,\ldots,g)$ is a 1:1 map of the isospectral class P into the g-dimensional torus formed by the product

---

* $\bullet = \partial/\partial\lambda$. $' = \partial/\partial x$.

** §6.1.

of these circles; in fact, the map is onto, as will be seen below.

FIG. 4

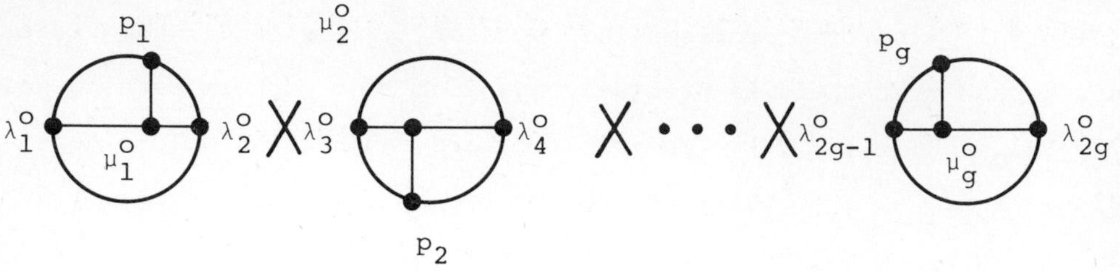

This seems to have been first proved by STEIMANN [1957];* see, also, AKHIEZER [1961].

6.3 Proof of Borg's Theorem; see LEVINSON [1949] for a proof of the full statement. The present article confirms that $q \to p_i$ $(i=1,\ldots,g)$ is a 1:1 map of P onto the torus of Fig. 4. The proof is adapted from McKEAN-MOERBEKE [1975] and TRUBOWITZ [1978].

STEP 1. Let q be translated by $0 \le x < 1$. This does not change the periodic or anti-periodic spectra of Q, but it does move μ_i^o $(i=1,\ldots,g)$. The first step is to prove that these numbers, as functions of $0 \le x < 1$, solve

$$X_1\mu_i^o = (\mu_i^o)' = \frac{2y(\mu_i^o)}{\prod_{j\neq i}(\mu_i^o - \mu_j^o)} \qquad (i=1,\ldots,g).$$

* F. Dyson pointed out this priority to me.

Let $\dot{q}$ be an infinitessimal change in q and let $f(x) = c_i y_2(x,\mu_i^o)$ be the normalized eigenfunction for $\mu = \mu_i^o$. Then $(Q-\mu)f^{\bullet} + q^{\bullet}f = \mu f^{\bullet} + \mu^{\bullet} f$, and taking the inner product with f yields * $\mu^{\bullet} = \int_0^1 f^2 q^{\bullet} df$, i.e., $\partial\mu_i^o/\partial q = f^2$. Now f^2 satisfies $Kf^2 = 2\mu_i^o (f^2)'$ with $K = qD + Dq - (1/2)D^3$, and this fact is now used to compute

$$X_1\mu_i^o = \int_0^1 \frac{\partial\mu_i^o}{\partial q} q' \, dx = \int_0^1 f^2 K1 \, dx$$

$$= -(f')^2\Big|_0^1 - \int_0^1 Kf^2 \, dx$$

$$= \frac{1-[y_2'(1)]^2}{\dot{y}_2 y_2'(1)}$$

$$= \frac{2\sqrt{\Delta^2-1}}{\dot{y}_2(1)}$$

$$= \frac{2y(\mu_i^o)}{\prod_{j\neq i}(\mu_i^o-\mu_j^o)} .$$

The identities $y_2'y_1 - y_2y_1' = 1$ and $y_1 + y_2' = 2\Delta$ are used in the next to the last step. The final step is achieved by factoring out the double spectrum.

STEP 2. The joint motion of μ_i^o $(i=1,\ldots,g)$ appears to have singularities: μ_i^o could come to rest at λ_{2i-1}^o or λ_{2i}. The fact is that this does not happen: <u>if μ_i^o hits λ_{2i-1}^o or λ_{2i}^o, then the signature of the radical flips and μ_i^o turns back, continuing its motion without rest</u>. This means that the motion of $\mu_i^o(x)$ $(i=1,\ldots,g)$ for $0 \leq x < 1$ is fully determined by p_i $(i=1,\ldots,g)$ for $x = 0$ which specifies $\mu_i^o(0)$ $(i=1,\ldots,g)$ and the initial directions of motion.

* $\int_0^1 f^{\bullet} f \, dx = 0$.

STEP 3. The 1:1 nature of the map $q \to p_i$ $(i=1,\ldots,g)$ is now confirmed by the trace formula:*

$$q(x) = \sum_{i=0}^{2g} \lambda_i^o - 2 \sum_{i=1}^{g} \mu_i^o(x) \qquad (0 \le x < 1).$$

PROOF of the trace formula for $x = 0$. Let p be the elementary solution of $\partial p/\partial t = -Qp$ in the doubled period interval $0 \le x < 2$, and note that**

$$p(t,x,x) = (4\pi t)^{-1/2}[1 - tq(x)] + O(t^{3/2}) \qquad (t \downarrow 0).$$

Now let p^o be the elementary solution in $0 \le x \le 1$ for Q restricted to functions vanishing at $x = 0$ and $x = 1$. Then $p^o(t,x,x)$ has the same development as $p(t,x,x)$ inside that interval, and if q is extended symmetrically to the left of $x = 0$, then, up to an exponentially small error $O[\exp(-c/t)]$, $p^o(t,x,x) = p(t,x,x) - p(t,x,-x)$ for $0 \le x < 1/2$, say. The upshot is that, for $t \downarrow 0$,

$$\int_0^2 p(t,x,x) - 2\int_0^1 p^o(t,x,x) = 2\int_0^1 [p(t,x,x) - p^o(t,x,x)] = \int_{-1/2}^{1/2} p(t,x,-x),$$

up to such errors; moreover, the final integral is well approximated if q is replaced by $q(0)$, i.e.,

$$\int_{-1/2}^{1/2} p(t,x,-x) \simeq e^{-tq(0)} \int \frac{e^{-x^2/t}}{\sqrt{4\pi t}}\, dx = [1 - tq(0)] + O(t^{3/2}).$$

The trace formula is now plain from the evaluation:

$$\int_0^2 p - 2\int_0^1 p^o = \sum_{i=0}^{\infty} e^{-\lambda_i t} - 2\sum_{i=1}^{\infty} e^{-\mu_i t}$$

upon noting that the double spectrum cancels out, leaving a finite sum.

---

* See, e.g., McKEAN-MOERBEKE [1975].

** See §2.4.

STEP 4 is to confirm that the map $q \to p_i$ (i=1,...,g) is onto. Let X_i (i=1,...,g) be the commuting vector fields $X_i: q \to D\, \partial H_i/\partial q$ (i=1,...,g). These preserve the isospectral manifold P in view of $X_i \int_0^2 p(t,x,x)\, dx = 0$ (i=1,...,g), which is proved just as in §2.4, and the evaluation $\int_0^2 p\, dx = \sum \exp(-\lambda_i t)$. Now a small translation places the auxiliary spectrum in general position $[\lambda^o_{2i-1} < \mu^o_i < \lambda^o_{2i}$ (i=1,...,g)], and a computation confirms

$$\det [X_i \mu^o_j : 1 \le i,j \le g] = \det [(2\mu^o_j)^{i-1}] \prod_{j=1}^{g} X_1 \mu^o_j \neq 0 .$$

This means that $X_1 q, \ldots, X_g q$ span the tangent space of P at such a point [dim P $\le$ g by Borg], and the commutation of the fields permits the same conclusion to be drawn even if μ^o_i (i=1,...,g) is in special position. The fact that the map $q \to p_i$ (i=1,...,g) is onto will now be plain; indeed, more is proved than was advertised: P is a g-dimensional torus and the commuting fields X_i (i=1,...,g) span its tangent bundle, so that the flows $\partial q/\partial t = X_i q$ (i=1,...,g) restricted to P are integrable. This will be seen in a more conventional format later on; see §8.7 for alternative classical formats.

6.4 Differential Equations

The fact that X_i (i=1,...,g) span the tangent bundle of P means that $X_{g+1} = c_1 X_1 + \cdots + c_g X_g$ on P. This represents a common nonlinear differential equation of degree 2g + 1 satisfied by q in the class P. It may be integrated g + 1 times with common constants of integration $c_0, c_{-1}, \ldots, c_{-g}$; the numbers c_i ($|i| \le g$) are (nonelementary) symmetric polynomials in λ^o_i (i=0,...,2g); see McKEAN-MOERBEKE [1975].

EXAMPLE 1: g = 0. $X_0 = 0$, so $X_1 q = q' = 0$.
This proves the theorem of BORG [1945] stating that, aside from λ_0 ,

the periodic and antiperiodic eigenvalues Q *are all double if and only if* q *is constant*.

EXAMPLE 2: $g = 1.$* $c_1X_1 + X_2 = 0$, so $c_1q' + 3qq' - (1/2)q''' = 0$ which leads to the cnoidal wave $q(x) = 2\,p(x + \omega_2) + c_{-2}$ of primitive real period $1/m$ $(m=1,2,3,\ldots)$; in particular, if $m = 1$ and $c_{-2} = 0$, then the simple spectrum of Q is $\lambda_0 = -e_1 < \lambda_1 = -e_2 < \lambda_2 = -e_3$, the numbers $e_1 > e_2 > e_3$ being the roots of the cubic $y^2 = 4(x - e_1)(x - e_2)(x - e_3)$ determining the associated elliptic curve.

6.5 Open Questions

It would be interesting to form some idea as the the general shape of the functions q from a fixed isospectral class P, e.g., *how many peaks and valleys can such a function have*? The only information to date is numerical; see HYMAN [1975]. I do not even know if P can contain an *odd* function. The question is: *If* q *is odd and of primitive period* 1, *must* Q *have infinite simple spectrum*? The situation with respect to *even* functions is simple: q is even if and only if $\mu_i^o = \lambda_{2i-1}^o$ or λ_{2i}^o $(i=1,\ldots,g)$, so P contains precisely 2^g such functions. The maximum and minimum of $q(0)$ on P are easily determined: by the trace formula, the maximum $\lambda_0^o + \sum (\lambda_{2i}^o - \lambda_{2i-1}^o)$ is achieved by placement of $\mu_i^o = \lambda_{2i-1}^o$ $(i=1,\ldots,g)$; similarly, the minimum is $\lambda_0^o - \sum (\lambda_{2i}^o - \lambda_{2i-1}^o)$.

---

* HOCHSTADT [1965].

7. KDV ON THE CIRCLE: PERIODICITY AND THE JACOBI VARIETY

7.1 Periodicity

Let f_i^o $(i=0,\ldots,2g)$ be the simple eigenfunctions of Q with $\int_0^1 (f_i^o)^2 dx = 1$. $\partial\lambda_i^o/\partial q(x) = [f_i^o(x)]^2$ is easily computed in the style of §6.3, and it follows from $\Delta(\lambda_i^o) = \pm 1$ that

$$\partial\Delta(\lambda)/\partial q(x) \text{ at } \lambda = \lambda_i^o \quad = -\Delta^{\bullet}(\lambda_i^o)\ [f_i^o(x)]^2 \qquad (i=0,\ldots 2g).$$

Now $\partial\Delta(\lambda)/\partial q(x) = y_2(x+1)y_1(x) - y_1(x+1)y_2(x)$ is the function $y_2(1)$ computed for q translated by $0 \le x < 1$; as such it is the product of $\prod_{i=1}^{g} [\lambda - \mu_i^o(x)]$ and a trivial factor coming from the double spectrum, so $[f_j^o(x)]^2$ is proportional to $\prod_{i=0}^{g} [\lambda_j^o - \mu_i^o(x)]$ $(j=0,\ldots,2g)$, with the self-evident corollary that $\mu_i^o(x)$ <u>hits both</u> <u>λ_{2i-1}^o and λ_{2i}^o m_i times in a full period</u> $0 \le x < 1$, m_i <u>being the common number of roots per period of f_{2i-1}^o and f_{2i}^o</u>. [f_0 is root-free, f_1 and f_2 have 1 root period, f_3 and f_4 have 2, <u>etc</u>.]
Now the differential equations of step 1, §6.3 imply *

$$\sum_{i=1}^{g} \frac{(\mu_i^o)^{j-1}}{y(\mu_i^o)} d\mu_i^o = 2 \sum_{i=1}^{g} (\mu_i^o)^{j-1} \prod_{k\neq i} (\mu_i^o - \mu_k^o)^{-1}\ dx$$

$$= 0 \qquad (j < g)$$

$$= 2\,dx \qquad (j = g),$$

and integration over $0 \le x < 1$ yields the <u>periodicity condition</u>:

$$\sum_{i=1}^{g} 2m_i \int_{\lambda_{2i-1}^o}^{\lambda_{2i}^o} \mu^{j-1}\, d\mu\ /y(\mu) = 0 \quad (j < g) \qquad = 2 \quad (j = g).$$

---

* The second sum is evaluated as $(2\pi\sqrt{-1})^{-1} \int \lambda^{j-1} \prod_{k=1}^{g} (\lambda-\mu_k^o)^{-1}\, d\lambda$ taken about a large circle.

The purpose of this article is to prove that* <u>this condition is not only necessary but also sufficient for</u> λ_i^o $(i=0,\ldots,2g)$ <u>to be the simple spectrum of a Hill's operator</u> Q <u>of period</u> 1.

PROOF. The differential equations step 1 of §6.3 are solved for $\mu_i^o(x)$ $(i=1,\ldots,g)$, subject to the understanding of step 2 of that article. The periodicity condition is now used to prove that this motion is of period 1. Let the motion begin at $o_i = [\lambda_{2i-1}^o, 0]$ $(i=1,\ldots,g)$ and end at p_i $(i=1,\ldots,g)$. Then

$$\sum_{i=1}^{g} \int_{o_i}^{p_i} \lambda^{j-1}\, d\lambda\, /y(\lambda) = \sum_{i=1}^{g} 2m_i \int_{\lambda_{2i-1}^o}^{\lambda_{2i}^o} \lambda^{j-1}\, d\lambda\, /y(\lambda) \qquad (j=1,\ldots g)$$

or, what is better for the purpose,

$$\sum_{i=1}^{g} \int_{o_i}^{p_i} \lambda^{j-1}\, d\lambda\, /y(\lambda) = 0 \qquad (j=1,\ldots,g)$$

by choice of paths, and it is easy to see that if the paths $o_i p_i$ are trivial for, <u>e.g.</u>, $m < i \le g$ and non-trivial for $1 \le i \le m$, then

$$\det \left[\int_{o_i}^{p_i} \lambda^{j-1}\, d\lambda\, /y(\lambda) : 1 \le i,j \le m \right] \ne 0 \ ,$$

which is contradictory, <u>i.e.</u>, $p_i = o_i$ $(i=1,\ldots,g)$. The function

$$q(x) = \sum_{i=0}^{2g} \lambda_i^o - 2 \sum_{i=1}^{g} \mu_i^o(x)$$

is now seen to be infinitely differentiable and of period 1; it is to be proved that λ_i^o $(i=0,\ldots,2g)$ <u>is the simple spectrum of</u> $Q = -D^2+q$. A tiresome but elementary computation shows that the product $F = \prod_{i=1}^{g} [\lambda - \mu_i^o(x)]$ satisfies $KF = 2\lambda F'$ with the customary

---

* McKEAN-MOERBEKE [1975]; the present proof is simpler.

$K = qD + Dq - (1/2)\,D^3$. Now for most values of λ, this equation is easily seen to have just one periodic solution:

$$\partial\Delta(\lambda)/q(x) = \prod_{i=1}^{\infty} (i^2\pi^2)^{-1} [\mu_i(x) - \lambda] ,$$

μ_i $(i=1,2,\ldots)$ being the full auxiliary spectrum of Q, so $\mu_i^o(x)$ $(i=1,\ldots,g)$ agree with some g of the functions $\mu_i(x)$ $(i=1,2,\ldots)$, and the rest of the auxiliary spectrum of Q is fixed under translation, as is easily verified by considering the motion of $\mu_i(x)$ $(i=1,2,\ldots)$ in a full period $0 \leq x < 1$. The periodicity conditions are now employed to prove that $\mu_i^o(x)$ hits λ_{2i-1}^o and λ_{2i}^o exactly m_i times per period. This proves that Q has $2g + 1$ simple eigenvalues: λ_i^o $(i=1,\ldots,2g)$ and a ground state someplace to the left. The latter is identified with λ_0^o by means of the formula for $X_1\mu_1^o$. The proof is finished.

<u>7.2 The Jacobi Variety;</u> see SIEGEL [1971] for general information on this subject.

The ubiquity of the radical $y(\lambda) = \sqrt{-(\lambda-\lambda_0^o)\cdots(\lambda-\lambda_{2g}^o)}$ suggests the introduction of the nonsingular hyperelliptic curve K of genus g with points $p = [\lambda, y(\lambda)]$. This is a sphere with g handles formed by cutting the λ-plane along $(-\infty,\lambda_0^o]$ and the lacunae $[\lambda_{2i-1}^o,\lambda_{2i}^o]$ $(i=1,\ldots,g)$, opening up the cuts, and pasting two copies together as in Fig. 5 [g = 2]. The nomenclature Hill's curve is reserved for such K with real branch points ∞ and λ_i^o $(i=0,\ldots,2g)$ satisfying the periodicity condition of §7.1. Let a_i $(i=1,\ldots,g)$ and b_i $(i=1,\ldots,g)$ be the standard homology basis of the curve depicted in Fig. 6 and let $\omega_j = \lambda^{j-1}\,d\lambda/y(\lambda)$ be the customary differentials of the first kind [DFK]; in this language, the periodicity condition of §7.1 states that $\sum m_i a_i(\omega_j) = 0$ $(j < g)$ $= 2$ $(j = g)$. The Jacobi variety of K is defined as follows. Let p_i $(i=1,\ldots,g)$ be a

FIG. 5

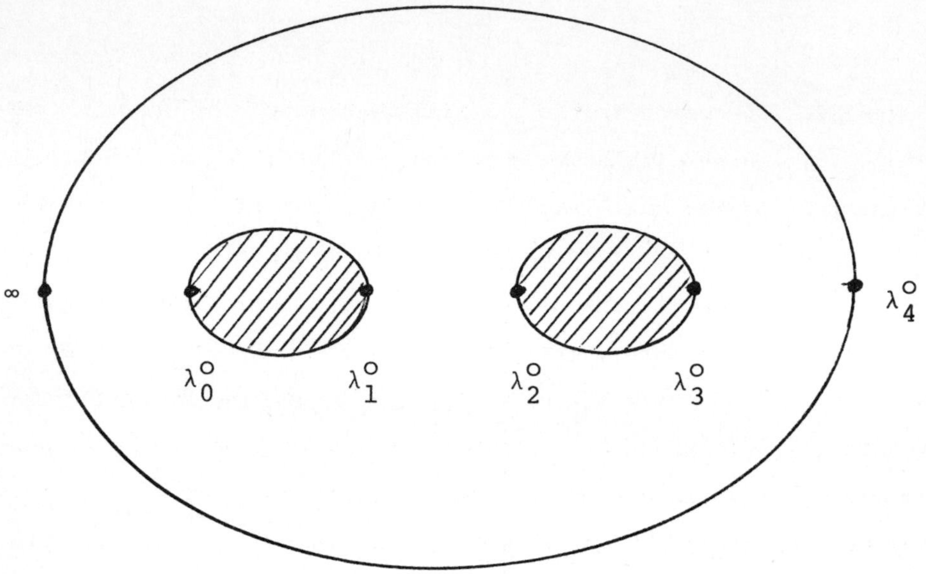

FIG. 6

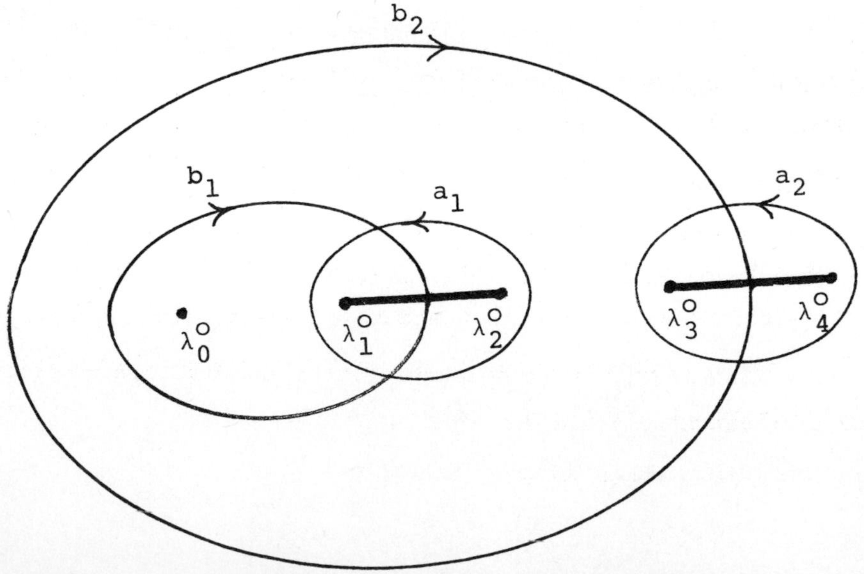

general divisor on the curve. Then the sums *

$$\sum_{i=1}^{g} \int_{o_i}^{p_i} \omega_j = x_j \qquad (j=1,\ldots g)$$

determine a point $\mathbf{x} = (x_1,\ldots,x_g) \in C^g$; in fact, the map is onto but ambiguous, the paths of integration $o_i p_i$ $(i=1,\ldots,g)$ being unspecified. The ambiguity is removed by factoring C^g by the lattice L_C of periods $\mathbf{x}$ arising from closed paths. The quotient space C^g/L_C is the Jacobi variety J_C of K; it is a g-dimensional complex torus. Now let p_i $(i=1,\ldots,g)$ be in *real position*, meaning that the projection $\lambda(p_i)$ lies in the lacuna $[\lambda^o_{2i-1},\lambda^o_{2i}]$ $(i=1,\ldots,g)$. Then the map is onto R^g provided the path $o_i p_i$ is restricted to the cycle covering $[\lambda^o_{2i-1},\lambda^o_{2i}]$ $(i=1,\ldots,g)$; also, the period lattice L_R obtained from closed paths is now real, and the quotient space $J_R = R^g/L_R$ is a g-dimensional *real* torus. The latter is denoted by J for brevity and is termed the *real part* of the Jacobi variety. The map from P to $\mathbf{x} \in J$ via the divisor p_i $(i=1,\ldots,g)$ is easily seen to be 1:1. The upshot is an identification of the isospectral class P with the real part J of the Jacobi variety of K.

7.3 Straightening out the Flows

This is where Isaiah 40:3-4 comes in. The isospectral flows $\partial q/\partial t = X_i q$ $(i=1,\ldots,g)$ commute on P and provide it with a nice addition which is not at all transparent in the auxiliary coordinates p_i $(i=1,\ldots,g)$; see §3.2. The point of introducing the Jacobi variety is *to identify the addition with that of* J, *i.e., with the addition theorem of hyperelliptic integrals*. The task is to prove that if $\mathbf{x}$ is the point of J assigned to Q and if, *e.g.*, $q^{\bullet} = X_2 q$, then $\mathbf{x}^{\bullet} = \mathbf{x}_2$ is independent of $\mathbf{x}$, so that $\mathbf{x}$ *moves in a straight line at constant speed* $[\mathbf{x} \to \mathbf{x} + t\,\mathbf{x}_2]$.

* $o_i = [\lambda^o_{2i-1},0]$ $(i=1,\ldots,g)$.

PROOF for X_1: Let $\mathfrak{x} \in J$ be the point corresponding to Q. Then *

$$X_1 x_j = X_1 \sum_{i=1}^{g} \int_{o_i}^{p_i} \omega_j$$

$$= \sum_{i=1}^{g} \frac{(\mu_i^o)^{j-1}}{y(\mu_i^o)} \frac{2y(\mu_i^o)}{\prod_{k\neq i}(\mu_i^o-\mu_k^o)}$$

$$= 0 \quad (j < g) \qquad = 2 \quad (j = g)$$

$$= \sum_{i=1}^{g} m_i a_i(\omega_j) \qquad (j = 1,\ldots,g) ,$$

i.e., $\mathfrak{x}_1 = (0,\ldots,2)$ is the co-tangent vector at $\mathfrak{x} \in J$ corresponding to the infinitessimal translation $X_1: q \to q'$. The proof is finished, $\mathfrak{x}_1$ being independent of the place $\mathfrak{x} \in J$.

PROOF for X_2: This is only a little harder. $X_2\mu_i^o = [2\mu_i^o+q(0)]X_1\mu_i^o$ is computed much as in step 1, §6.3, and it follows that **

$$X_2 x_j = \sum_{i=1}^{g} [2\mu_i^o + q(0)](\mu_i^o)^{j-1} \prod_{k\neq i} (\mu_i^o - \mu_k^o)^{-1}$$

$$= \frac{1}{2\pi\sqrt{-1}} \int [2\lambda + q(0)]\lambda^{j-1} \prod_{k=1}^{g} (\lambda - \mu_k^o)^{-1} d\lambda$$

$$= 0 \qquad (j=1,\ldots,g-2)$$

$$= 2 \qquad (j=g-1)$$

$$= q(0) + 2\sum_{i=1}^{g} \mu_i^o = \sum_{i=0}^{2g} \lambda_i^o \qquad (j=g) .$$

The cotangent vector $\mathfrak{x}_2 = \mathfrak{x}^{\bullet} = (0,\ldots,2,\sum\lambda_i^o)$ so defined is independent of the place $\mathfrak{x} \in J$, as before.

---

* See footnote, §7.1.

** The integral is taken about a big circle.

8. KDV ON THE CIRCLE: THE THETA FUNCTION AND BAKER'S FORMULA

8.1 The Theta Function

ITS-MATVEEV [1975] found that the map $P \to J$ has an elegant inversion in terms of the Riemann theta function. The formula stems from BAKER [1897]. The theta function is defined for $\mathfrak{x} \in C^g$ by the sum

$$\vartheta(\mathfrak{x}) = \sum e^{2\pi\sqrt{-1}\mathfrak{x}\cdot n} e^{-\pi C[n]} ,$$

in which the sum is taken over the dual lattice L_R' of points $n \in R^g$ such that $\mathfrak{x}\cdot n \in Z$ for periods $\mathfrak{x}$ L_R and $C[n]$ is the quadratic form based upon the period matrix $C = BA^*$ formed from $A = [a_i(\omega_j)]$ and $B = -\sqrt{-1}\,[b_i(\omega_j)]$. Riemann's period relations tstate that C is symmetric and positive, and Jacobi's transformation of the theta function shows that $\vartheta(\mathfrak{x}) > 0$ on J; see BAKER [1897] or SIEGEL [1971] for such matters.

8.2 Baker's Formula

Let prime denote differentiation on J in the direction $\mathfrak{x}_1 = (0,\ldots,2)$ corresponding to infinitessimal translation and let *

$$\omega = \frac{1}{2}\Delta^{\bullet}\, d\lambda \Big/ \sqrt{\Delta^2 - 1} = \frac{1}{2}\prod_{i=1}^{g}(\lambda - \lambda_i^{\bullet})\, d\lambda \,/\, y(\lambda) .$$

Baker's formula states that

$$\sum_{i=1}^{g}\int_{o_i}^{p_i}\omega = \frac{\vartheta'(\mathfrak{x})}{\vartheta(\mathfrak{x})} .$$

PROOF of Baker's Formula. Let p_i $(i=1,\ldots,g)$ be fixed in, e.g., real position, and let p be variable. Riemann's vanishing theorem states that if $\mathfrak{x}_\infty = [\int_\infty^p \omega_j : 1 \le j \le g]$ and if $\mathfrak{x} = [\sum_{i=1}^{g}\int_{o_i}^{p_i}\omega_j : 1 \le j \le g]$,

---

* $\lambda_i^{\bullet}$ $(i=1,\ldots,g)$ are the nontrivial roots of $\Delta^{\bullet}(\lambda) = 0$. ω is the differential figuring in Hochstadt's formula in §6.1.

then $\vartheta(\mathfrak{x} - \mathfrak{x}_\infty)$, considered as a function $f(p)$ of p alone, vanishes simply at $p = p_i$ $(i=1,\ldots,g)$ and no place else. Now fix $o = [\lambda_0, 0]$, and compute

$$\frac{1}{2\pi\sqrt{-1}} \int \left(\int_o^p \omega \right) d\, \ell g\, f(p) ,$$

the integral being taken as p runs over the boundary $a_1 b_1 a_1^{-1} b_1^{-1} \ldots$ $\ldots a_g b_g a_g^{-1} b_g^{-1}$ of a dissection of the curve. The sum $\sum \int_o^{p_i} \omega$ is produced by the poles of $d\, \ell g\, f(p)$ coming from the roots p_i $(i=1,\ldots,g)$ of $f(p) = \vartheta(\mathfrak{x} - \mathfrak{x}_\infty)$. The pole $-\sqrt{-\lambda}$ of $\int_o^p \omega$ produces a residue at ∞: in fact, in the local parameter $k = 1/\sqrt{-\lambda}$, $\mathfrak{x}_\infty \sim k(0,\ldots,2)$, and as this is proportional to the direction $\mathfrak{x}_1 = (0,\ldots,2)$, the integral may be evaluated as

$$\sum_{i=1}^{g} \int_o^{p_i} \omega - \frac{\vartheta'(\mathfrak{x})}{\vartheta(\mathfrak{x})} .$$

The integral is now re-evaluated via periods: $a_i(\omega) = 0$ $(i=1,\ldots,g)$ by inspection; also, $\vartheta(\mathfrak{x})$ is real periodic, i.e., the addition of a real period $[a_i(\omega_j): 1 \le j \le g]$ to $\mathfrak{x}$ leaves it unchanged, while the addition of an imaginary period $[b_i(\omega_j): 1 \le j \le g]$ simply multiplies it by an exponential * $\exp[2\pi\sqrt{-1}\ (A^{-1*}x)_i]$. The upshot is that, with a change of base points,

$$\sum_{i=1}^{g} \int_{o_i}^{p_i} \omega - \frac{\vartheta'(\mathfrak{x})}{\vartheta(\mathfrak{x})}$$

is a constant independent of p_i $(i=1,\ldots,g)$. The latter is seen to vanish by taking $p_i = o_i$ $(i=1,\ldots,g)$ $[\mathfrak{x} = 0]$ and noting that $\vartheta(0)$ is the biggest value of $\vartheta(\mathfrak{x})$ on J so that $\vartheta'(0) = 0$.

---

* $A = [a_i(\omega_j): 1 \le i,j \le g]$ as before.

8.3 Its-Matveev's Formula

The formula of ITS-MATVEEV [1975] is obtained from Baker's formula: a further differentiation in the direction $\mathfrak{x}_1$ yields

$$[\lg \vartheta(\mathfrak{x})]'' = \sum_{i=1}^{g} \prod_{j=1}^{g} (\mu_i^o - \lambda_j^o) \prod_{j\neq i} (\mu_i^o - \mu_j^o)^{-1}$$

$$= \frac{1}{2\pi\sqrt{-1}} \int \prod_{j=1}^{g} \frac{\lambda - \lambda_j^\bullet}{\lambda - \mu_j^o}\, d\lambda$$

$$= \sum_{j=1}^{g} (\mu_j^o - \lambda_j^\bullet) ,$$

the integral being taken about a big circle. The trace formula of §6.3 is now applied to identify the sum as $-\frac{1}{2}[q(0) - c]$ with $c = \sum \lambda_i^o - 2 \sum \lambda_i^\bullet$, whence the final result:

$$q(x) = -2[\lg \vartheta(\mathfrak{x} + x\mathfrak{x}_1)]'' + c \qquad (0 \leq x < 1);$$

in particular, if $\mathfrak{x}_2$ is the vector co-tangent to J corresponding to $X_2: q \to 3qq' - (1/2)q'''$, then

$$\exp(tX_2)\, q(\mathfrak{x}) = -2\, [\lg \vartheta(\mathfrak{x} + x\mathfrak{x}_1 + t\mathfrak{x}_2)]'' \qquad (0 \leq x < 1).$$

BAKER [1897] presents another formula to the effect that the jth simple eigenfunction of Q is proportional to $\vartheta(\mathfrak{x} + x\mathfrak{x}_1 + \mathfrak{x}_\infty) \times \vartheta^{-1}(\mathfrak{x} + x\mathfrak{x}_1)$ with $\mathfrak{x}_\infty = \left[\int_\infty^o \omega_i : i=1,\dots,g\right]$ and $o = [\lambda_j^o, 0]$. This may be employed for $j = 0$ to prove that <u>if f_0 is the ground state of Q, then the map</u> $Q \to Q - 2(\lg f_0)''$ <u>is an isospectral involution of C_1^∞, expressible in Jacobi language by the addition of half the sum of the real periods</u>:

$$\mathfrak{x} \to \mathfrak{x} + [\tfrac{1}{2} \sum_{i=1}^{g} a_i(\omega_j) : 1 \leq j \leq g] .$$

This is the analogue of the Bäcklund transformation of §5.5.

8.4 Solitons and Singular Curves

The formula of Its-Matveev is similar to that of Hirota [§2.3] and to that of Dyson [§4.4]. MATVEEV [1976] and McKEAN [1979] clarified this point. Let the spectrum of Q be shifted so that $\lambda_{2g}^{o} = 0$ and let the intervals $[\lambda_{2i-2}^{o}, \lambda_{2i-1}^{o}]$ $(i=1,\ldots,g)$ shrink to points $-k_1^2 < \ldots < -k_g^2 < 0$. The final singular curve $y(\lambda) = \sqrt{-\lambda}(\lambda + k_1^2)\ldots$ $\ldots(\lambda + k_g^2)$ so produced is a sphere with parameter $k = \sqrt{-\lambda}$ with the points $\pm\sqrt{-1}\, k_i$ $(i=1,\ldots,g)$ identified in pairs. The behavior of the period matrix $C = BA^*$ is most readily computed relative to a new basis of DFK: $\omega_j = \prod_{g>j\neq i} (\lambda - \lambda_{2j}^{o})\, d\lambda / y(\lambda)$ $(i=1,\ldots,g)$. The result is achieved only with tears: *

$$a_{ij} = a_i(\omega_j) = \lg \frac{1}{\varepsilon_i} + O(1) \qquad (i=j)$$

$$= \lg \left|\frac{k_i + k_j}{k_i - k_j}\right| + o(1) \qquad (i\neq j)$$

$$b_{ij} = b_i(\omega_j) = 1 + o(1) \qquad (i=j)$$

$$= o(1) \qquad (i\neq j)$$

with $\varepsilon_i = \lambda_{2i-1}^{o} - \lambda_{2i-2}^{o}$ $(i=1,\ldots,g)$. The behavior of $\vartheta(x)$ may now be seen by means of Jacobi's transformation: up to an inessential factor,

$$\vartheta(\mathfrak{x}) = \sum e^{-\pi(\mathfrak{x}-\ell)AB^*(\mathfrak{x}-\ell)},$$

the sum being taken over the lattice AZ^g. Let $1/2 = (1/2,\ldots,1/2)$, $\mathfrak{x} = \mathfrak{x}' + A(1/2)$, and $\ell = An$ with $n \in Z^g$. Then the general term of the sum is

---
* See FAY [1973] for such matters.

$$\exp\ [o(1) + 2\pi(n - \tfrac{1}{2})\not{x}' - \pi(n - \tfrac{1}{2})\ A[1 + o(1)]\ (n - \tfrac{1}{2})]\ ,$$

and the principal contribution is produced by terms with
$(n - 1/2)\ A(n - 1/2) = \sum (n - 1/2)^2\ \ell g(1/\varepsilon_i) + O(1)$ as small as
possible, i.e., $n_i = 0$ or 1 $(i=1,\ldots,g)$. The upshot is that with $\not{x}$
in place of $2\pi\not{x}'$ and an inessential factor $\exp(a\not{x} + b)$ omitted, the
chief part of the sum is

$$\not{\theta}\ _0(\not{x}) = \sum e^{\not{x}\cdot n} \prod_{i<j} \left(\frac{k_i - k_j}{k_i + k_j}\right)^{2n_i n_j}$$

with $n_i = 0$ or 1 $(i=1,\ldots,g)$. This is just Hirota's sum of §2.3,

now identified as the theta function of the singular curve. It is
also easy to estimate the directions $\not{x}_j$ co-tangent to J corresponding
to the fields X_j $(j=1,\ldots,g)$: in fact, $\not{x}_j$ is proportional to
$k^{2j-1} = (k_1^{2j-1},\ldots,k_g^{2j-1})$, or approximately so, the upshot being that
if $\not{\theta}_1 = \not{\theta}_0(\not{x} + xk + t_1 k + t_2 k^3 + \cdots + t_g k^{2g-1})$, then
$q = - 2(\ell g\ \not{\theta}_1)''$ satisfies $\partial q/\partial t_j = X_j q$ $(j=1,\ldots,g)$ up to
inessential scalings, as it should.

8.5 Dyson's Formula: Reprise

McKEAN [1979] noticed that, in the absence of bound states, [*]
$\not{\theta}\ (\not{x}) = \det\ [I + L(\xi+\eta): 0 \le \xi,\eta]$ can also be construed as the
theta function of a (nonclassical) singular curve like
$\dot{y}(\lambda) = \sqrt{-\lambda}\ (\lambda + k_1^2) \cdots (\lambda + k_g^2)$ but with additional singularities
arising from the nontrivial part of the continuous spectrum $[0,\infty)$ of
Q, the adjective nontrivial meaning $s_{11}(k) \neq 1$: in fact, the
substitution of

$$L(\xi + \eta) = \frac{1}{2\pi} \int e^{\sqrt{-1}k(\xi+\eta)}\ s_{21}(k)\ dk$$

[*] See §4.4; $\not{x}$ = phase s_{21}.

into

$$\mathfrak{D}(\not{x}) = 1 + \sum_{n=1}^{\infty} \frac{1}{n!} \int_{[0,\infty)^n} \det [L(x_i + x_j): 1 \leq i,j \leq n]\, d^n x$$

yields

$$\mathfrak{D}(\not{x}) = 1 + \sum_{n=1}^{\infty} \left(\frac{\sqrt{-1}}{2}\right)^n \int_{k_1<\dots<k_n} \frac{|s_{21}(k_1)|}{k_1} \cdots \frac{|s_{21}(k_n)|}{k_n} d^n k$$

$$\times\, e^{\sqrt{-1}[\not{x}(k_1)+\cdots+\not{x}(k_n)]} \prod_{i<j} \left(\frac{k_i - k_j}{k_i + k_j}\right)^2 ,$$

in which the k's downstairs are construed as $k + \sqrt{-1}\,0+$. The self-evident analogy with Hirota's sum requires further investigation.

8.6 Infinite Genus

See McKEAN-TRUBOWITZ [1976 & 1978]. The assumption $g < \infty$ is now removed: in fact, the periodic and antiperiodic spectrum of Q is commonly all simple, and it will be most instructive to place ourselves in that case. The case of mixed simple and double spectra does not offer any additional difficulty. Now $y(\lambda) = \sqrt{\Delta^2-1}$, the curve K being of infinite genus with handles piling up at $p = \infty$. They do not get much in the way, being most of them very small, i.e., K is doing its best to be compact. The isospectral class P is specified by Δ as before, and Borg's theorem remains in force, i.e., the map from Q to the divisor $p_i = [\mu_i, y(\mu_i)]$ $(i=1,2,\dots)$ is 1:1 and onto the now ∞-dimensional torus depicted in Fig. 4, §6.2. The tangent bundle to P is more delicate: in general, the local fields X_j $(j=1,2,\dots)$ do not span it, and it is necessary to use the nonlocal fields $X: q \to D[\partial\Delta(\lambda)/\partial q]$ which are technically nicer anyhow; see §8.9, below, for more information on this point. The differentials of the first kind [DFK] also pose technical problems. The map *

$$\mathcal{X}: \omega \to \sum_{i=1}^{\infty} \int_{o_i}^{p_i} \omega$$

associated with the divisor p_i $(i=1,2,\dots)$ in real position is viewed as an element of the dual space (DFK)', and after dividing out the appropriate lattice L_R, the quotient $J = J_R = (DFK)'/L_R$ is required to be a faithful copy of the isospectral manifold Q and to have a sufficiently large tangent space to accommodate the fields X_j $(j=1,2,3,\dots)$. The correct balance is struck by taking for DFK the differentials $\omega = \phi(\lambda)\, d\lambda \,/\, y(\lambda)$ with integral transcendental $\phi(\lambda)$ subject to 1) $|\phi(\lambda)| \le \exp\left[\sqrt{|\lambda|} \times [1 + o(1)]\right]$, 2) $\phi(\lambda) = [\phi(\lambda^*)]^*$, 3) $\int_0^{\infty} |\phi(\lambda)|^2 \lambda^{3/2} d\lambda < \infty$, and 4) $\phi(\lambda)$ vanishing rapidly as $\lambda \uparrow +\infty$ via the lacunae $[\lambda_{2i-1}, \lambda_{2i}]$ $(i=1,2,\dots)$. The direction cotangent to J corresponding to the infinitessimal translation X_1 is

$$\mathcal{X}_1(\omega) = 2 \lim_{\lambda \downarrow -\infty} (-\lambda)^{3/2} \phi(\lambda) \Big/ y(\lambda) ,$$

and the periodicity condition of §7.1 is now restated as

$$\mathcal{X}_1(\omega) = \sum_{j=1}^{\infty} j a_j(\omega)$$

The map from P to J straightens out the flows; in particular, the Korteweg-de Vries flow $\partial q/\partial t = X_2 q$ appears as a straight-line motion at constant speed on the compact ∞-dimensional torus J, explaining the quasi-periodicity of the numerical experiments of HYMAN [1975]; see, also, TAPPERT [197]. The picture also clarifies the stability of the flow; see McKEAN [1977]. The map is inverted as before $[q = -2(\lg \vartheta)'']$ with a (nonclassical) theta function on (DFK)'. ϑ has a nice vanishing theorem and behaves pleasantly

---

* $o_i = [\lambda_i, 0]$ $(i=1,2,\dots)$.

in every respect. The *complex* Jacobi variety J_C awaits proper investigation; it is always unbounded. The function theory of K requires further development, too: for example, a) What is the correct function field? b) What is the correct form of Abel's theorem? c) How is Riemann-Roch to be construed? See McKEAN [1979] for additional information and comment. The periodicity condition can be dropped so long as some control over the ramifications of K is maintained. Then $q = -2(\ell g\, \vartheta)''$ is merely *almost periodic*. The question as to whether every such function arises via the theta function from a (nonclassical) curve of infinite genus is *open* despite intimations to the contrary in the Soviet literature.

8.7 The Foliation of C_1^∞.

The whole space C_1^∞ is now foliated* by isospectral classes P identified as the real Jacobi varieties J of curves $y = \sqrt{\Delta^2-1}$, properly desingularized; the identification imputes to the leaves of the foliation an addition reflecting the addition theorem of hyperelliptic integrals. It is a remarkable fact that there do not appear any singular leaves such as are familiar in the classical mechanics of, *e.g.*, the simple pendulum. MARCENKO-OSTROVSKII [1975] proved that finite-dimensional leaves are dense; see §11.5. The flows may be reduced, leaf-wise, to conventional Hamiltonian form by use of the coordinates $q_i = \mu_i^o$ and $p_i = \ell g\,[\Delta(\mu_i^o) + \sqrt{\Delta^2(\mu_i^o) - 1}\,]$ $(i=1,\ldots,g)$; see FLASCHKA-McLAUGHLIN [1976], and, for a different approach, DIKII-GELFAND [1975]. The naive statement that *the existence of the foliation and the commutativity of flows confirms the integrability of KDV* requires further comment. The number of degrees of freedom is infinite and it is not clear how to count $\infty/2$ involutive integrals of the motion; in particular, the integrals H_j $(j=0,1,2,\ldots)$ are generally too few, as in §5.3. The values of the discriminant $\Delta(\lambda)$

* The nomenclature is abused in that a local product structure is absent.

for real λ are better: They are involutive; the gradient $\partial\Delta/\partial q$ is normal to Q, while the field X: $q \to D\partial\Delta/\partial q$ is tangent; and, what is decisive, span$[\partial\Delta/\partial q] \oplus$ span$[D\partial\Delta/\partial q]$ fills up the whole of C_1^∞ so that no direction of the ambient space is left unclassified; see McKEAN-TRUBOWITZ [1976]. X_j $(j=1,2,...)$ do span the tangent bundle if $\lambda_{2i} - \lambda_{2i-1} \leq a\ e^{-bi}$ $(i\uparrow\infty)$ which is the condition for real analyticity of q; see McKEAN-TRUBOWITZ [1976] and TRUBOWITZ [1977] for more information.

8.8 A Geometrical Comment

Let the spectrum of Q be simple. Then J_R is compact, but the unit ball of its tangent space (DFK)' is not. The anomaly is explained by the fact that the lattice L_R contains arbitrarily small periods so that the exponential map (DFK)' $\to$ J is not locally 1:1 at any point. Now J_R is an actual variety in any sense you like, though infinite-dimensional and transcendental: for example, it embeds in projective space via 3^∞ theta functions in the manner of LEFSCHETZ [19] * and may be defined by an infinite number of equatioss of degree ≤ 4 . ** I would draw the moral: that the conventional insistence upon locally 1:1 exponential maps is only finite-dimensional prejudice and need not be respected in infinite-dimensional geometry.

---

* See LANG [n972] for the case of finite genus.

** See §10.6.

9. THE VIEWPOINT OF ALGEBRAIC GEOMETRY

KDV on the circle is re-examined from the viewpoint of algebraic curves. The present development was initiated by BAKER [192 & 193] and BURCHNELL-CHAUNDY [1922,1928, & 1931], forgotten, and rediscovered in an improved form by KRICHEVER [1976]. The present account follows him.

9.1 Transcendental Singularities; see BAKER [1897].

Let K be any nonsingular algebraic curves of genus $g < \infty$ with standard homology basis a_i, b_i $(i=1,\ldots,g)$ and fix a nonspecial divisor q_i $(i=1,\ldots,g)$. The adjective nonspecial means that no nonconstant function of rational character on K has poles q_i $(i=1,\ldots,g)$ or softer. The Riemann-Roch theorem implies that every curve has such divisors; for example, if K is a Hill curve, any divisor in real position* will do, A new point p_∞ is now fixed and a local parameter $1/k$ is introduced in its vicinity $[k(p_\infty) = \infty]$. BAKER [1897] proved that if $E(k)$ is any small polynomial, then K carries a unique function $f(p)$, of rational character away from p_∞, with poles q_i $(i=1,\ldots,g)$ and transcendental singularity $\exp(E)$ at p_∞; more precisely, $f(p) = \exp(E) \times [1 + c_1 k^{-1} + c_2 k^{-2} + \ldots]$ in the vicinity of p_∞.

PROOF. $f(p) \neq 0$ near p_∞ and $\int d\, \ell g\, f(p) = 0$ on a small circuit about p_∞, so $f(p)$ must have g roots p_i $(i=1,\ldots,g)$. Now it is a standard fact ** that K carries 1) a unique differential $\omega_\infty \sim dE(k)$ near p_∞ with no other poles and $a_i(\omega_\infty) = 0$ $(i=1,\ldots,g)$, and 2) a unique differential ω_{qp} with poles at q and p, residues -1 and +1, and $a_i(\omega_{qp}) = 0$ $(i=1,\ldots,g)$. Now $\omega = d\, \ell g\, f(p)$ can be expressed, up to

---

* $\lambda(q_i) \in [\lambda^o_{2i-1}, \lambda^o_{2i}]$ $(i=1,\ldots,g)$.

** SIEGEL [1971].

a possible differential of the first kind, as $\omega_\infty + \sum \omega_{q_i p_i}$ with unspecified p_i $(i=1,\ldots,g)$. The periods $a_i(\omega)$ $(i=1,\ldots,g)$ vanish automatically, so

$$f(p) = \exp\left(\int_{p_o}^{p} \omega\right)$$

is a <u>bona fide</u> single-valued function on K if $b_i(\omega) \equiv 0$ <u>modulo</u> $2\pi\sqrt{-1}$ $(i=1,\ldots,g)$. The task is to prove that this can be achieved by choice of p_i $(i=1,\ldots,g)$. Let ω_i $(i=1,\ldots,g)$ be a basis of DFK such that $a_i(\omega_j) = 2\pi\sqrt{-1}$ or 0 according as $i = j$ or not. Then *

$$b_j(\omega_{qp}) = \int_p^q \omega_j \qquad (j=1,\ldots,g),$$

so

$$b_j(\omega) = b_j(\omega_\infty) + \sum_{i=1}^{g} \int_{p_i}^{q_i} \omega_j \qquad (j=1,\ldots,g),$$

and since the map

$$p_i\ (i=1,\ldots,g) \to \sum_{i=1}^{g} \int_{p_o}^{p_i} \omega_j \qquad (j=1,\ldots,g)$$

is <u>onto</u> C^g, $b_j(\omega)$ $(j=1,\ldots,g)$ can, in fact, be made to vanish by choice of p_i $(i=1,\ldots,g)$. The smallness of E(k) now comes into play. The effect is to make ω_∞ small and p_i $(i=1,\ldots,g)$ close to q_i $(i=1,\ldots,g)$, so that the former inherits the nonspecial character of the latter. The uniqueness of f(p) is now plain.

---

* SIEGEL [1971].

9.2 Differential Operators

Now let $f(p)$ have the singularity $\exp[xk + t_1E_1(k) + t_2E_2(k)+\ldots]$ with distinguished parameter x. The next step is to prove the existence of differential operators L_i, acting upon x and independent of p, such that $\partial f/\partial t_i = L_i f$ $(i=1,2,\ldots)$

PROOF for $t_1E_1(k) + \cdots = t\,k^3$. The proof may be carried out near p_∞ and the result continued over the rest of K. Let $^\bullet = \partial/\partial t$ and $' = \partial/\partial x$. Then

$$f^\bullet = [k^3 + (1+c_1/k+\cdots)^{-1}(c_1^\bullet/k+\cdots)]f = [k^3+O(k^{-1})]f$$

$$f' = [k + O(k^{-1})]f$$

$$f'' = [k^2 + 2c_1' + O(k^{-1})]f$$

$$f''' = [k^3 + 3c_1'k + 3(c_2'-c_1c_1'+c_1'') + O(k^{-1})]f ,$$

by elementary calculations, which makes it plain that there is a differential operator L of degree 3 such that $Lf = f^\bullet + O(k^{-1})f$ in the vicinity of p_∞. The function $h = f^{-1}(Lf - f^\bullet)$ is of rational character on K with poles p_i $(i=1,\ldots,g)$ or softer, and as this divisor is nonspecial, h vanishes identically, being constant and vanishing at p_∞. The proof is finished.

COMMUTATORS. The differential operators L so produced satisfy $0 = [L_i,L_j] + \partial L_i/\partial t_j - \partial L_j/\partial t_i$ $(i,j=1,2,\ldots)$; in fact, the operator in question annihilates $f(p)$ and it is easy to see from the singularity $f(p) \sim \exp(xk)$ that the operator itself vanishes identically.

PROJECTION AS EIGENVALUE. Let K carry a function $\lambda(p)$ with pole at p_∞ only: $\lambda(p) = E_0(k) + O(k^{-1})$ near p_∞. Then the function f_o with singularity $\exp[xk + t_0E_0(k) + t_1E_1(k) + \ldots]$ satisfies $\partial f_0/\partial t_0 = L_0 f_0$. But f_0 is just the product of the function f with singularity $\exp[xk + t_1E_1(k) + \ldots]$ and the global function $\exp[t_0\lambda(p)]$, so $L_0 f = \lambda f$; in particular, L_0 is independent of the parameter t_0, and $[L_i, L_0] = \partial L_0/\partial t_i$ $(i=1,2,\ldots)$. This defines flows of the coefficients of L_0; plainly, they commute.

9.3 Examples.

Let $f(p) = f_x(p) \sim \exp[xk + t_1k + t_2k^2 + t_3k^3 + \ldots]$ in the vicinity of p_∞ and note the evaluations

$$L_1 = D\,,\quad L_2 = D^2 - q,\quad L_3 = D^3 - (3/2)qD - p,$$

$$[L_2, L_3] = q''' - (3/2)qq' + [(3/2)q'' - 2p']D - p''$$

with $q = 2c_1'$ and $p = 3(c_2' - c_1c_1' + c_1'')$; see §9.2.

EXAMPLE 1: KDV. Let $\lambda(p) \sim k^2$. Then K is hyperelliptic, $\lambda(p)$ having valence 2, and p_∞ is a Weierstrass point if $g \geq 2$. $L_0 = L_2 = D^2 - q$, $f'' - qf = \lambda f$, and the flows are computed as follows:

a) $\partial q/\partial t_1 = [L_0, L_1] = q' = X_1 q$; b) $\partial q/\partial t_2 = [L_0, L_2] = 0$;

c) $\partial q/\partial t_3 = [L_0, L_3] = q''' - (3/2)qq' + [(3/2)q'' - 2p']D - p''$,

in which $(3/2)q'' = 2p'$, D being absent from the left-hand side, so that $\partial q/\partial t_3 = -(3/2)qq' + (1/4)q''' = (-1/2)X_2 q$; in fact, the present choice of $\lambda(p)$ trivializes the <u>even</u> flows [$L_4 = L_0^2$, $L_6 = L_0^3$, <u>etc</u>.], while the <u>odd</u> flows coincide with $\partial q/\partial t = X_j q$ $(j=1,2,3,\ldots)$ up to inessential scaling. The association of X_j with the singularity $\exp(t\,k^{2j-1})$ is to be compared to the scattering recipe $s_{21}(k) \to s_{21}(k)\exp\left(\sqrt{-1}\,k^{2j-1}\,t\right)$ of §5.1. Now let the projections

$\lambda(p) = \lambda_0, \lambda_1, \ldots, \lambda_{2g}, \infty$ of the branch points of K be real and in order of increase. Then $q = 2c_1'$ is real if and only if the pole divisor q_i $(i=1,\ldots,g)$ is in real position. The projection of the root divisor p_i $(i=1,\ldots,g)$ of the function $f_1(p)$ plays the role of the nontrivial auxiliary spectrum of $Q = -D^2 + q = -L_0$. Q is of period 1 only if $f_{x+1}(p) = e(p)f_x(p)$ $(0 \le x < 1)$ with multiplier $e(p) \neq 0, \infty$ for $p \neq p_\infty$ and $e(p) \sim \exp(k)$ at p_∞. Now* $e(p)e(p')$ is an integral function of $\lambda = \lambda(p)$ and ~ 1 at p_∞, so $e(p)e(p') \equiv 1$, i.e., $e(p)$ is a unit of the function field of K; it is identical to the multiplier $\Delta \pm \sqrt{\Delta^2 - 1}$ of §6.1. The importance of such units was emphasized by MARCENKO-OSTROVSKII [1975]; see McKEAN [1979] for the case $g = \infty$.

EXAMPLE 2: BOUSSINESQ. Let $\lambda(p) \sim k^3$. K is now a 3-sheeted cover of its projection. $L_1 = 0$, $L_2 = D^2 - q$, $L_3 = D^3 - (3/2)qD - p$, as before, only now $L_0 = L_3$. The first flow is translation:

$$\frac{3}{2}\dot{q}D + \dot{p} = [L_0, L_1] = \frac{3}{2}q'\,D + p'$$

The second is a version of Boussinesq:

$$\frac{3}{2}\dot{q}D + \dot{p} = [L_0, L_2] = [\frac{3}{2}q'' - 2p']D + q''' - \frac{3}{2}qq' - p''$$

is equivalent to $\dot{q} = q'' - (4/3)p'$, $\dot{p} = q''' - (3/2)qq' - p''$, and that implies $\partial^2 q/\partial t^2 = (q^2 - (1/3)q'')''$. The third flow is trivial $[L_0 = L_3]$; indeed, the present choice of $\lambda(p)$ trivializes all the flows of index $\equiv 0$ modulo 3, while the others fall into two series to be described fully in §14.2.

EXAMPLE 3: KADMOTSEV-PETIASHVILII. Let K be any nonsingular curve and let $t_2 = a$, $t_3 = b$ for ease of writing. Then

---
* $p \to p'$ is the sheet map of K.

$$0 = [L_2, L_3] + \partial L_2/\partial b - \partial L_3/\partial a$$

$$= q''' - \frac{3}{2} qq' + \left(\frac{3}{2} q'' - 2p'\right) D - p'' - \frac{\partial q}{\partial b} + \frac{3}{2} \frac{\partial q}{\partial a} D + \frac{\partial p}{\partial a}$$

is equivalent to

$$\frac{\partial q}{\partial a} = \frac{4}{3} p' - q'' , \qquad \frac{\partial q}{\partial b} = q''' - \frac{3}{2} qq' - p'' + \frac{\partial p}{\partial a} .$$

The elimination of p by cross-differentiation yields the equation of KADOMTSEV-PETIASHVILII [1970]:

$$\frac{\partial^2 q}{\partial a^2} = \frac{4}{3} \frac{\partial}{\partial b} \frac{\partial q}{\partial x} + \frac{\partial^2}{\partial x^2} \left(q^2 - \frac{1}{3} \frac{\partial^2 q}{\partial x^2}\right) .$$

KDV results if q is independent of a, and Boussinesq if q is independent of b.

EXAMPLE 4: COMMUTING DIFFERENTIAL OPERATORS. Let $f(p) \sim \exp(xk)$ at p_∞ and let R be the ring of polynomials E(k) which are the principal parts at p_∞ of functions $\lambda(p)$ of rational character on K with no other poles. The structure of R is regulated by the Weierstrass gap theorem. The procedure of §9.2 associates to each $E \in R$ a differential operator L such that $Lf = \lambda f$. These form a commutative ring R' and $R \to R'$ is an isomorphism. The pleasant thing is that <u>any such commuting ring</u> R' <u>of ordinary differential operators containing</u> L_1 <u>and</u> L_2 <u>of coprime degree is obtained in this way</u>. This is the chief result of BURCHNELL-CHAUNDY [1922,1928 & 1931] rediscovered by KRICHEVER [1976]; see MUMFORD [1978] for additional information.

10. A SYSTEM OF C. NEUMANN.

The motion of translation of the simple eigenfunctions f_{2j}^{o} $(j=0,\ldots,g)$ of the Hill's operator Q is embedded in the integrable mechanical system of NEUMANN [1895], to wit, the motion of $g+1$ uncoupled harmonic oscillators $\ddot{x}_i + \omega_i^2 x_i = 0$ $(i=0,\ldots,g)$ with unequal frequencies $\omega_0 < \omega_1 < \ldots < \omega_g$, pulled back to the sphere S^g: $|x| = 1$. The motion goes on in the co-tangent bundle $T^*(S^g)$ of $|x| = 1$. T^* contains infinitely many g-dimensional tori of trajectories of period 1, these being copies of isospectral classes, i.e., they are real parts of Jacobi varieties.

10.1 A Remarkable Identity

The embedding of the Hill system into Neumann's is based upon a remarkable identity for the simple eigenfunctions of Q :*

$$1 = \sum_{i=0}^{g} [f_{2i}^{o}(x)]^2/e_i^2$$

with positive e_i^2 depending only upon the isospectral class. The derivation is indicated. $\partial\Delta(\lambda)/\partial q(0) = y_2(1,\lambda)$ was identified in §7.1 with $\prod_{i=1}^{g} (\lambda - \mu_i^{o})$, up to inessential factors coming from the double spectrum of Q; also, $\Delta^{\bullet}(\lambda) = -\prod_{i=1}^{g} (\lambda - \lambda_i^{\bullet})$ up to the same inessential factors, $\lambda_i^{\bullet} \in (\lambda_{2i-1}^{o}, \lambda_{2i}^{o})$ $(i=1,\ldots,g)$ being the nontrivial roots of $\Delta^{\bullet}(\lambda) = 0$. The formula

$$-\Delta^{\bullet}(\lambda_i^{o})[f_i^{o}(x)]^2 = \partial\Delta(\lambda)/\partial q(x) \text{ at } \lambda = \lambda_i^{o} \quad (i=0,\ldots,2g)$$

now implies

$$\prod_{j=1}^{g} (\lambda_i^{o} - \lambda_j^{\bullet})[f_i^{o}(x)]^2 = \prod_{j=1}^{g} [\lambda_i^{o} - \mu_j^{o}(x)] \quad (i=0,\ldots,2g)$$

---

* McKEAN-MOERBEKE [1975].

by cancellation of common factors. This is used to interpolate $\prod_{j=1}^{g} [\lambda - \mu_j^o(x)]$ off the g+1 points λ_{2i}^o $(i=0,\dots,g)$. The result is

$$\sum_{i=0}^{g} \frac{e_i^{-2}[f_{2i}^o(x)]^2}{\lambda - \lambda_{2i}^o} = \frac{\prod_{i=1}^{g} [\lambda - \mu_i^o(x)]}{\prod_{i=0}^{g} (\lambda - \lambda_{2i}^o)}$$

with positive

$$e_i^2 = \prod_{j=1}^{g} (\lambda_{2i}^o - \lambda_j^\bullet) \prod_{\substack{0\le j\le g \\ j\ne i}} (\lambda_{2i}^o - \lambda_{2j}^o)^{-1} \qquad (i=0,\dots,g).$$

The remarkable identity follows by appraisal near $\lambda=\infty$; indeed, expanding in inverse powers of λ yields a whole series of identities of which the first is the remarkable identity, cited above. The second is

$$q(x) = \sum_{i=0}^{g} 2\lambda_{2i}^o [f_{2i}^o(x)]^2/e_i^2 + \lambda_0^o + \sum_{i=1}^{g} (\lambda_{2i}^o - \lambda_{2i-1}^o) ;$$

it is derived using the trace formula of §6.3. These two identities form the starting point of the present story. The results are recorded in a more convenient format: with the substitution $x \to t$ and $f_{2i}^o(x)/e_i = x_i(t)$, $\mu_i^o(x) = a_i(t)$, $\lambda_{2i}^o = \omega_i^2$, $\lambda_0^o + \sum (\lambda_{2i}^o - \lambda_{2i-1}^o) = c$,

1) $$\ddot{x}_i + \omega_i^2 x_i = q x_i \qquad (i=0,\dots g),$$

2) $$|x|^2 = \sum_{i=0}^{g} x_i^2 = 1 ,$$

3) $$q = \sum_{i=0}^{g} 2\omega_i^2 x_i^2 + c ,$$

4) $$\sum_{i=0}^{g} \frac{x_i^2}{\lambda - \omega_i^2} = \frac{\prod_{i=1}^{g} (\lambda - a_i)}{\prod_{i=0}^{g} (\lambda - \omega_i^2)} = \frac{P(\lambda)}{Q_+(\lambda)} .$$

10.2 Neumann's System

The system 1), 2), 3) of §10.1 has a simple mechanical interpretation: qx is just the extra force required to pull back the system $\ddot{x}_i + \omega^2 x_i = 0$ $(i=0,\ldots,g)$ to the sphere S^g: $|x| = 1$; in fact, if $\ddot{x}_i + \omega_i^2 x_i = qx_i$ $(i=0,\ldots,g)$ and if $|x|^2 = 1$, then, $\dot{x}\cdot x = 0$ implies

$$-|\dot{x}|^2 = \ddot{x}\cdot x = q - \sum \omega_i^2 x_i^2 \quad \text{and} \quad \dddot{x}\cdot \dot{x} + \sum \omega_i^2 x_i \dot{x}_i = 0\,;$$

in particular, $|\dot{x}|^2 + \sum \omega_i^2 x_i^2 = -c$ is a constant of the motion, and consequently, 3): $q = \sum 2\omega_i^2 x_i^2 + c$. This is Neumann's system in the cotangent bundle $T^*(S^g)$: $|x|^2 = 1$, $x\cdot y = 0$ with $\dot{x} = y$; more precisely, the Hill system is a subsystem of Neumann's singled out by the particular choice of g constants of motion λ_{2i-1}^o $(i=1,\ldots,g)$ so as to achieve the periodicity condition of §7.1, as will be clarified below. Notice that the embedding map $Q \to xy \in T^*(S^g)$ is 1:1 since 4) permits the recovery of μ_i^o $(i=1,\ldots,g)$ from x and of $y(\mu_i^o))(i=1,\ldots,g)$ from y, by differentiation and the formula $X_1\mu_i^o = 2y(\mu_i^o)\prod_{j\neq i}(\mu_i^o-\mu_j^o)^{-1}$ of §6.3. The fact that the Hill system is contained in Neumann's was pointed out to me by J. Moser and E. Trubowitz; see D. & G. CHOODNOVSKY [1978], MOSER [1979], and, for more information, MOSER-TRUBOWITZ [1979]. NEUMANN [1859] proved the integrability of 1), 2), 3). The next three articles follow him.

10.3 New Coordinates

The system 1), 2), 3) in the full 2g-dimensional cotangent bundle is described by the ellipsoidal coordinates a_i $(i=1,\ldots,g)$ of 4) interlacing ω_i^2 $(i=0,\ldots,g)$, and by g more canonically conjugate coordinates b_i $(i=1,\ldots,g)$, so chosen as to make the restriction to the cotangent bundle of the customary symplectic form $\sum_{i=0}^{g} dx_i \wedge dy_i$ coincide with $\sum_{i=1}^{g} da_i \wedge db_i$. The computation is tiresome but elementary with the result that $b_i = -(1/4)P'(a_i)\dot{a}_i/Q_+(a_i)$ $(i=1,\ldots,g)$. Now the total energy of the motion is the sum $H_1(x,y)$ of $y^2/2$ and the equation can be expressed as

$$0 = -2(H_3 - \frac{1}{2}\eta_1) = \sum_{i=1}^{g} [4Q_+(a_i)b_i^2 + Q_-(a_i)]\,[P'(a_i)]^{-1}$$

with $b = \partial G/\partial a$ and $Q_-(\lambda) = \lambda^g + \eta_1\lambda^{g-1} + \eta_2\lambda^{g-2} + \ldots + \eta_g$, the numbers η_i $(i=2,\ldots,g)$ being arbitrary. Now let $G(a,\eta) = \sum_{i=1}^{g} G_i(a_i)$ [separation of variables]. Then a solution is obtained by choice of $G_i' = (1/2)\sqrt{-Q_-/Q_+}$ $(i=1,\ldots,g)$, or, what is the same,

$$G = \sum_{i=1}^{g} \frac{1}{2}\int^{a_i} \sqrt{-Q_-(\lambda)/Q_+(\lambda)}\,d\lambda\ .$$

The upshot is that

$$\xi_j = \partial G/\partial\eta_j = \frac{1}{4}\sum_{i=0}^{g}\int_{a_i(0)}^{a_i(t)} \frac{\lambda^{g-j}}{\sqrt{-Q(\lambda)}}\,d\lambda \qquad (j=1,\ldots,g)$$

with $Q = Q_-Q_+$. Neumann's system is solved thereby in terms of hyperelliptic integrals; in particular, the co-tangent bundle is foliated by invariant leaves indexed by η_i $(i=1,\ldots,g)$. The form of these leaves will be clarified presently. The individual leaf arises from a Hill system if and only if the roots λ_{2i-1}^o $(i=1,\ldots,g)$ of $Q_-(\lambda) = 0$ fill out the odd part of a simple Hill spectrum, <u>i.e.</u>, $\lambda_0^o = \omega_0^2 < \lambda_1^o < \lambda_2^o = \omega_2^2 < \lambda_3^o < \ldots < \lambda_{2g}^o = \omega_g^2$ and

$$\sum_{i=1}^{g} m_i \int_{\lambda_{2i-1}^o}^{\lambda_{2i}^o} \frac{\lambda^{j-1}}{\sqrt{-Q(\lambda)}}\,d\lambda = 0 \quad (j < g) \qquad = 1 \quad (j = g)$$

for some integral m_i $(i=1,\ldots,g)$. Meiman's theorem [§12.1] asserts that this can be done uniquely for every choice of $m_1 < \ldots < m_g$, so one sees, among the leaves of the foliation, a countable infinity of real parts of Jacobi varieties on which every trajectory is of period 1; indeed, there appear such tori for <u>every</u> period, and they potential energy $U = (1/2) \sum_{i=0}^{g} \omega_i^2 x_i^2$, and it follows from the variational form of mechanics that $a^{\bullet} = [a,H_2]$ and $b^{\bullet} = [b,H_2]$ with the classical Poisson bracket

$$[A,B] = \sum_{i=1}^{g} \left[\frac{\partial A}{\partial a_i}\frac{\partial B}{\partial b_i} - \frac{\partial A}{\partial b_i}\frac{\partial B}{\partial a_i}\right]$$

and $H_2(a,b) = H_1(x,y)$. H_2 is computed from 4) with tears:

$$H_2(a,b) = -2 \sum_{i=1}^{g} \frac{Q_+(a_i)}{P'(a_i)} b_i^2 - \frac{1}{2} \sum_{i=1}^{g} a_i + \frac{1}{2} \sum_{i=0}^{g} \omega_i^2 .$$

The equivalent Hamiltonian $H_3 = H_2 - (1/2) \sum_{i=0}^{g} \omega_i^2 = (1/2)\eta_1$ is an integral of the motion.

10.4 Hamilton-Jacobi Theory

The plan is to make a further canonical substitution $ab \to \xi\eta$ by means of a function $G = G(a,\eta)$ $[b = \partial G/\partial a,\ \xi = \partial G/\partial \eta]$. G is determined by the partial differential equation of Hamilton-Jacobi: $H_3(a,b) = H_3(a,\partial G/\partial a) = \eta_1/2$; it depends upon g-1 more constants of integration η_i (i=2,...g) representing additional integrals of the motion, and the statement of the Hamilton-Jacobi theory is that the flow is equivalent to $\xi_1^{\bullet} = 1$, $\xi_i^{\bullet} = 0$ (i=2,...,g), $\eta_i^{\bullet} = 0$ (i=1,...g). The solution is achieved by a favorite device of Jacobi:

$$\begin{aligned}\sum_{i=1}^{g} a_i^n [P'(a_i)]^{-1} &= \frac{1}{2\pi\sqrt{-1}} \int \frac{\lambda^n \, d\lambda}{P(\lambda)} \\ &= 0 \qquad (n = 0,1,\ldots,g-2) \\ &= 1 \qquad (n = g-1) \\ &= \sum_{i=1}^{g} a_i \quad (n = g) ,\end{aligned}$$

the integral being taken about a big circle, so the Hamilton-Jacobi are ubiquitous but sparse in the full co-tangent bundle; see McKEAN-MOERBEKE [1979]. This seems quite remarkable.

10.5 Integrals of the Motion

NEUMANN [1859] found $g+1$ involutive integrals of the motion:

$$F_i = x_i^2 + \sum_{j\neq i} \frac{(x_i y_j - x_j y_i)^2}{\omega_i^2 - \omega_j^2} \qquad (i=0,\ldots,g),$$

of which any g are independent, confirming integrability. Their relation with the previous integrals η_i $(i=1,\ldots,g)$ is expressed by

$$\sum_{i=0}^{g} (\lambda - \omega_i^2)^{-1} F_i = \frac{Q_-(\lambda)}{Q_+(\lambda)} .$$

The identifications $Q_-(\lambda) = \prod_{i=1}^{g} (\lambda - \lambda_{2i-1}^{o})$ and $Q_+(\lambda) = \prod_{i=0}^{g} (\lambda-\lambda_{2i}^{o})$ in the Hill case make it plain that all the flows $\partial q/\partial t = X_j q$ $(j=1,2,3,\ldots)$ extend from the Hill-type leaves to the whole cotangent bundle in a natural (involutive) way; in fact, $H_1 = \sum_{i=0}^{g} \lambda_{2i}^{o} F_i$, $H_2 = \sum_{i=0}^{g} (\lambda_{2i}^{o})^2 F_i$ + a multiple of H_1, <u>etc</u>. The formula also permits the identification of the Hill tori in the cotangent bundle as the intersection of g quartics.

10.6 Infinite Genus

The extension of all this to the case of simple spectra as in §8.6 is easy. The Hamilton-Jacobi function is

$$G = \sum_{i=1}^{\infty} \int_{o_i}^{p_i} \sqrt{\prod_{i=1}^{\infty} \frac{\lambda-\lambda_{2i-1}}{\lambda-\lambda_{2i}}} \; \frac{d\lambda}{\sqrt{\lambda-\lambda_0}} .$$

Now λ_{2i-1} $(i=1,2,\ldots)$ play the role of integrals of the motion, and since

$$\omega_j = (\lambda-\lambda_{2j-1})^{-1} \prod_{i=1}^{\infty} (i^2\pi^2)^{-1} (\lambda_{2i-1} - \lambda)\, d\lambda/\sqrt{\Delta^2(\lambda)-1}$$

is a basis of DFK, the variables

$$x_j = -2 \frac{\partial G}{\partial \lambda_{2j-1}} = \sum_{i=1}^{\infty} \int_{o_i}^{p_i} \omega_j \qquad (j=1,2,\ldots)$$

form coordinates of a point on the Jacobi variety moving, as the Hamilton-Jacobi theory predicts, along a straight line. To my knowledge this is the first explicit example of Hamilton-Jacobi theory with infinitely many degrees of freedom. It should be emphasized that the integrals F_i $(i=0,\ldots,g)$ carry over to $g = \infty$, permitting the identification of the ∞-dimensional Jacobi variety J_R as the intersection of $\infty = g$ quartics. If that is not an infinite-dimensional algebraic variety, what is? D. Mumford informs me of the following beautiful description of the Jacobi variety J_C of the curve $y^2(\lambda) = \prod_{i=0}^{2g+1} (\lambda - \lambda_i)$ due to REID [1972]; see, also, DONAGI [1978]. J_C is isomorphic to the variety of g-dimensional subspaces of P^{2g+1} on which both the forms $\sum x_i^2$ and $\sum \lambda_i x_i^2$ vanish identically. This must be connected to the preceding development.

10.7 A Little History

The tricks employed in §§10.2-4 have been used for other purposes:

$$\sum_{i=1}^{g} \int_{o_i}^{p_i} \lambda^{g-j} \, d\lambda \,/\, \sqrt{-Q(\lambda)} = c_1 t \quad (j = 1) \qquad = 0 \quad (j > 1)$$

represents 1) straight line motion on R^g expressed in ellipsoidal coordinates if $\deg Q = 2g-1$; 2) simple harmonic motion similarly expressed if $\deg Q = 2g$; 3) the present system of Neumann if $\deg Q = 2g + 1$; and 4) the motion of the Toda lattice if $\deg Q = 2g + 2$. JACOBI [1884] treated 1) and 2); in particular, he used 1) to give a beautiful proof of the addition theorem for hyperelliptic integrals. 4) is due to MOERBEKE [1978]; see §11. The same ideas enter into KOVALEVSKAYA's celebrated memoir on the top [1889]; see ARNOLD [1974], GOLUBEV [1953], and MANAKOV [1976].

11. THE TODA LATTICE

The lattice was introduced by TODA [1967]; it is quite similar to Korteweg-de Vries. The infinite lattice is solved in FLASCHKA [1974] and MANAKOV [1974]; see, also, FLASCHKA [1975], HÉNON [1974], HIROTA [1973], MOSER [1975], and TODA [1974]. The periodic lattice is solved in KAC-MOERBEKE [1975]; see, also, DATE-TANAKA [1976], MOERBEKE [1978], and MOERBEKE-MUMFORD [1978]. The next article deals with the allied algebraic curves, the actual lattice makes its entrance later. The material was prepared in collaboration with P. van Moerbeke.

11.1 Toda Curves

Let K be a nonsingular hyperelliptic curve of genus $g < \infty$, with real branch points $\lambda_0^o < \lambda_1^o < \dots < \lambda_{2g+1}^o$, defined by the radical $y(\lambda) = \sqrt{(\lambda-\lambda_0^o) \dots (\lambda-\lambda_{2g+1}^o)}$. The following statements about K are equivalent: 1) K has a real division point p_∞ , meaning that the projection $\lambda(p_\infty)$ is real and that $\int_{p_\infty}^{p'_\infty}$ is a fractional real period:*

$$n \int_{p_\infty}^{p'_\infty} \omega_j = \sum_{i=1}^{g} m_i a_i(\omega_j) \qquad (j=1,\dots,g)$$

for some integral n, m_i (i=1,...,g).** p_∞ and p'_∞ have the same projection $\lambda(p_\infty)$ which is taken to be ∞. 2) K carries a unit† $e(p) = a(\lambda) + y(\lambda)\, b(\lambda)$ with rational integral a,b satisfying the reality condition†† and having an n-fold pole at p_∞ and an n-fold root at p'_∞. 3) K is the desingularization of a curve

---

* $p \to p'$ is the sheet map. a_i is the counterclockwise loop about the lacuna $[\lambda_{2i-1}^o, \lambda_{2i}^o]$ (i=1,...,g). $\omega_j = \lambda^{j-1}\, d\lambda / y(\lambda)$ (j=1,...,g), as usual.

** K has infinitely many division points if g = 1. Mumford conjectured that curves of genus g > 1 can have only a finite number; see, e.g., LANG [1965] and LIARDET [1975] for such matters.

† $e(p)$ is a unit if $e(p)e(p') = a^2(\lambda) - y^2(\lambda)b^2(\lambda)$ is unity; compare example 1, §9.3.

†† $c^*(\lambda^*) = c(\lambda)$.

$e^2(p) - 2a(\lambda)e(p) + 1 = 0$ <u>expressed by the vanishing of a determinant</u>

$$\det \begin{bmatrix} b_1-\lambda & a_1 & 0 & \cdots & & 0 & a_n/e \\ a_1 & b_2-\lambda & a_2 & & & 0 & 0 \\ \vdots & & & & & \vdots & \\ 0 & 0 & 0 & & a_{n-2} & b_{n-1}-\lambda & a_{n-1} \\ a_n e & 0 & 0 & \cdots & 0 & a_{n-1} & b_n-\lambda \end{bmatrix} = \prod_{i=1}^{n} a_i \times (e + e^{-1} - 2a)$$

with real $a_i \neq 0$, b_i $(i=1,\ldots,n)$.

PROOF: see MOERBEKE-MUMFORD [1978] for more details. 1) implies that K carries a function e(p) of rational character with a single n-fold pole at p_∞ and a single n-fold root at p'_∞. e(p)e(p') has no roots or poles, so it is constant aand may be adjusted to be unity. Now $2a(\lambda) = e(p) + e(p')$ is rational with pole only at ∞, i.e., it is integral, and $e = a + yb$ with rational b. $e(p)e(p') = a^2 - y^2b^2 = 1$ now implies that b is also integral, the roots of $y^2(\lambda) = 1$ being simple. e is the unit of 2). The reality of a and b follow from a self-evident uniqueness. The opposite implication is even easier. The existence of the unit implies that $n(p'_\infty - p_\infty)$ is a period; its reality follows from that of a and b.

PROOF that 2) implies 3). Let q_i $(i=1,\ldots,g)$ be a nonspecial divisor in real position * on K. Then for $j = 0,1,2,\ldots$, K carries a function $f_j(p)$ of rational character, unique up to a constant multiplier, with poles q_i $(i=1,\ldots,g)$, j more poles at p_∞, j roots at p'_∞, and necessarily g more roots p_i $(i=1,\ldots,g)$ some place else; compare §9.1. Now the function $\lambda(p)f_j(p)$ has one more pole at p_∞ and one less root at p'_∞, so it can be expressed as

---

* $\lambda(q_i) \in [\lambda^o_{2i-1}, \lambda^o_{2i}]$ $(i=1,\ldots,g)$; such a divisor is automatically nonspecial.

$a_{j-1}f_{j-1}(p) + b_j f_j(p) + a_j f_{j+1}(p)$ with real $a_j \neq 0$, b_j $(j=0,1,2,\ldots)$, by a suitable disposition of the multiplicative constants, and the existence of the unit e of 2) implies a periodicity: $f_{j+n} = e(p)f_j(p)$, $a_{j+n} = a_j$, $b_{j+n} = b_j$ $(j=0,1,2,\ldots)$. The matrix of 3) is now seen to annihilate $(f_1,\ldots,f_n)$, so its determinant vanishes, producing a quadratic equation $e(p) + e^{-1}(p) - 2a(\lambda) = 0$ with rational integral $a(\lambda)$ of degree n. The curve $e^2 - 2ae + 1 = 0$ so produced is evidently a singularization of K: in fact, $e(p) = a(\lambda) \pm \sqrt{a^2(\lambda)-1}$ with $2a(\lambda) = e(p) + e(p')$ and $\pm\sqrt{a^2(\lambda)-1} = y(\lambda)b(\lambda)$, so the singular curve is simply K with the points p, p' covering the roots of $b(\lambda) = 0$ identified in pairs.

PROOF that 3) implies 2). Let K be the desingularization of the curve defined by the vanishing of the determinant of 3): $e^2 - 2ae + 1 = 0$. The branch points of K are real: indeed, $e = a \pm \sqrt{a^2-1}$ so $a^2 - 1 = 0$ if and only if $e = \pm 1$, in which case the matrix of 3) is real and symmetric and has real eigenvalues. e is the unit of 2): $e(p)e(p') = (a + \sqrt{a^2-1})(a - \sqrt{a^2-1}) = 1$.

11.2 Isospectral Classes

The discussion of §11.1 touches upon a number of important issues; see MOERBEKE [1978] for more information. The roots of $y^2(\lambda) = 0$ and $b^2(\lambda) = 0$ comprise the simple, resp., double periodic and anti-periodic spectra of

$$Q = \begin{bmatrix} b_1 & a_1 & 0 & \cdots & 0 & a_n \\ a_1 & b_2 & a_2 & & 0 & 0 \\ \vdots & & & & & \\ 0 & 0 & 0 & a_{n-2} & b_{n-1} & a_{n-1} \\ a_n & 0 & 0 & 0 & a_{n-1} & a_n \end{bmatrix}.$$

Let $\lambda_0^o < \lambda_1^o < \ldots < \lambda_{2g+1}^o$ be the simple spectrum of Q and let

μ_i^o (i=1,...,g) be the points of the auxiliary spectrum * which fall in the lacunae $[\lambda_{2i-1}^o, \lambda_{2i}^o]$ (i=1,...,g). These are the projections of the roots p_i (i=1,...,g) of the function $f_0(p)$, and the map $Q \to p_i$ (i=1,...,g) is 1:1, as is the map from this divisor to the sums **

$$x_j = \sum_{i=1}^{g} \int_{o_i}^{p_i} \omega_j \qquad (j=1,...,g)$$

construed modulo periods. The isospectral class P of Q is identified thereby with the real part J of the Jacobi variety of K, just as in §7.2. Now † $Q^\bullet = [Q,(Q^j)^+]$ (j=1,...,g) is an isospectral motion; in fact, the fields $X_j: Q \to [Q,(Q^j)^+]$ (j=1,...,g) commute, span the tangent bundle of Q, and the induced motions on J proceed in straight lines at constant speed; compare §§7.3 and 3.4. For example, if n = 4,

$$Q^+ = \begin{bmatrix} 0 & a_1 & 0 & -a_4 \\ 0 & 0 & a_2 & 0 \\ 0 & 0 & 0 & a_3 \\ 0 & 0 & 0 & 0 \end{bmatrix},$$

and the motion $Q^\bullet = X_1 Q = [Q,Q^+]$ is equivalent to $a_i^\bullet = a_i(b_i - b_{i+1})$, $b_i^\bullet = 2(a_i^2 - a_{i-1}^2)$ (i=1,...,4) with $a_0 = a_4$ and $b_5 = b_1$. Now comes the trick: †† if $a_i = (1/2)\exp[(1/2)(x_i - x_{i+1})]$ and $b_i = (1/2)\, y_i$ (i=1,...,4), then $(x_i - x_{i+1})^\bullet = y_i - y_{i+1}$ and $y_i^\bullet = \exp(x_{i-1} - x_i) - \exp(x_i - x_{i+1})$ (i=1,...,4), and this is just the Toda lattice of period 4 if you identify $x_i^\bullet = y_i$ (i=1,...,4); see §1.2. The same

---

* i.e., the spectrum of Q with its nth row and column crossed out.

** $o_i = [\lambda_{2i-1}^o, 0]$ (i=1,...,g).

† $(Q^j)^+$ is the upper triangular part of Q^j with the diagonal left out and an extra minus sign in the upper corner.

†† FLASCHKA [1974]. $0 < a_i$ (i=1,...,g) if this is so initially, as is now assumed.

recipe works for any n. The inverse map from $\mathfrak{x} \in J$ to Q may be expressed by means of the Riemann theta function much as in §8.3. The analogue of Baker's formula is *

$$\sum_{i=1}^{g} \int_{o_i}^{p_i} \prod_{i=1}^{g} (\lambda - \lambda_i^{o})\, y^{-1}(\lambda)\, d\lambda = \ell g \frac{\vartheta(\mathfrak{x}^{+} + j\mathfrak{x}_1)}{\vartheta(\mathfrak{x}^{-} + j\mathfrak{x}_1)},$$

in which p_i $(i=1,\ldots,g)$ is the auxiliary divisor obtained by crossing out the jth row and column of Q, $\mathfrak{x}_1$ is the element of J corresponding to the shift, and

$$x_j^{\pm} = \sum_{i=1}^{g} \int_{o}^{p_i} \omega_j \qquad (j=1,\ldots,g)$$

with $o = p_\infty$ or p'_∞ according to the choice of the sign. The analogue of Its-Matveev's formula is $f_j = (1/2)[c_j - c_{j+1}]$ $(j=1,\ldots,n)$, in which c_j is the right-hand side of Baker's formula; see DUBROVIN-MATVEEV-NOVIKOV [1976] for fuller explanations.

11.3 Spike Domains and Moduli

Another criterion for K to be a Toda curve is

4) $$\sum_{i=1}^{g} m_i a_i(\omega_j) = 2n \int_{\lambda_{2g+1}^{o}}^{p_\infty} \omega_j \qquad (j=1,\ldots,g),$$

i.e., this is equivalent to 1), 2), and/or 3) of §11.1. The idea is adapted from MARCENKO-OSTROVSKII [1975]; see below for more information.

PROOF. Let K be a Toda curve and e the unit of 2), §11.1. Then $e = a + yb$ with $a = \cosh\left(\int_0^p \omega\right)$ and $\omega = d \cosh^{-1} a = a^{\bullet}(a^2 - 1)^{-1/2}\, d\lambda \sim nk^{-1}\, dk$ in terms of the local parameter $k = \lambda$ in the vicinity of p_∞ or p'_∞. ω is a differential of the second kind

---

* $\lambda_i^{\bullet}$ $(i=1,\ldots,g)$ are the nontrivial roots of $a^{\bullet}(\lambda) = 0$ falling in the lacunae $[\lambda_{2i-1}^{o}, \lambda_{2i}^{o}]$ $(i=1,\ldots,g)$.

with * $a_i(\omega) = 0$ and $b_i(\omega) \equiv 0$ modulo $2\pi\sqrt{-1}$ $(i=1,\ldots,g)$, seeing as $a(\lambda_i^o) = \pm 1$ in pairs $(i=0,\ldots,2g+1)$, so by the period relations for differentials of the first and second kinds, **

$$\sum_{i=1}^{g} 2\pi\sqrt{-1}\, m_i a_i(\omega_j) = 4\pi\sqrt{-1}\, n \int_{\lambda^o_{2g+1}}^{p_\infty} \omega_j \qquad (j=1,\ldots,g)$$

with the required integral m_i $(i=1,\ldots,g)$. The opposite implication is even easier. 4) implies

$$4') \qquad -\sum_{i=1}^{g} m_i a_i(\omega_j) = 2n \int_{\lambda^o_{2g+1}}^{p'_\infty} \omega_j \qquad (j=1,\ldots,g)$$

by change of sheet, and subtraction of 4') from 4) yields 1).

The geometrical picture can now be clarified in the manner of MARCENKO-OSTROVSKII [1975]: in general, $\omega = d \cosh^{-1} a$ is the differ-differential $\prod_{i=1}^{g} (\lambda - \lambda_i^{\bullet})\, d\lambda \,/\, y(\lambda)$ of the second kind with $\lambda_i^{\bullet}$ so chosen in the lacunae $[\lambda^o_{2i-1}, \lambda^o_{2i}]$ $(i=1,\ldots,g)$ as to make $a_i(\omega) = 0$ $(i=1,\ldots,g)$. Now † $\lambda \to k = (\sqrt{-1}/2) \int_0^{\lambda} \omega$ maps the λ-plane cut along $[0,\infty)$ onto the spike domain of Fig. 7. The corner πx_i- $[\pi x_i+]$ corresponds to λ^o_{2i-1} $[\lambda^o_{2i}]$, the ith spike to the intervening interval

FIG. 7

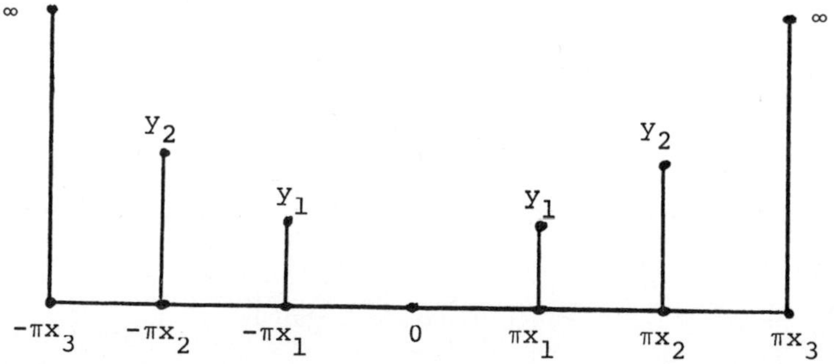

* b_i $(i=1,\ldots,g)$ are the customary clockwise cycles about $[\lambda_0^o, \lambda^o_{2i-1}]$ $(i=1,\ldots,g)$.

** SIEGEL [1971].

† λ_0^o is taken to be 0 for simplicity.

$[\lambda^o_{2i-1}, \lambda^o_{2i}]$, and

$$\sum_{i=1}^{g} x_i a_i(\omega_j) = 2x_{g+1} \int_{\lambda^o_{2g+1}}^{p_\infty} \omega_j \qquad (j=1,\dots,g)$$

by the period relations previously invoked. The map is normalized by $k' \sim (\sqrt{-1}/2)\, \ell g\, \lambda$ for $\lambda \to \infty$, and the inverse Schwarz-Christoffel map to the cut plane reproduces λ^o_i $(i=1,\dots,2g+1)$ from the places x_i $(i=1,\dots,g+1)$ and the heights y_i $(i=1,\dots,g)$; in particular, $xy \in R^{2g+1}$ is a set of moduli for K, and 5) K is a Toda curve if and only if $x_i/m_i = x_{g+1}/n$ $(i=1,\dots,g)$ with integral m_i $(i=1,\dots,g)$ and n.

11.4 Hill Curves

The same ideas apply to Hill curves; see MARCENKO-OSTROVSKII [1975]. Now λ^o_{2g+1} is removed to ∞, $k = (\sqrt{-1}/2) \int_0^\lambda \omega \sim \sqrt{\lambda}$,

$$y(\lambda) = \sqrt{-\lambda(\lambda - \lambda^o_1) \dots (\lambda - \lambda^o_{2g})},^{*}$$

Fig.7 is changed by the removal of x_{g+1} to ∞, and

$$\sum_{i=1}^{g} x_i a_i(\omega_j) = 0 \quad (j < g) \qquad = 2 \quad (j = g);$$

in particular, K is a Hill curve of period 1 if and only if x_i is integral $(i=1,\dots,g)$. MARCENKO-OSTROVSKII [1975] used the present spike picture to prove that the Hill operators $Q = -D^2 + q$ having finitely many lacunae are dense; see, also, MEIMAN [1977], and, for additional information about Hill and Toda curves, McKEAN-MOERBEKE [1979].

---

* $\lambda^o_0 = 0$, as in §11.4.

11.5 Spectral Rigidity in More Dimensions

The content of §11.2 can be summed up, so: Let Q_e denote the action of the periodic map $f \to a_{j-1}f_{j-1} + b_j f_j + a_j f_{j+1}$ in the class of functions $f\colon Z \to R$ with multiplier $e \neq 0$ $[f_{j+n} = e\, f_j \ (j \in Z)]$ and determine a <u>curve</u> K by the vanishing of $\det (Q_e - \lambda)$. Then the fibers of the map $Q \to K$ are real parts of Jacobi varieties. Now let Q be a difference operator on Z^2 with two independent periods and let Q_e denote its action on the class of functions $f\colon Z^2 \to R$ with multipliers $e_1, e_2 \neq 0$, . Then $\det (Q_e - \lambda) = 0$ determines a <u>surface</u> K. MUMFORD [1979] proved the striking fact that now the fibers of the map $Q \to K$ are generally points. This type of spectral rigidity seems to be typical of several dimensions; compare §3.4.

12. THE TODA LATTICE CONTINUED

12.1 Meiman's Theorem*

Let K be a Hill or a Toda curve. Then $\lambda_0^o < \lambda_2^o < \ldots < \lambda_{2g}^o$ [$< \lambda_{2g+1}^o$], the integers m_i $(i=1,\ldots,g)$, and the period T [n] determine the interlacing branch points $\lambda_1^o < \lambda_3^o < \ldots < \lambda_{2g-1}^o$ of K.

PROOF for Hill's curves. Let $k \to k'$ be the map of the upper half of the $k = \sqrt{\lambda}$ plane to the spike domain of Fig. 7, §11.3, with x_{g+1} removed to ∞, and let K' be the image of an interval K of $R \times 0$. The harmonic measure of K, as viewed from $k = \sqrt{-1}\, b$,** is

$$\frac{b}{\pi} \int_K (x^2 + b^2)^{-1} \, dx \sim \frac{|K|}{b\pi} \qquad (b \uparrow \infty).$$

This quantity is also the harmonic measure $H_{\sqrt{-1}b'}(K')$ of K' viewed from the point $k' = \sqrt{-1}\, b'$ of the spike domain, so

$$H_{\sqrt{-1}\, b'}(K') = \frac{b' - 1}{\pi} \int \frac{H_{x+\sqrt{-1}}(K')}{x^2 + (b'-1)^2} \, dx \sim \frac{|K|}{b\pi} \qquad (b \uparrow \infty) \quad ,$$

assuming for simplicity that $y_i < 1$ $(i=1,\ldots,g)$; in particular,

$$\int H_{x+\sqrt{-1}}(K') \, dx = |K|$$

by the increase of $b^2(x^2 + b^2)^{-1}$ to 1 as $x \uparrow \infty$ and the fact that $b' \sim b$. Now let λ_{2i}^o $(i=0,\ldots,g)$ and $x_i = m_i/T$ $(i=1,\ldots,g)$ be fixed, let λ'_{2i-1}, resp., λ''_{2i-1} $(i=1,\ldots,g)$ be two determinations of the odd branch points of K, and let y'_i, resp., y''_i $(i=1,\ldots,g)$ be the

---

* MEIMAN [1977].

** $|K|$ is the length of K.

corresponding spike heights. Let K', resp., K" be the sum of the elbows joining πx_{j-1}, πx_j, and $\pi x_j+\sqrt{-1}y_j'$, resp., $\sqrt{-1}\, y_j''$ for such values of j as have $y_j' > y_j''$. The principle of CARLEMAN [1927] * now asserts that $H_{x+\sqrt{-1}}(K') > H_{x+\sqrt{-1}}(K'')$, as is intuitively evident. The common pre-image of K' and K" is a sum of lacunae $[\lambda^o_{2j-1}, \lambda^o_{2j}]$, and now a contradiction arises between this fact and Carleman's inequality. The only way out is to have $y_j' = y_j''$ $(j=1,\dots,g)$ and consequently $\lambda'_{2j-1} = \lambda''_{2j-1}$ $(j=1,\dots,g)$. The proof is finished.

12.2 Approximating Hill by Toda Curves; see McKEAN-MOERBEKE [1979] for more details.

The possibility is suggested by the spike pictures of §§11.3-4. Let λ^o_i $(i=0,\dots,2g)$ be the branch points of a Hill's curve of period 1 so that ** $\sum m_i a_i(\omega_j) = 0$ $(j < g)$ $= 2$ $(j = g)$, let n be a large integer, c an unspecified positive number, and use Meiman's theorem† to determine from $\lambda^o_0 < \lambda^o_2 < \dots < \lambda^o_{2g} < \lambda'_{2g+1} = cn^2$, m_i $(i=1,\dots,g)$, and n a Toda spectrum $\lambda^o_0 < \lambda'_1 < \lambda^o_2 < \lambda'_3 < \dots < \lambda^o_{2g} < \lambda'_{2g+1}$ of period n. Let

$$\omega_j' = \lambda^{j-1}[(\lambda - \lambda^o_0)(\lambda-\lambda'_1)(\lambda - \lambda^o_2)(\lambda-\lambda'_3)\dots(\lambda - \lambda^o_{2g})(\lambda-\lambda'_{2g+1})]^{-1/2}\, d\lambda.$$

Then

$$\sum_{i=1}^{g} 2m_i \int_{\lambda'_{2i-1}}^{\lambda^o_{2i}} \omega_j' = 2n \int_{\lambda'_{2g+1}}^{\infty} \omega_j' \qquad (j=1,\dots,g)$$

is well approximated by

---

* See NEVANLINNA [1970].

** $\omega_j = \lambda^{j-1}[-(\lambda - \lambda^o_0)\ \dots\ (\lambda - \lambda^o_{2g})]^{-1/2}\, d\lambda$ $(j=1,\dots,g)$, as usual.

† §12.1.

$$\frac{1}{n\sqrt{c}} \sum_{i=1}^{g} 2m_i \int_{\lambda'_{2i-1}}^{\lambda^o_{2i}} \lambda^{j-1}[(\lambda - \lambda^o_0)(\lambda - \lambda'_1)\ldots(\lambda - \lambda^o_{2g})]^{-1/2}\, d\lambda$$

$$\sim 0 \qquad (j < g)$$

$$\sim 2n \int_{cn^2}^{\infty} \frac{d\lambda}{\lambda^{3/2}\sqrt{\lambda - cn^2}} \qquad (j = g),$$

and for this to go over into the Hill periodicity condition, it is necessary to have

$$\frac{4}{\sqrt{c}} = n\sqrt{c} \times 2n \int_{cn^2}^{\infty} \frac{d\lambda}{\lambda^{3/2}\sqrt{\lambda - cn^2}} = 2, \quad \underline{\text{i.e.}}, \quad c = 4.$$

Now λ'_{2i-1} must tend to λ^o_{2i-1} $(i=1,\ldots,g)$, while the division points p_∞, p'_∞ of the approximating curve coalesce over ∞; see Fig. 8. Let k be a local parameter in the vicinity of $\lambda'_{2g+1} = 4n^2$ so chosen as to make $k(p_\infty) = k_\infty = 1/2n$, $k(p') = -k(p)$, and

$$\lambda(p) = n[(k + k_\infty)^{-1} - (k - k_\infty)^{-1}] \sim -1/k^2, \quad \underline{\text{i.e.}}, \quad k \sim 1/\sqrt{-\lambda}\ .$$

FIG. 8

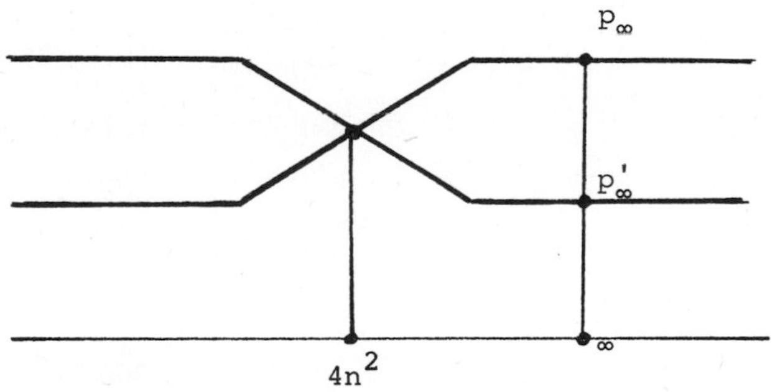

This is done to facilitate appraisal of the functions $f_j(p)$ $(j=1,\ldots,n)$ of §11.1. The pole divisor q_i $(i=1,\ldots,g)$ is fixed, and, in the vicinity of ∞, $f_j(p)$ is the product of a local unit and

$$\left(\frac{k+p_\infty}{k-p_\infty}\right)^j = \exp\left[j\ \ell g\left(\frac{k+p_\infty}{k-p_\infty}\right)\right] = \exp\left[(2k_\infty j/k)(1+k_\infty^2/3k^2+\ldots)\right]$$

$$= \exp\ (x/k)[1+o(1)]$$

with $x = j/n$. This indicates that $f_j(p)$ approximates the function $f_x(p) \sim \exp(x/k)$ of §9.3 and that the corresponding divisor of roots of $f_j(p)$ for $j = n$ approximates the root divisor of $f_x(p)$ for $x = 1$. Use is now made of trace formulas to see what is happening in more detail, <u>esp</u>., to make explicit the approximation of the Hill's operator $-D^2+q$ by the enlarged $(n+1)\times(n+1)$ matrix $Q = \begin{bmatrix} b_n & a_n & 0 & \ldots 0 \\ a_n & b_1 & a_1 & 0 \end{bmatrix}$. Let Q^+, Q^-, Q^∞ denote the latter acting upon f_i $(i=0,\ldots,n)$ with a) $f_n = f_0$ (periodic spectrum), b) $f_n = -f_0$ (antiperiodic spectrum), c) $f_n = f_0 = 0$ (auxiliary spectrum). Then

$$2b_n = sp\left(Q^- + Q^+ - 2Q^o\right) = \lambda_0^o + \sum_{i=1}^{g}\left(\lambda'_{2i-1} + \lambda^o_{2i}\right) + \lambda'_{2g+1} - 2\sum_{i=1}^{g}\lambda(p_i)$$

$$= \lambda_0^o + \sum_{i=1}^{g}\left(\lambda^o_{2i-1} + \lambda^o_{2i}\right) + 4n^2 - 2\sum_{i=1}^{g}\lambda(p_i) + o(1),$$

so $2b_n = q(0) + 4n^2 + o(1)$ by the trace formula of §6.3, and $2b_j = q(j/n) + 4n^2 + o(1)$ by applicationof the shift; similarly,

$$2b_n^2 + 4a_{n-1}^2 + 4a_n^2 = sp[(Q^-)^2 + (Q^+)^2 - 2(Q^o)^2] = 16n^4 + O(1)\ ,$$

so $a_{n-1}^2 + a_n^2 = -n^2 q(0) + 2n^4 + o(n^2)$, and, hopefully, $a_j = (1/4)q(j/n) - n^2 + o(1)$. The upshot is that

$$\lambda f_j = a_{j-1}f_{j-1} + b_j f_j + a_j f_{j+1} = -n^2(f_{j-1} - 2f_j + f_{j+1}) + q(j/n)f_j + o(1),$$

as it should be.

12.3 The Kac-Moerbeke Lattice

The system KVM: $x_i^\bullet = \exp(x_{i+1}) - \exp(x_{i-1})$ $(i=1,\ldots n,\ x_0 = x_n,\ x_{n+1} = x_1)$ of KAC-MOERBEKE [1975] is closely related to the Toda lattice in several ways. 1) KVM sits inside the second Toda flow Toda-2: $Q^\bullet = [Q,(Q^2)^+]$. The latter is equivalent to

$$a_i^\bullet = a_i(a_{i+1}^2 - a_{i-1}^2 + b_{i+1}^2 - b_i^2)$$

$$(i=1,\ldots,n),$$

$$b_i^\bullet = 2b_i(a_i^2 - a_{i-1}^2) + 2b_{i+1}a_i^2 - 2b_{i-1}a_{i-1}^2$$

it restricts to the locus $b_i = 0$ $(i=1,\ldots,n)$, and the restricted flow may be identified with KVM via the substitution $\sqrt{2}\, a_i = \exp(x_i/2)$ $(i=1,\ldots,n)$. 2) KVM also sits inside the first Toda flow Toda-1: $Q^\bullet = [Q,Q^+]$. The substitution $x_{i-1} - x_i \to x_i$ converts the latter into $x_i^{\bullet\bullet} = \exp(x_{i+1}) - 2\exp(x_i) + \exp(x_{i-1})$ $(i=1,\ldots,n)$, while the substitution $x_i + x_{i+1} \to x_i$ converts KVM into $x_i^{\bullet\bullet} = \exp(x_{i+2}) - 2\exp(x_i) + \exp(x_{i-2})$ $(i=1,\ldots,n)$ which is equivalent to Toda-1 if n is odd and splits into 2 uncoupled copies of Toda in half as many degrees of freedom if n is even. 3) KVM is also birationally equivalent to Toda-1 as it stands. The substitution*

$$b_1 = c_{2n} + c_1\ ,\quad b_2 = c_2 + c_3\ ,\ldots,\ b_n = c_{2n-2} + c_{2n-1}$$

$$a_1 = \sqrt{c_1 c_2}\ ,\quad a_2 = \sqrt{c_3 c_4}\ ,\ldots,\ a_n = \sqrt{c_{2n-1}c_{2n}}$$

from $c \in R^{2n}$ with $c_{2i-1}c_{2i} > 0$ $(i=1,\ldots,n)$ to $ab \in R^{2n}$ with $a_i > 0$ $(i=1,\ldots,n)$ is nice on an open top-dimensional figure, avoiding the vanishing of

$$c_1, c_3 = b_2 - \frac{a_1^2|}{|c_1}\ ,\quad c_5 = b_3 - \frac{a_2^2|}{|b_2} - \frac{a_1^2|}{|c_1}\ ,\ \underline{\text{etc}}.,$$

---

* STIELTJES [1918] suggested this.

and requiring the matching of $b_1 = c_{2n} + c_1$ with

$$c_{2n} = \frac{a_n^2|}{|b_n} + \cdots + \frac{a_2^2|}{|b_2} + \frac{a_1^2|}{|c_1} ,$$

and in that figure, the Toda-l flow of ab is equivalent to the KVM flow of c via the substitution $c_i = \exp(x_i/2)$ $(i=1,\ldots,2n)$. The significance of all these intricate inter-relations is not fully understood.

12.4 Tied Lattices

The tied Toda lattice $[a_n = 0, x_0 = -\infty, x_n = \infty]$ was integrated by KAC-MOERBEKE [1975] and MOSER [1975]; see §15.1 for details and McKEAN [1979] for the connection with singular curves. The analogous connection between the scattering theory of §4 and singular curves is an attractive subject for investigation. The KVM lattice, similarly tied, was integrated by KAC-MOERBEKE [1975] and MOSER [1975].

12.5 Approximation by Partial Differential Equations

M. Kac verified that the front of the soliton in the infinite KVM lattice approximates the KDV soliton; the scaling $x_j(t) \to n^2 x_{[j-nt]}(n^3 t)$ is employed. SCHWARTZ [1979] extended this to the general wave in the infinite lattice. The question of approximating the middle of the wave, presumably by Boussinesq, awaits careful investigation; compare §1.3.

12.6 Solitons

HIROTA [1973] obtained a many-soliton formula for the infinite Toda lattice: $4a_j^2 = 1 + [\ell g\, \phi(\chi - jk + ct)]$ with $0 < k_1 < \ldots < k_g$, $c_i = \pm 2\, \mathrm{sh}(k_i/2)$ $(i=1,\ldots,g)$, and

$$\phi(\chi) = \sum e^{\chi\cdot n} \prod_{i<j} \left[\frac{\mathrm{ch}[(k_i-k_j)/2]-1}{\mathrm{ch}[(k_i+k_j)/2]-1}\right]^{2n_i n_j} ,$$

the sum being taken over $n_i = 0$ or 1 ($i=1,\ldots,g$), much as in §2.3; presumably, ϑ is the theta function of a singular curve, as in §8.4.

13. KDV: SOLUTIONS OF RATIONAL CHARACTER FOR SPHERE, CYLINDER, ELLIPTIC CURVE

13.1 Solutions of Rational Character

Let E be the sphere P^1, the cylinder $R \times [0,1)$, or an elliptic curve with primitive periods 1, $2\omega_1 = 1$ and $2\omega_2 \in \sqrt{-1}\, R$, and let p be the function x^{-2}, or the simply periodic function

$$\pi^2[\sin^{-2}(\pi x) - 1/3] = x^{-2} + \sum_{n\neq 0} [(x-n)^{-2} - n^{-2}] ,$$

or else the doubly periodic Weierstrassian function

$$x^{-2} + \sum_{\omega = n_1\omega_1 + n_2\omega_2 \neq 0} [(x-2\omega)^{-2} - (2\omega)^{-2}] ,$$

according to the choice of E. AIRAULT-McKEAN-MOSER [1977] studied the classes C of functions of rational character on E such as are 1) <u>invariant under the flows</u> $\partial q/\partial t = X_j q$ $(j=1,2,3,\ldots)$ and 2) <u>minimal in this regard</u>; see also, CHOODNOVSKY [1977/78] and KRICHEVER [1978]. Let $\partial q/\partial t = X_2 q$ be a curve in P. The poles of q move smoothly, preserving their multiplicity in open intervals filling up most of the time axis, and inserting the principal part* $a(x-c)^{-m} + b(x-c)^{-m+1} + \ldots$, it develops that $a = 2$, $m = 2$, $b = 0$, <u>i.e.</u>, the general function of Q is of the form $q(x) = \sum_{i=1}^{n} 2p(x-x_i)+c$ with common n and c and movable poles x_i $(i=1,\ldots,n)$; indeed, more detailed computation shows that <u>the rational character of</u> $q(x)$ <u>can be maintained under the flow</u> $\partial q/\partial t = X_2 q$ <u>only if the poles</u> x_i $(i=1,\ldots,n)$ <u>lie in the closure of the locus</u>

$$x_i \neq x_j \quad (i \neq j) , \qquad \sum_{j\neq i} p'(x_i - x_j) = 0 \quad (i=1,\ldots,n)$$

<u>and that the flow is equivalent in the open locus to</u>

---
* a,b,c depend upon t.

$$\dot{x}_i = -6 \sum_{j\neq i} p(x_i - x_j) \qquad (i=1,\ldots,n).$$

13.2 The Sphere

The situation is simplest for the sphere $E = P^1$; see AIRAULT-McKEAN-MOSER [1977]. *The locus*

$$x_i \neq x_j \quad (i \neq j), \qquad \sum_{j\neq i} (x_i - x_j)^{-3} = 0 \quad (i=1,\ldots,n)$$

is either void, or else $n = g(g+1)$ *for some* $g = 1,2,3,\ldots,$ *and the closed locus may be identified with* C^g *via the map*

$$(x_1,\ldots,x_n) \to \pi_1 = \sum x_i\,, \quad \pi_3 = \sum x_i^3\,, \;\ldots,\; \pi_{2g-1} = \sum x_i^{2g-1}\;;$$

in fact, under the fields X_j $(j=1,\ldots,g)$, the π_i $(i=1,3,\ldots,2g-1)$ move in straight lines at constant speed, and $X_{g+1}x = 0$. The origin of C^g corresponds to the function $2g(g-1)x^{-2}$ of Q, and the whole of Q flows out from this point: Fig. 9 depicts some special figures found in the locus; see AIRAULT-McKEAN-MOSER [1977] and §13.4, below, for more information. The matter is treated in less detail in CHOODNOVSKY [1977/78]; see, also. THICKSTUN [1976].

FIG. 9

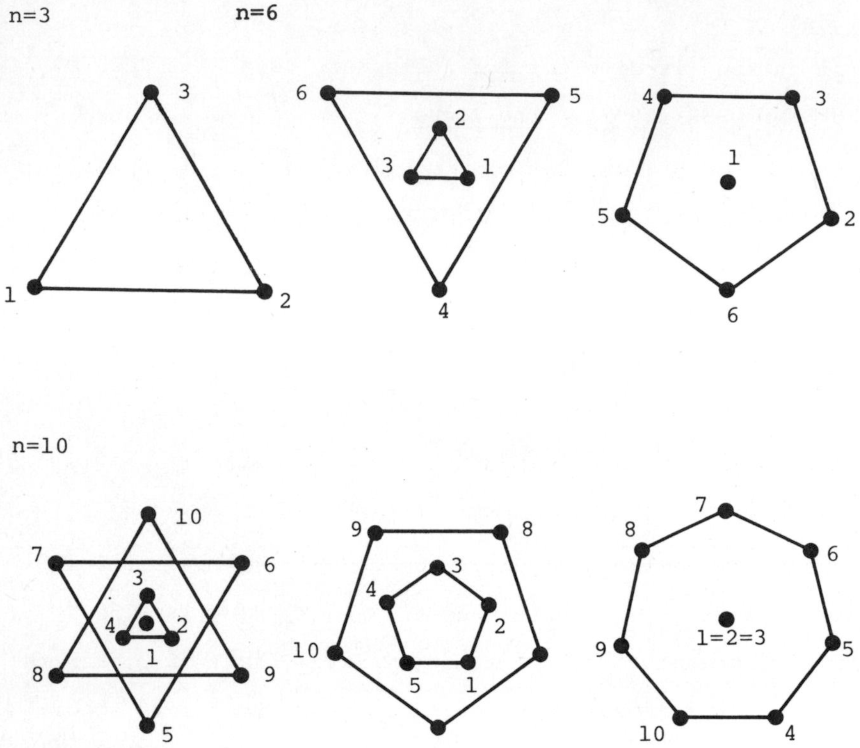

13.3 The Calogero Lattice

CALOGERO [1971] noticed that the n-dimensional eigenvalue problem $(-\Delta + U)f = \lambda f$ could be solved explicitly for $U = \sum_{i<j} (x_i - x_j)^{-2}$. Now it is a part of the conventional wisdom that in such a case, the classical system with Hamiltonian $H = y^2/2 + U$ is integrable. MOSER [1975] confirmed this fact.* The phase space is $M = [xy \in R^{2n}: x_1 < \ldots < x_n]$ with $y = x^{\bullet}$. The estimate $x_i(t) = x_i' t + y_i' + O(t^{-1})$ $(i=1,\ldots,n,\ t \uparrow \infty)$ defines a map $\sigma(xy) = x'y'$ of M into itself. This remarkable map is a) onto, b) 1:1, c) canonical with multiplier -1, and d) involutive; moreover, the motion $x^{\bullet} = \partial H/\partial y = y$, $y^{\bullet} = -\partial H/\partial x$ can be expressed as

---

* See OLSHANETSKY-PERELOMOV [1976] and ADLER [1977] for related systems.

$xy = \sigma(x', x't+y)$. The numbers $x_1' < \ldots < x_n'$ are the eigenvalues of the complex symmetric matrix L with y_i $(i=1,\ldots,n)$ on diagonal and $\sqrt{-1}\,(x_i - x_j)^{-1}$ $(1 \le i \ne j \le n)$ off diagonal. Now x_i' $(i=1,\ldots,n)$ are involutive integrals of the motion, and it is a fact of classical mechanics that if H_1 and H_2 are two involutive Hamiltonians, then the 1-flow restricts to the submanifold where the 2-flow is trivial [grad $H_2 = 0$], and <u>vice versa</u>. Let $H_2 = y^2/2 + U = (1/2)\,\mathrm{sp}(L^2)$ be the Calogero Hamiltonian and let

$$H_3 = \frac{1}{3}\,\mathrm{sp}(L^3) = \frac{1}{3}\sum_{i=1}^{n} y_i^3 + \sum_{1\le i\ne j\le n} y_i (x_i - x_j)^{-2} .$$

Then the locus grad $H_2 = 0$ is the product of $[x \in R^n\colon x_1 <\ldots< x_n, \sum_{j\ne i}(x_i - x_j)^{-3} = 0 \ (i=1,\ldots,n)]$ and the point $y = 0$, and is <u>void</u> but need not be so if you move into the complex as in §13.2, and it is a remarkable fact * that if $n = g(g + 1)$ $(g=1,2,3,\ldots)$ so that the <u>complex</u> locus is nonvoid, then <u>the 3-flow reduces there to $\dot{x}_i = 6\sum_{j\ne i}(x_i-x_j)^{-2}$ $(i=1,\ldots,n)$ which is just the KDV motion of poles described before, up to a sign</u>. This is quite unexpected, as also the fact that <u>the 2-flow restricted to the complex locus</u> grad $(H_3 - H_1) = 0$ <u>with $H_1 = \mathrm{sp}(L)$</u> <u>is equivalent to the pole motion of rational solutions of Boussinesq:</u> $\partial^2 q/\partial t^2 = (\partial^2/\partial x^2)[q - (1/2)q^2 + (1/2)q'']$.
The latter locus:

$$y_i^2 = 1 - \sum_{j\ne i}(x_i-x_j)^{-2}, \quad \sum_{j\ne i}(y_i+y_j)(x_i-x_j)^{-3} = 0 \ (i=1,\ldots n)$$

awaits investigation; for $n = 2$, it is nonvoid and produces the elementary solution $q = [x-x(-)]^{-2} + [x-x(+)]^{-2}$ with $x(\pm) = x_0 \pm \sqrt{t^2 + \frac{1}{4}}$. The 2-3 numerology is to be compared to that of §9.3. The significance of this coincidence is not yet clear, but see §15.4. KRICHEVER [1978] has computed the solutions of rational character

* AIRAULT-McKEAN-MOSER [1977].

of Kadomtsev-Petiashvilli * by the methods of §9 applied directly to a singular curve; rational KDV and Boussinesq are special cases.

13.4 Theta Functions

ADLER-MOSER [1978] described the locus grad $H_2 = 0$ at one stroke by computing the theta function $\vartheta_g(x = t_1, t_2, \ldots, t_g) = \prod_{i=1}^{n} (x - x_i)$ in which $n = g(g+1)$ and x_i $(i=1,\ldots,n)$ are the poles of the functions flowing out from the origin $2nx^{-2}$ under $\exp(t_1X_1+\ldots+t_gX_g)$ ** $[q = -2(\ell g\,\vartheta)'']$. ϑ is computed from the recipe

$$\vartheta'_{g+1}\vartheta_{g-1} - \vartheta'_{g-1}\vartheta_{g+1} = (2g+1)\vartheta_g^2 ,$$

the parameters t_i $(i=1,\ldots,g)$ appearing successively as constants of integration; for example, up to an inessential triangular substitution in t_i $(i=1,\ldots,n)$,

$$\vartheta_0 = 1$$
$$\vartheta_1 = x$$
$$\vartheta_2 = x^3 + t_2$$
$$\vartheta_3 = x^6 + 5t_2x^3 + t_3x - 5t_2^2$$
$$\vartheta_4 = x^{10} + 15t_2x^7 + 7t_3x^5 - 35t_2t_3x^2 + 175t_2^3x - (7/3)t_3^2 + t_4x^3 + t_4t_2.$$

13.5 The Cylinder

The case of the cylinder $E = R \times [0,1)$ is easily settled with the help of the many-soliton formula of §2.3; see McKEAN [1979]. The theta function

---

* See §9.3.

** See KRICHEVER [1978] for an improved derivation in the manner of §9, starting from a singular curve.

$$\vartheta_1 = \vartheta(\mathfrak{x} + 2xk) = \sum e^{\mathfrak{x}\cdot n + 2xk\cdot n} \prod_{i<j} \left(\frac{k_i - k_j}{k_i + k_j}\right)^{2n_i n_j}$$

with arbitrary $\mathfrak{x} = (x_1, \ldots, x_g)$ and integral $k = (k_1, \ldots, k_g)$ leads to functions $q = -2(\ell g\, \vartheta_1)''$ which are rational in $y = e^x$ and so of rational character on E, ϑ_1 being a polynomial in y of total degree $n = k_1 + \ldots + k_g$. The roots of this polynomial necessarily lie in the closure of the locus

$$y_i \neq y_j \quad (i \neq j), \quad \sum_{j \neq i} (y_i + y_j)(y_i - y_j)^{-3} = 0 \quad (i = 1, \ldots, n)$$

by §13.1; in fact, <u>for fixed</u> $n = 1, 2, 3, \ldots$, <u>the closed locus breaks up into subloci identifiable as the root loci of</u> ϑ_1 <u>for the several solutions of</u> $n = k_1 + \ldots + k_g$ <u>in unequal integers</u> $0 < k_1 < \ldots < k_g$. The proof employs the scattering theory of §4. McKEAN [1979] connects the present case to that of the sphere by observing that if $k = (1, 2, \ldots, g)$ and if

$$\vartheta_2 = \vartheta(\mathfrak{x} + \sqrt{-1}\, k + 2xk/T + t_1 k/T + t_2 k^3/T^3 + \ldots + t_g k^{2g-1}/T^{2g-1}),$$

then $T^n \vartheta_2 \to \vartheta_3$ as $T \uparrow \infty$, ϑ_3 being the theta function of ADLER-MOSER described in §13.4. ϑ_3 is of degree $n = g(g+1) = 2(1+2+\ldots+g)$ in x, so it is a not unexpected but still curious fact that $(1, 2, \ldots, g)$ is the only choice of k for which such a limit can be obtained. The proof involves a curious little vanishing theorem for ϑ_2.

13.6 The Elliptic Curve

The case of an elliptic curve E is more complicated. The Weierstrassian function * $q(x) = g(g+1)\, p(x + \omega_2)$ of INCE [1946]

---

* The primitive periods of p are $2\omega_1 = 1$ and $2\omega_2 \in \sqrt{-1}\, \mathbb{R}$.

provides an example. The simple spectrum of $Q = -D^2 + q$ comprises λ_i $(i=0,\ldots,2g)$, the rest of the spectrum being double. The corresponding isospectral manifold P is of the required kind, as is seen from the fact that, more generally, * <u>if one point of a finite-dimensional isospectral class</u> P <u>comes from an elliptic function, then the whole class comes from a single elliptic function field</u>.

PROOF. The idea is that if q is an elliptic function with nonreal period, $2\omega_2$, say, and if x traces a complex path from $x_0 \in [0,1)$ to $x_0 + 2\omega_2$, then the elementary symmetric functions of the projected divisor $\mu_i^o = \lambda(p_i)$ $(i=1,\ldots,g)$ corresponding to q translated by x must come back to their original values after 1 circuit, so that the divisor itself must do likewise after $m \le 2^g$ circuits. It follows** from

$$(\mu_i^o)' = 2y(\mu_i^o) \prod_{j\neq i} (\mu_i^o - \mu_j^o)^{-1} \qquad (i=1,\ldots,g)$$

that if c is the closed circuit traveled by the divisor, then

$$c(\omega_j) = 0 \quad (j < g) \qquad = 4m\omega_2 \quad (j = g)$$

is a period of the complex variety J_C. Now $\not{x}_1 = (0,\ldots,2) \in T^*(J_R)$ points in direction of infinitessimal translation and the Its-Matveev formula of §8.3 serves to finish the proof.

The stated condition on the circuit c is equivalent to

$$\sum_{i=1}^{g} m_i' b_i(\omega_j) = 0 \quad (j < g) \qquad = 4m\omega_2 \quad (j = g)$$

---

* McKEAN-MOERBEKE [1975] stated this with inadequate attention to the proof; see AIRAULT-McKEAN-MOSER [1977] for details.

** See §6.3.

with integral m_i' $(i=1,\ldots,g)$, which is the same as to say that $-\lambda_{2g}^{o} < \ldots < -\lambda_1^{o} < -\lambda_0^{o}$ is the simple periodic and antiperiodic spectrum of a Hill's operator of period $2m\omega_2/\sqrt{-1}$. McKEAN-MOERBEKE [1979] proved that such leaves are dense but rare in the foliation of C_1^{∞}. The condition is void for $g = 1$: every isospectral class comes from an elliptic curve. For $g = 2$ $[n = 3]$, the pole locus is the closure of *

$$[x_1 \neq x_2 \neq x_3 \ , \ p_{12}' + p_{13}' = p_{21}' + p_{23}' = p_{31}' + p_{32}' = 0] \subset S^3(E) \ ,$$

which splits into 9 isolated copies of the elliptic curve, plus the sublocus $x_1 \neq x_2 \neq x_3$, $p_{12} + p_{23} + p_{31} = 0$ with its diagonal adjoined. The second piece is a copy of the complex variety J_C associated to the isospectral class of 6p, alias the Jacobi variety of the quintic curve K: $y^2 = -(\lambda^2 + 12\sigma_2)(\lambda^3 + 9\lambda\sigma_2 - 27\sigma_3)$, σ_i $(i=1,2,3)$ being the elementary symmetric functions of the invariants e_i $(i=1,2,3)$. J_C fibers over E via the projection $(x_1,x_2,x_3) \to x_1 + x_2 + x_3$, the fiber F being also an elliptic curve; see NOVIKOV [1974] and AIRAULT-McKEAN-MOSER [1977]. Nothing more is known about the elliptic case, though it may be conjectured, if $g = 2$ is any guide, that J_C always fibers in a pleasant way over E. D. and G. Choodnovsky concur in this conjecture.

---

* $p_{12} = p(x_2 - x_1)$, etc. $S^3(E)$ is the 3-fold symmetric product of E.

14. BOUSSINESQ'S EQUATION ON THE CIRCLE

The equation is taken in the form $\partial^2 q/\partial t^2 = (\partial^2/\partial x^2)[(4/3)q^2 \pm (1/3)\, \partial^2 q/\partial x^2]$. The plus sign appears unfavorable to the existence of solutions, though it leads to a simpler spectral problem. The minus sign is adopted below. The integrability of the system is conjectured but not proved; see ZAKHAROV [1973]. The state of the art is quite primitive.

14.1 Commutator Formalism

The equation is equivalent to $Q^\bullet = [Q,K_2]$ with $Q = -D^3+qD+Dq+p$ and $K_2 = -D^2 + (4/3)q$; in fact, $[Q,K_2] = p'D + Dp' + [(4/3)q^2 - (1/3)q'']'$, so $q^\bullet = p'$ and $3p^\bullet = (4q^2 - q'')'$. The Dikii-Gelfand recipe of §3.4 has its counterpart in the following rule: if $J = \sqrt[3]{-Q} = D + c_0 + c_1 D^{-1} + c_2 D^{-2} + \ldots$, then $c_0 = 0$, $c_1 = (-2/3)q$, $c_2 = (2/3)q - (1/3)p$, and

$$(J^2)^+ = [(D - \tfrac{2}{3}\, q\, D^{-1})^2]^+ = D^2 - \tfrac{4}{3}\, q = K_2 \;;$$

see ADLER [1979] and SYMES [1978] for additional information.

14.2 Hamiltonian Formalism

The phase space is pairs of functions qp of class C_1^∞. Let $J = \begin{pmatrix} 0 & D \\ D & 0 \end{pmatrix}$. Then the Poisson bracket *

$$[A,B] = \int_0^1 \text{grad } A\; J\; \text{grad } B\, dx = \int_0^1 \left[\frac{\partial A}{\partial q}\, D\, \frac{\partial B}{\partial p} - \frac{\partial B}{\partial q}\, D\, \frac{\partial A}{\partial p}\right] dx$$

satisfies Jacobi's identity, and $q^\bullet = p'$, $3p^\bullet = (4q^2 + q'')'$ is equivalent to $q^\bullet p^\bullet = J \text{ grad } H_2$ with

---

* grad C signifies the pair $[\partial C/\partial q,\ \partial C/\partial p]$.

$$H_2 = \int_0^1 [\tfrac{1}{2} p^2 + \tfrac{4}{9} q^3 + \tfrac{1}{6} (q')^2]\, dx .$$

$H_{-1} = \int_0^1 q\, dx$, $H_0 = \int_0^1 (1/2)\, p\, dx$, $H_1 = \int_0^1 qp\, dx$, and H_2 are involutive integrals of the motion, as may easily be checked by hand. The corresponding vector fields $X: (q,p) \to J \text{ grad } H$ are $X_{-1} = 0$, $X_0 = 0$, $X_1(qp) = (q',p')$, and $X_2(qp) = (p',(8/3)qq' - (1/3)q''')$.

14.3 More Integrals

H_{-1}, H_1 and H_0, H_2 are only the first of two series of involutive integrals produced by a recipe similar to that of §2.4 for KDV:

$$X_{j+2}(qp) = J \text{ grad } H_{j+2} = K \text{ grad } H_j \qquad (j=-1,0,2,\dots)$$

with

$$K = \begin{bmatrix} -D^3+qD+Dq & 3pD+2p' \\ 3pD+p' & \frac{1}{3} D^5 - \frac{5}{3}(qD^3+D^3q)+q''D+Dq''+ \frac{16}{3} qDq \end{bmatrix} ;$$

see ADLER [1979] or McKEAN [1978] for the proof. K, like J, is skew-symmetric relative to the inner product $\int_0^1 f_1 \cdot f_2 dx$ for functions $f: [0,1) \to R^2$. The integrals have an isobaric character if degrees are ascribed as follows: deg ' = 1, deg q = 2, deg p = 3; for instance, deg $H_2 = 6$. The next two integrals are

$$H_3 = \int_0^1 [\tfrac{1}{2} (p')^2 + 2p^2q + \tfrac{1}{6} (q'')^2 + \tfrac{8}{9} q^4 + 2q(q')^2]\, dx$$

$$H_4 = \int_0^1 [\tfrac{5}{6} p^3 + \tfrac{20}{9} pq^3 + \tfrac{5}{6} p(q')^2 + \tfrac{1}{3} p''q'' - \tfrac{5}{3} q^2p'']\, dx .$$

14.4 The Multiplier Curve

The eigenvalue problem $Qf = \lambda f$ now plays the central role. The 3×3 monodromy matrix $M = [D^{i-1}y_j(1,\lambda): 1 \le i,j \le 3]$ is formed from the solutions of $Qy = \lambda y$ with $D^{i-1} y_j(0) = 1$ on diagonal and 0 off. The multiplier m is its eigenvalue, viewed as a 3-valued function of λ determined by the cubic $m^3 - am^2 + bm - c = 0$, in which $a = m_1 + m_2 + m_3$ is the trace $\operatorname{sp} M = y_1(1,\lambda) + y_2'(1,\lambda) + y_3''(1,\lambda)$, $b = m_1m_2 + m_2m_3 + m_3m_1$ is the corresponding trace for Q^*, and $c = m_1m_2m_3$ is the determinant $\det M = 1$. The cubic determines a 3-sheeted curve M of genus $g \le \infty$ on which the multiplier lives: the discriminant Δ of the cubic is an integral function of order 1/3 and type 1 and so has an infinite number of isolated roots; in particular, M has an infinite number of ramifications and/or singular points at which 2 or 3 multipliers coincide. M is also <u>irreducible</u>, <u>i.e.</u>, no single branch of the cubic is analytic; indeed, Cardano's formula for the roots of the cubic gives an estimate $|m(\lambda)| \le \exp(|k|)$ with $-k^3 = \lambda$, so an analytic branch $m_1(\lambda)$ of m would be an integral function of order $\le 1/3$ without roots $[m_1m_2m_3 = 1]$ and so constant. Then $a = m_1 + m_2 + m_3$ and $b = m_1(m_2 + m_3) + 1/m_1$ would have the <u>same</u> growth at ∞, contradicting the elementary estimates

$$a \sim \sum_0^2 \exp(\omega^j k) ,$$

$$b \sim \sum_0^2 \exp(-\omega^j k)$$

for $\omega = \exp(2\pi\sqrt{-1}/3)$ and $k \to \infty$. The same estimates show that the triple points of M $[a = b = 3]$ are finite in number. The distant critical points of M are <u>either</u> double ramifications covering conjugate pairs of simple roots of $\Delta(\lambda) = 0$ in the vicinity of $\pm(2\pi n/\sqrt{3})^3$ $(n \uparrow \infty)$, <u>or else</u> they are double points covering double roots of $\Delta(\lambda) = 0$, representing the coalesence of such a pair of simple roots. The point (λ,m) of M is declared to belong to the auxiliary spectrum of Q if the eigenfunction f satisfying $Qf = \lambda f$, $f(x+1) = mf(x)$ $(0 \le x \le 1)$, and $f(0) = 1$ exhibits a pole at that place.

14.5 Examples: Genus 0 and 1

GENUS 0. Let q and p be constant. Then $Q = -D^3 + 2qD + 1$ and the multiplier is $m = \exp(k)$ with $-k^3 + 2qk + p = \lambda$. This cubic defines a curve K projecting onto M via the exponential. $p = 0$ entails no loss of generality.

CASE 1: $q = 0$. K is a sphere triply ramified over 0 and ∞. M inherits these ramifications; it exhibits a double point over λ if $k^3 = (k + 2\pi\sqrt{-1}\,n)^3$ for some $n \neq 0$, which is to say $\lambda = \pm (2\pi n/\sqrt{3})^3$, and has no triple points at all; in particular, the desingularization of M is a sphere.

CASE 2: $q \neq 0$. K is triply ramified over ∞ and doubly ramified over $\pm (4/3)\sqrt{-2q^3/3}$. M inherits these ramifications and has also double points over $\pm (4/3)\sqrt{-2q^3/3}$ if $q = -2\pi^2 n^2/3$ for some $n \neq 0$. The other double or triple points of M cover

$$\pm \frac{4}{3}\, 3\sqrt{3}\ \sqrt{\pi^2 n^2 + 2q}\ (2\pi^2 n^2 + q) \sim \pm (2\pi n/\sqrt{3})^3 .$$

The number of triple points is equal to the number of representations[†]

[†] See DIRICHLET [1871] for such matters.

of $-9q/2\pi^2$ by the quadratic form $n^2 + nm + m^2$ with $n, m, n+m \neq 0$; in particular, M can have an indefinitely large number of triple points. The desingularization of M is a sphere, as before: indeed, its ramification index is $4 = 1 \times (3-1) + 2\times(2-1) = 2\times(3$ sheets $+ g - 1)$, i.e., $g = 0$. The fact is that the desingularization of M is a sphere only if q and p are constant. This is easy to prove upon observing that $k = \lg m$ would be 1-valued on M and 3-valued over the projection, and using the estimate $\lg^+|m| = O(\sqrt[3]{\lambda})$.

GENUS 1. A traveling wave for Boussinesq is obtained by equating X_2 and cX_1, c being the speed of propagation. Now $X_2(qp) = (p', (8/3)qq' - (1/3)q''')$, so $X_2 = cX_1$ means that $p' = cq'$ and $(8/3)qq' - (1/2)q''' = cp'$, whence 1) $cq = p + h_1$, 2) $c^2q = (4/3)q^2 - (1/3)q'' + h_2$, and 3) $(c^2/2)q^2 = (4/9)q^3 - (1/6)(q')^2 + h_2q + h_3$ with constants of integration h_1, h_2, h_3. The solution of 3) is $q = (3/8)c^2 + (3/2) \times$ the Weierstrassian function with primitive periods $2\omega_1 = 1$ and $2\omega_2 \in \sqrt{-1}\,R$, evaluated at $x + \omega_2$ $(0 \le x < 1)$ for reality. The fields X_3, X_4, etc. are also proportional to X_1; in fact, $X_2 = cX_1$ is maintained under the corresponding flows, so it suffices to identify h_1, h_2, h_3 as integrals of the motion. Now by 1), $h_1 = c\int_0^1 q - \int_0^1 p = cH_{-1} - H_0$ is an integral; similarly, $c\int_0^1 q^2 = \int_0^1 qp + h_1 = H_1 + h_1$, so by 2), h_2 is also an integral; as to h_3, $\int_0^1 p^2 = cH_1 - 2h_1H_0$ is an integral by 1), $\int_0^1 [(4/3)q^3 + (1/6)(q')^2]$ is an integral by 2), so $\int_0^1 q^3$ and $\int_0^1 (q')^2$ are also integers by $H_2 = \int_0^1 [(1/2)p^2 + (4/9)q^3 + (1/6)(q')^2]$, and h_3 likewise, by 3). The upshot is that 1) and 3) determine a 1-dimensional compact leaf [= circle] of the foliation of $C_1^\infty \times C_1^\infty$ under the flows $\dot{q}\dot{p} = X(qp) = J\,\text{grad}\,H$. Now let $c = 0$ and permit me to re-employ the letter p to denote the Weierstrassian function evaluated at $x + \omega_2$ $(0 \le x < 1)$, so that $q = (3/2)p$. The multiplier curve M is covered by the elliptic curve E; indeed, viewing E as the quotient by the period lattice of the plane with parameter $\mathfrak{x} = 1/k$, the function $f(\mathfrak{x})$ of §9.3 with singularity $\exp(xk)$ at $\mathfrak{x} = 0$ $[k = \infty]$ is expressed as *

$$f(\mathfrak{x}) = e^{x\zeta(\mathfrak{x})+} \quad \frac{\sigma(\mathfrak{x}-\mathfrak{x}_0)}{\sigma(\mathfrak{x}-\mathfrak{x}_\infty)} \frac{\sigma(-\mathfrak{x}_\infty)}{\sigma(-\mathfrak{x}_0)}$$

with fixed pole $\mathfrak{x}_\infty \neq 0$ and movable root $\mathfrak{x}_0$ depending upon $0 \leq x < 1$. The double periodicity of $f(\mathfrak{x})$ requires ** the vanishing of $\omega_1 c + \eta_1(x + x_\infty - x_0)$ and $\omega_2 c + \eta_2(x + x_\infty - x_0)$, as follows from the transformation rule of the sigma function under addition of periods, x being taken halfway up the fundamental cell for reality. Now $QF = \lambda g$ with $\lambda(\mathfrak{x}) = (1/2)p'(\mathfrak{x})$, exhibiting E as a 3-sheeted cover of the λ-plane; also the multiplier

$$m(\mathfrak{x}) = e^{\zeta(\mathfrak{x}) \quad 2\eta_1\mathfrak{x}}$$

is exhibited as a doubly periodic function† which is 3-valued over λ. M is now seen to be covered by E and to inherit a triple ramification over $\lambda = \infty$ $[\mathfrak{x} = 0]$ and 4 double ramifications over the points $\lambda = (1/2)p'(\mathfrak{x})$ corresponding to the 4 roots of $p''(\mathfrak{x}) = 0$. The ramification index $1\times(3-1)+4\times(2-1) = 6 = 2(3+g-1)$ shows that the genus of M is unity; in fact, the desingularization of M is isomorphic to E. The singular points of M are fairly complicated but need not be examined.

---

* $\zeta(\mathfrak{x})$ $[\sigma(\mathfrak{x})]$ is the customary odd [even] Weierstrassian function with simple poles [roots] on the period lattice.

** $2\eta_1$ and $2\eta_2$ are, resp., the real and imaginary periods of $\zeta(\mathfrak{x})$.

† $\eta_1\omega_2 - \eta_2\omega_1 = \pi\sqrt{-1}/2$.

14.6 Finite Genus

The general curve M awaits investigation. The case of finite genus should be accessible with function-theoretic aid from BAUR [1895], HENSEL-LANDSBERG [1962], and BLISS [1933], but even the existence of such curves for $\infty > g \geq 2$ is not confirmed. The latter is equivalent to the existence of g-dimensional leaves of the foliation.

14.7 The Line

To date, there is no scattering theory for $Q = -D^3 + qD + Dq + p$ with, e.g., $qp \in C_\downarrow^\infty$, and so no method of solution of Boussinesq on the line. HIROTA [1973] found a many-soliton formula in the style of §2.3:

$$q(t,x) = \frac{3}{2}\,[\ell g\, \vartheta\, (\mathfrak{x} + xk + tc)]'' - \frac{1}{12}$$

with $k_1 < \ldots < k_g$, $c = \pm\sqrt{3}\, k\, \sqrt{1 + k^2}$ componentwise, and

$$\vartheta(\mathfrak{x}) = \sum e^{\mathfrak{x}\cdot n} \prod_{i<j} \left[\frac{(c_i-c_j)^2 - (k_i-k_j)^2 - (k_i-k_j)^4}{(c_i+c_j)^2 - (k_i+k_j)^2 - (k_i+k_j)^4}\right]^{2n_in_j},$$

the sum being taken over $n_i = 0$ or 1 $(i=1,\ldots,g)$; presumably, this is connected with a singular curve, as in §8.4. CHU-SCOTT-McLAUGHLIN [1973] present a whole series of formulas of this type awaiting such geometrical interpretation.

15. THE VIEWPOINT OF GROUPS

The present section contains a brief account of the group-theoretical picture introduced by KOSTANT [1978 & 1979]. J. Moser, T. Ratiu, and D. Sims clarified the material for me.

15.1 The Tied Toda Lattice

The tied Toda lattice induces an isospectral motion of the tridiagonal matrix

$$Q = \begin{bmatrix} b_1 & a_1 & 0 & \dots & 0 & 0 \\ a_1 & b_2 & a_2 & \dots & 0 & 0 \\ \vdots & & & & & \\ 0 & 0 & 0 & & a_{n-1} & b_n \end{bmatrix}$$

with $a_i > 0$ $(1 \le i < n)$ and $\sum b_i = 0$, say. The equations of motion are $a_i^{\bullet} = a_i(b_i - b_{i+1})$, $b_i^{\bullet} = 2(a_i^2 - a_{i-1}^2)$ $(i=1,\dots,n)$ with $a_n = a_0 = b_{n+1} = 0$. The substitution $a_i = (1/2)\exp[(1/2)(x_i - x_{i+1})]$, $b_i = (1/2)x_i^{\bullet}$ of §11.2 converts them into the Toda lattice $x_i^{\bullet\bullet} = \exp(x_{i-1} - x_i) - \exp(x_i - x_{i+1})$ $(i=1,\dots,n)$ with $x_0 = -\infty$, $x_{n+1} = +\infty$. The system was integrated by MOSER [1975] in the following way. Let λ_i $(i=1,\dots,n)$ be the spectrum of Q. Then

$$(\lambda - Q)^{-1}_{11} = \sum_{i=1}^{n} c_i^2(\lambda - \lambda_i)^{-1}$$

with positive c_i^2 $(i=1,\dots,n)$ summing to 1. Now it is easy to see that the spectrum is fixed under the Toda flow and that c_i^2 $(i=1,\dots,n)$, viewed projectively, satisfy $(c_i^2)^{\bullet} = \lambda_i c_i^2$, i.e., $c_i^2(t) = \exp(\lambda_i t) \times c_i^2(0)$ $(i=1,\dots,n)$ in RP^{n-1}, which is to say

$$c_i^2(t) = e^{\lambda_i t} c_i^2(0) \times \Big[\sum_{j=1}^{n} e^{\lambda_j t} c_j^2(0)\Big]^{-1} \qquad (i=1,\dots n).$$

The integration is completed by use of the continued fraction of STIELTJES [1918]:

$$(\lambda - Q)^{-1}_{11} = \frac{1}{\lambda - b_1} \Big| - \frac{a_1^2}{\lambda - b_2} \Big| - \cdots - \frac{a_{n-1}^2}{\lambda - b_n} \Big| .$$

15.2 Kostant's Picture

KOSTANT [1979] explained the tied Toda lattice from the viewpoint of groups. Let G be a Lie group, L its Lie algebra, and L' the dual of L relative to the Killing form. Let M be the orbit of a fixed point of L' under the dual of the adjoint action of G on L'. M is always even dimensional and carries a natural symplectic structure; in fact, if (ad x)' is the dual of the action ad $x = [x,\cdot]$ of $x \in L$ on L, then the general co-tangent vector to M at a point y is (ad x)'y, and the symplectic form is $[(\text{ad } x_1)'y, (\text{ad } x_2)'y] = y[x_2, x_1]$ with the Killing pairing. This makes M into a symplectic manifold. Now M is equipped with a natural set of integrals: the Chevalley invariants, and it will sometimes happen that <u>they are equal in number to</u> $(1/2) \times \dim M$, <u>and, miraculously, that they are involutive</u>. Then the flows induced upon M by the coadjoint action of $\exp(t\mathfrak{x})$ ($\mathfrak{x} \in G$) will commute and be integrable on the leaves of the foliation of M obtained by fixing the values of the Chevalley invariants.

How does the Toda lattice fit into this scheme? Let G be the group of $n \times n$ lower triangular real matrices with positive diagonal. Then L [L'] is the class of lower [upper] triangular matrices with unrestricted diagonal, and the Killing form is sp(xy). The co-adjoint action of $\mathfrak{x} \in G$ upon $y \in L'$ is * $y \to (\mathfrak{x} y \mathfrak{x}^{-1})^+$. The orbit M passing through

* + means take the upper triangular part, i.e., project onto L'.

$$y_0 = \begin{bmatrix} 0 & 1 & 0 & 0 & \cdots \\ 0 & 0 & 1 & 0 & \\ 0 & 0 & 0 & 1 & \\ \vdots & & & & \end{bmatrix}$$

consists of all $n \times n$ matrices

$$y = \begin{bmatrix} b_1 & a_1 & 0 & 0 & \cdots \\ 0 & b_2 & a_2 & 0 & \\ 0 & 0 & b_3 & a_3 & \\ \vdots & & & & \end{bmatrix}$$

with $a_i > 0$ $(1 \leq i < n)$ and $\sum b_i = 0$, and the symplectic form is expressed as

$$\omega = - \sum_{j=1}^{n-1} \left[\sum_{i=j}^{n-1} a_i^{-1} \, da_i \right] \wedge db_j \, .$$

Chevalley's invariants are just the elementary symmetric functions $H_j = sp(y^j)$ $(j=1,\ldots,n)$ in the eigenvalues of y; they are equal in number to $n = (1/2) \dim M$ and involutive, and up to inessentials, the motion produced by the Hamiltonian $H_2 = sp\,(y^2)$ is the Toda flow of the allied symmetric matrix

$$\begin{bmatrix} b_1 & a_1/2 & 0 & 0 & \cdots \\ a_1/2 & b_2 & a_2/2 & 0 & \\ 0 & a_2/2 & b_3 & a_3/2 & \\ \vdots & & & & \end{bmatrix} ;$$

in fact, more remarkably still, it lifts to a 1-dimensional subgroup $\exp(tx/2)$ of G. The moral is that <u>the Toda lattice is merely a complicated projection of a simple motion upstairs</u>.

15.3 Other Groups

KOSTANT [197] associates similar Toda-like systems with any Dynkin diagram; for example, the system $\dot{q}_i = \partial H/\partial p_i$, $\dot{p}_i = -\partial H/\partial q_i$ ($i=1,2,3$) with $H = (1/2)p^2 + \exp(q_1 - q_2) + \exp(q_2 - q_3) + \exp(q_3)$ is associated with $G = SO(7)$; compare ADLER [1979], CHOODNOVSKY [197], and OLSHANETSKY-PERELOMOV [1976]. Kostant conjectures that the periodic Toda lattice of §§11-12 fits into a similar scheme of infinite-dimensional groups associated with the Kac-Moody algebras, but this remains to be seen; an exciting part of such a project would be to see how the curves and Jacobi varieties come in.

15.4 The Calogero Lattice

KAZHDAN-KOSTANT-STERNBERG [1978] found a group theoretical interpretation of the Calogero lattice of §13.3 involving $G = U(n)$; see MOSER [1979] for a simplified account. The relation of the lattice to solutions of rational character of KDV, Boussinesq, and Kadomtsev-Petiashvilii prompts the conjecture that KDV and Boussinesq are the restrictions of group actions $\exp(tX_2)$ and $\exp(tX_3)$ to coadjoint orbits of the dual of an infinite-dimensional Lie algebra , the orbit having the interpretation as the locus where the other flow is trivial. This would be a remarkable development. The plan is closely connected to the group-theoretical methods employed by ARNOLD [1966] and EBIN-MARSDEN [1969] to integrate the Eulerian hydrodynamical equations; see, also, ADLER [1979], SYMES [1978], and, for possibly related infinite-dimensional Lie algebras, ESTABROOK-WAHLQUIST [1975] and MORRIS [1976].

REFERENCES

ADLER, M.: Some finite-dimensional systems and their scattering behavior. Comm. Math. Phys. 55 (1977) 195-230.

_________: Completely integrable systems and symplectic action. J. Math. Phys. (to appear).

_________: On a trace functional for formal pseudo-differential operators and the symplectic structure of the Korteweg-de Vries equation. Inventiones Math. (1979) - .

ADLER, M., and J. MOSER: On a class of polynomials connected with the Korteweg-de Vries equation. Comm. Math. Phys. 61 (1978) 1-30.

AGRANOVICH, Z., and V. MARCENKO: The Inverse Scattering Theory. Gordon and Breach, New York, 1963.

AIRAULT, H., H. P. McKEAN, and J. MOSER: Rational and elliptic solutions of the Korteweg-de Vries equation and a related many-body problem. CPAM 30 (1977) 95-148.

AKHIEZER, N.: A continuous analogue of orthogonal polynomials on a system of intervals. Dokl.Akad. Nauk 141 (1961) 263-266; Sov. Math. 2 (1961) 1409-1412.

ARNOLD, V.: Sur la géométrie différentielle des groupes de Lie de dimension infini.et ses spplications à l'hydrodynamique des fluides parfaits. Ann. Inst. Grenoble 16 (1966) 319-361.

_________: Mathematical Methods of Classical Mechanics. Nauka, Moscow, 1974; Springer Verlag, New York, 1978.

BAKER, H.: Abel's Theorem and the Allied Theory, Including the Theory of the Theta Functions. Cambridge U. Press, Cambridge, 1897.

_________: Note on the foregoing paper "Commutative ordinary differential operators" by J. L. BURCHNALL and T. W. CHAUNDY. Proc. Royal Sco. London 118 (1928) 584-593.

BARGMANN, V.: On the connection between phase shifts and scattering potentials. Rev. Mod. Phys.. 21 (1949) 488-493.

BAUR, L.: Aufstellung eines vollständigen System von Differentialen erster Gattung in einem cubischen Funktionenkörpers. Math. Ann. 46 (1895) 31-61.

BLISS, G.: Algebraic Functions. AMS Publ. 16, New York, 1933.

BORG, G.: Eine Umkerhrung der Sturm-Liouvilleschen Eigenwertaufgabe. Acta Math. 78 (1945) 1-96.

BOUSSINESQ, J.: Théorie des ondes et des remous qui se propagent le long d'un canal rectangulaire horizontal, en communiqant au liquide contenu dans ce canal des vitesses sensiblement pareilles de la surface au fond. J. Math. Pures Appl. 7 (1872) 55-108.

BURCHNELL, J., and T. CHAUNDY: Commutative ordinary differential operators. Proc. London Math. Soc. 21 (1922) 420-440; Proc. Royal Soc. London 118 (1928) 557-583; 134 (1931) 471-485.

BUSLAEV, V., and L. FADDEEV: On trace formulas for singular Sturm-Liouville operators. Dokl. Akad. Nauk. 132 (1960) 13-16.

CALOGERO, F.: Solution of the n-body problems with quadratic and/or inversely quadratic pair potentials. J. Math. Phys. 12 (1971) 419-436.

CARLEMAN, T.: Sur les fonctions inverses des fonctions entières. Ark. Mat. Astron. Phys. 15 (1927).

CHOODNOVSKY, D., and G. CHOODNOVSKY: Seminaire sur les Équations Non-linéaires. Centre de Math., École Poly., Palaisseau, 1977/78.

________: Completely integrable class of mechanical systems connected with the Korteweg-de Vries and multicomponent Schrödinger equations (1). Sem. Diophantienne, École Poly., Palaisseau, 10 June 1978.

________: Multi-dimensional two-particle problem with non-central potential. Sem. Diophantienne, École Poly., 22 Nov. 1978.

CHU, F., D. McLAUGHLIN, and A. SCOTT: The soliton: a new concept in applied science. Proc. IEEE 61 (1973) 1443-1483.

DATE, E., and S. TANAKA: Analogue of inverse scattering theory for the discrete Hill's equation and exact solutions for the periodic Toda Lattice. Progress Theor. Phys. 55 (1976) 457-465.

DEIFT, P., and E. TRUBOWITZ: Inverse scattering on the line. CPAM (to appear).

DIKII, L., and I. M. GELFAND: Asymptotic behaviour of the resolvent of Sturm-Liouville equations and the algebra of the Korteweg-de Vries equations. Uspekhi Mat. Nauk 30 (1975) 67-100; Russ. Math. Surveys 30 (1975) 77-113.

________: Fractional powers of operators and Hamiltonian systems. Funkt. Anal. Pril. 10 (1976) 13-29.

DIRICHLET, P.: Vorlesungen über Zahlentheorie. F. Vieweg, Braunschweig, 1871.

DONAGI, R.: Group law on the intersection of two quadrics. (to appear).

DUBROVIN, B., and S. NOVIKOV: Periodic and conditionally periodic analogues of multi-soliton solutions of the Korteweg-de Vries equation. Dokl. Akad. Nauk 6 (1974) 2131-2144.

DUBROVIN, B., V. MATVEEV, and S. NOVIKOV. Non-linear equations of Korteweg-de Vries type, finite-zone linear operators, and abelian varieties. Uspekhi Mat. Nauk 31 (1976) 55-136; Russ. Math. Surveys 31 (1976) 59-146.

DYM, H., and H. P. McKEAN: Fourier Series and Integrals. Academic Press, New York, 1972.

DYSON, F.: Fredholm determinants and inverse scattering problems. Comm. Math. Phys. 47 (1976) 171-183.

EBIN, D., and J. MARSDEN: Groups of diffeomorphisms and the solution of the classical Euler equations for a perfect fluid. BAMS 75 (1969) 962-967 , Ann. of Math, 92 (1970) 102-163.

ESTABROOK, F., and H. WAHLQUIST: Bäcklund transformations for solutions of the Korteweg-de Vries equation. Phys. Rev. Letters 31 (1973) 1386-1390.

________: Prolongation structures of non-linear evolution equations. J. Math. Phys. 17 (1976) 1-7.

FADDEEV, L.: On the relation between the S-matrix and the potential of the one-dimensional Schrodinger equation. Tr. Mat. Inst. V.A. Steklova 73 (1964) 314-336; TAMS 65 (1966) 139-166.

FADDEEV, L., and V. ZAKHAROV: Korteweg-de Vries equation: a completely integrable Hamiltonian system. Funkt. Anal. Pril. 5 (1971) 280-287.

FAY, J.: Theta Functions on Riemann Surfaces. Lect. Notes in Math. 352, Springer-Verlag, New York, 1973.

FERMI, E., J. PASTA, and S. ULAM: Studies of non-linear problems. Collected Papers of Enrico Fermi 2, p.978 , Chicago, 1965.

FLASCHKA, H.: The Toda lattice, (1) and (2). Phys. Rev. 9 (1974) 1924-1925; Prog. Theor. Phys. 51 (1974) 703-706.

________: Discrete and periodic illustrations of some aspects of the inverse method. Lect. Notes in Phys. 38 (1975) 441-466.

FLASCHKA, H., and D. McLAUGHLIN: Canonically conjugate variables for the Korteweg-de Vries equation and the Toda lattice with periodic boundary conditions. Prog. Theor. Phys. 55 (1976) 438- .

FORNBERG, G., and G. WHITHAM: A numerical and theoretical study of certain non-linear wave phenomena. Phil. Trans. Roy. Soc. London 289 (1978) 373-404.

GARDINER, C.: Korteweg-de Vries equation and generalizations (4). The Korteweg-de Vries equation as a Hamiltonian system.

J. Math. Phys. 12 (1971) 1548-1557.

GARDINER, C., J. GREENE, M. KRUSKAL, and R. MIURA: Method for solving the Korteweg-de Vries equation. Phys. Rev. Letters 19 (1967) 1095-1097.

________: Korteweg-de Vries equation and generalizations (6). Methods for exact solution. CPAM 27 (1974) 97- .

GARDINER, C., M. KRUSKAL, and R. MIURA. Korteweg-de Vries equation and generalizations (2). Existence of conservation laws and integrals of the motion. J. Math. Phys. 9 (1968) 1204-1209.

GELFAND, I. M., and B. LEVITAN: On the determination of a differential equation from its spectral function. Izvest. Akad. Nauk 15 (1951) 309-360.

GOLUBEV, V. V.: Lectures on Integration of the Equations of Motion of a Rigid Body about a Fixed Point. Nauka, Moscow, 1953; Monsin, Jerusalem, 1960.

GUILLEMIN, V., and D. KAZHDAN: Spectral rigidity of the Schrödinger operator on certain negatively curved 2-manifolds, (to appear).

HÉNON, M.: Integrals of the Toda lattice. Phys. Rev. 9 (1974) 1421-1423.

HENSEL, K., and G. LANDSBERG: Theorie der algebraischen Fonktionen einer Variabeln. Teubner, Leipzig, 1902; Chelsea Publ. Co., New York, 1965.

HIROTA, R.: Exact solution of the Korteweg-de Vries equation for multiple collisions of solitons. Phys. Rev. Letters 27 (1971) 1192-1194.

________: Exact n-soliton solution of the wave equation of long waves in shallow water and non-linear lattices. J. Math. Phys. 14 (1973) 810-815.

________: Exact n-soliton solutions of a non-linear lumped network equation. J. Phys. Soc. Japan 35 (1973) 286-288.

HOCHSTADT, H.: Function-theoretic properties of the discriminant of Hill's equation. Math. Zeit. 82 (1963) 237-242.

________: On the determination of Hill's equation from its spectrum. Arch. Rat. Mech. Anal. 19 (1965) 353-362.

HYMAN, M., appendix to P. LAX: Almost periodic solutions of the KdV equation. SIAM Rev. 18 (1976) 351-375.

INCE, E.: The periodic Lamé functions. Proc. Roy. Soc. Edinburgh 60 (1940) 47-63; 83-99.

ITS, A., and V. MATVEEV: Hill's operator with a finite number of lacunae. Funkt. Anal. Pril. 9 (1975) 69-70.

JACOBI, C.: Gessamelte Werke, Bd. 7, supplement: Vorlesungen über Dynamik. Reimer, Berlin, 1884.

KAC, M., and P. van MOERBEKE: On some periodic Toda lattices. PNAS USA 72 (1975) 1627-1629.

________: The solution of the periodic Toda lattice. PNAS USA 72 (1975) 2879-2880.

________: On an explicitly soluble system of nonlinear differential equations related to certain Toda lattices. Adv. Math. 16 (1975) 160-169.

KADOMTSEV, B. B., and V. I. PETIASHVILII: On the stability of solitary waves in weakly dispersing media. Dokl. Akad. Nauk 192 (1970) 753-756; Sov. Phys. Dokl. 15 (1970) 539-541.

KAY, I., and H. MOSES: Reflectionless transmission through dielectrics and scattering potentials. J. Appl. Phys. 27 (1956) 1503-1508.

KAZHDAN, D., B. KOSTANT, and S. STERNBERG: Hamiltonian group actions and dynamical systems of Calogero type. CPAM 31 (1978) 481-508.

KORTEWEG, D., and G. de VRIES: On the change of form of long waves advancing in a rectangular canal, and on a new type of long stationary waves. Phil. Mag. 39 (1895) 422-443.

KOSTANT, B.: Whittaker vectors and representation theory. Invent. Math. 48 (1978) 101-104.

_________: Ann. Math. (to appear).

KOVALEVSKAYA, S.: Sur le problème de la rotation d'un corps autour d'un point fixe. Acta Math. 12 (1889) 177-232; Navehnye Raboty, Acad. Sci. USSR, 1948.

KRICHEVER, I. M.: Algebraic-geometric construction of the Zakharov-Shabat equations and their solutions. Dokl. Akad. Nauk 227 (1976) 291-294; Sov. Math. Dokl. 17 (1976) 394-397.

_________: The method of algebraic geometry in the theory of non-linear equations. Uspekhi Mat. Nauk 32 (1977) 183-208.

_________: On rational solutions of the Kadmotsev-Petiashvilii equation and integrable systems of n particles on the line. Funkt. Anal. Pril. 12 (1978) 76-78.

KPUSKAL, M.: Non-linear wave equations. Lect. Notes in Phys. 38 (1975) 310-354.

KRUSKAL, M., and N. ZABUSKY: Interactions of solitons in a collision-less plasma and the recurrence of initial states. Phys. Rev. Letters 15 (1965) 240-243.

LAMB, G.: Bäcklund transformations at the turn of the century. Lect. Notes in Math. 515 (1976) 69-79.

LANG, S.: Division points on curves. Ann. Math. Pure Appl. 70 (1965) 229-234.

_________: Introduction to Algebraic and Abelian Functions.

Addison-Wesley Publ. Co., Reading, 1972.

LAX, P.: Integrals of non-linear equations of evolution and solitary waves. CPAM 21 (1968) 467-490.

_________: Periodic solutions of the Koretweg-de Vries equation. CPAM 28 (1975) 141-188.

LEVINSON, N.: On the uniqueness of the potential in the Schrödinger equation for a given asymptotic phase. Danske Vid. Selsk. Math. Fys. Medd. 25 (1944) 1-29.

_________: The inverse Sturm-Liouville problem. Math. Tidsskr. (1949) 25-30.

_________: Certain explicit relations between phase shift and scattering potential. Phys. Rev. 89 (1953) 755-757.

MAGNUS, W., and W. WINKLER: Hill's Equation. Interscience-Wiley, New York, 1966.

MANAKOV, S.: On complete integrability and stochastization in discrete dynamical systems. Z. Eksp. Teor. Fiz. 67 (1974) 543-555.

_________: Note on the integration of Euler's equations of the dynamics of an n-dimensional rigid body. Funkt. Anal. Pril. 10 (1976) 93-94.

MARCENKO, V., and I. V. OSTROVSKII: A characterization of the spectrum of Hill's operator. Sov. Math. Dokl. 16 (1975) 761-765; Mat. Sbornik 139 (1975) 540-606.

MATVEEV, V.: 373, U. Wroclaw, June 1976.

McKEAN, H. P.: Selberg's trace formula as applied to a compact Riemann surface. CPAM 25 (1972) 225-246.

_________: Stability for the Korteweg-de Vries equation. CPAM 30 (1977) 347-353.

_________: Boussinesq's equation as a Hamiltonian System. Adv. Math. Suppl. 3 (1978) 217-226.

_________: Theta functions, solitons, and singular curves. (to appear).

_________: Units of Hill's curves. (to appear).

_________: Int. Cong. Math. Helsinki (to appear).

_________: Toda lattice, tridiagonal matrices, and singular curves. (to appear).

McKEAN, H. P., and P. van MOERBEKE: The spectrum of Hill's equation. Inventiones Math. 30 (1975) 217-274.

_________: Hill and Toda curves. CPAM (to appear).

McKEAN, H. P., and E. TRUBOWITZ: Hill's operator and hyperelliptic function theory in the presence of infinitely many branch points. CPAM 29 (1976) 143-226.

_________: Hill's surfaces and their theta functions. BAMS (1978) -

MEIMAN, N.: The theory of 1-dimensional Schrödinger operators with periodic potential. J. Math. Phys. 18 (1977) 834-848.

MIURA, R.: Korteweg-de Vries equation and generalizations (1). A remarkable explicit non-linear transformation. J. Math. Phys. 9 (1968) 1202-1204.

_________, ed.: Bäcklund Transformations. Lect. Notes in Math. 515 Springer-Verlag, New York, 1976.

MOERBEKE, P. van: The spectrum of Jacobi matrices. Inventiones Math. 37(1976), 45-81.

MOERBEKE, P. van, and D. MUMFORD: The spectrum of difference operators and the theory of curves. (to appear).

MORRIS, H.: Prolongation structures and a generalized inverse scattering problem. J. Math. Phys. 17 (1976) 1867-1869.

MOSER, J.: Three integrable Hamiltonian systems connected with isospectral deformations. Adv. Math. 16 (1975) 197-220.

________: Finitely many mass points on the line under the influence of an exponential potential — an integrable system. Lect. Notes in Phys. 38 (1975) 467-497.

________: Various aspects of integrable Hamiltonian systems. CIME, Bressanone (1978).

MOSER, J., and E. TRUBOWITZ: (to appear).

MUMFORD, D.: An algebraic-geometric construction of commuting operators and of solutions to the Toda lattice equation, Korteweg-de Vries equation and related nonlinear equations. (to appear).

MURRAY, A.: Existence and regularity for solutions of the Korteweg-de Vries equation. Arch. Rat. Mech. (1979) -

________: Solutions of the Korteweg-de Vries equation evolving from a "box" and other irregular initial functions. Duke Math. J. 45 (1978) 149-181.

NEUMANN, C.: De problemate quodam mechanica, quod ad primam integralium ultra-ellipticorum classem revocatur. J. Reine U. Angew Math. 56 (1859) 54-66.

NEVANLINNA, F.: Analytic Functions. Springer-Verlag, Berlin, 1970.

NOVIKOV, S.: The periodic problem for the Korteweg-de Vries equation.

Funkt. Anal. Pril. 8 (1974) 54-66.

OLSHANETSKY, M., and A. PERELOMOV: Completely integrable Hamiltonian systems concerned with semi-simple Lie algebras. Inventiones Math. 37 (1976) 93-108.

POINCARÉ, H.: Les Méthodes Nouvelles de la Méchanique Céleste. Gauthier-Villars Cie., Paris, 1892; Dover Publ. Co., New York, 1957.

RAYLEIGH, : Phil. Mag. 1 (1876) 257.

SCHWARTZ, M.: (to appear).

SCOTT-RUSSELL, J.: Report on waves. Proc. Roy. Soc. Edinburgh (1844) 319-320.

SIEGEL, C. L.: Topics in Complex Function Theory (2). Interscience-Wiley, New York, 1971.

SINGER, I. M.: (to apper).

STEIMANN, U.: Äquivalente periodische potentiale. Helv. Phys. Acta 30 (1957) 515-520.

STIELTJES, T. J.: Sur la réduction en fraction continue d'une série procedant suivant les puissances descendentes d'une variable. Oeuvres Complètes 2, 184-206. Noordhoff, Groningen, 1918.

SYMES, W.: A Poisson structure on the space of symbols. MRC Technical Summary Report, Madison, Wisc., April 1978.

TANAKA, S.: On the n-tuple wave solutions of the Korteweg-de Vries equation. Publ. RIMS Kyoto 8 (1972/73) 419-427.

________: Korteweg-de Vries equation: construction of solutions in terms of scattering data. Osaka J. Math. 11 (1974) 49-59.

TAPPERT, F.: (to apper).

THICKSTUN, W.: A system of particles equivalent to solitons. J. Math. Anal. Appl. 55 (1976) 335-346.

TODA, M.: Wave propagation in anharmonic lattices. J. Phys. Soc. Japan 23 (1967) 501-506.

________: Studies on a nonlinear lattice. Ark. for Det Fysiske Sem. Trondheim 2 (1974).

TRUBOWITZ, E.: The inverse problem for periodic potentials. CPAM 30 (1977) 321-337.

WHITHAM, G.: Linear and Nonlinear Waves. Interscience-Wiley, New York, 1974.

ZAKHAROV, V.: On stochastization of one-dimensional chains of nonlinear oscillators. Zh. Eksp. Teor. Fiz. 65 (1973) 219-225.

ON THE EIGENVALUES OF A CLASS OF

HYPOELLIPTIC OPERATORS II

A. Menikoff* and J. Sjostrand**

0. INTRODUCTION

We shall study the asymptotic distribution of eigenvalues of a class of self-adjoint operators with double characteristics. The hypoellipticity of these operators has been investigated by Melin [3], Boutel de Monvel, Grigis, Helffer [1] and Hörmander [2].

Let X be a compact n-dimensional C^∞ manifold, let dx be a fixed smooth positive density on X and let $P \in L^m_c(X)$ be a classical pseudo-differential operator whose complete symbol is $\sim p(x,\xi) + p_{m-1}(x,\xi) + \cdots$ where $p = p_m$ and p_j is positively homogenous of degree m. We shall assume that P is formally self-adjoint

$$\int Pu \cdot \overline{v}\, dx = \int u \cdot \overline{Pv}\, dx , \qquad u,v \in C^\infty(X), \tag{0.1}$$

that $m > 1$ and that the principal symbol of P is non-negative

$$p(x,\xi) \geq 0. \tag{0.2}$$

We shall also suppose that

$$\Sigma = p^{-1}(0) \subset T^*X\backslash 0 \tag{0.3}$$

is a smooth submanifold of codimension $2d$ (d may be a half integer) and that p

---

* Partly supported by NSF Grant No. MCS77-21538.

** Both authors also supported by NSF Grants at the Institute for Advanced Study.

p vanishes exactly to second order on Σ, i.e., for every $K \subset T^*X\setminus 0$ there is a constant $C_K > 0$ such that

$$p(x,\xi) \geq C_K\,(d(x,\xi))^2 \qquad (x,\xi) \in K$$

where $d(x,\xi)$ denotes the distance of (x,ξ) to Σ. The subprincipal symbol $S_P(x,\xi)$ is invariantly defined on Σ by the formula

$$S_P(x,\xi) = p_{m-1}(x,\xi) + \frac{i}{2}\sum_{j=1}^{n} \frac{\partial^2 p}{\partial x_j \partial \xi_j} \qquad (x,\xi) \in \Sigma, \tag{0.4}$$

for any choice of local coordinates x in X. For $\rho \in \Sigma$ we denote by $p''_\rho(t,t)$ the Hessian quadratic form of p on $T_\rho(T^*X)$. Let $p''_\rho(t,s)$ be the corresponding symmetric bi-linear form. The Fundamental map of p is the linear map on $T_\rho(T^*X)$ determined by

$$\sigma(t,F_\rho s) = p''_\rho(t,s) \tag{0.5}$$

where $\sigma = \sum d\xi_j \wedge dx_j$ is the symplectic 2-form. It is easy to see the F_ρ is skew symmetric with respect to σ and that its eigenvalues are of the form $\pm i\mu_j$, $j = 1,\dots,n$, $\mu_j \geq 0$. Let $\tilde{\mathrm{tr}}\, F_\rho = \sum_1^n \mu_j$. In [1] and [2] it is proved that P is hypoelliptic with loss of one derivative if and only if

$$S_P(\rho) + \sum_{j=1}^{n} \left(\frac{1}{2} + \alpha_j\right)\mu_j + p''_\rho(v,v) \neq 0 \tag{0.6}$$

for all $(\alpha_1,\dots,\alpha_n) \in \mathbb{N}^n$, $\mathbb{N} = \{0,1,2,\dots\}$ and $v \in V_0$ the generalized null space of F_ρ at every point $\rho \in \Sigma$. We shall make the stronger assumption that

$$S_P(\rho) + \frac{1}{2}\tilde{\mathrm{tr}}\, F_\rho > 0 \qquad \rho \in \Sigma \tag{0.7}$$

From the hypoellipticity with loss of one derivative, it follows that P is an unbounded self-adjoint operator on the domain $\mathcal{D}_P = \{u \in L^2(X) : Pu \in L^2(X)\}$ and that the spectrum of P is discrete. From Melin's Theorem [3], we have that the spectrum of P is bounded from below. Let $N(\lambda)$ be the number of eigenvalues $\leq \lambda$. Our goal is to obtain an asymptotic formula for $N(\lambda)$.

Note that if (x,ξ) are symplectic coordinates then $dxd\xi = dx_1 \ldots dx_n d\xi, \ldots d\xi_n$ is an invariant density on T^*X. If $\theta = (\theta',\theta'')$ are local coordinates such that Σ is given by $\theta'' = 0$ and $dxd\xi = d\theta$, then

$$\omega'(d\theta') = (\det\ p''_{\theta''\theta''})^{-1/2}\, d\theta'$$

is an invariant density on Σ. When $md = n$, $\omega'(d\theta')$ is homogeneous of degree 0. If $(r,\tilde{\theta}')$ are "polar coordinates" on Σ, (so that $\tilde{\theta}'$ are local coordinates on $\tilde{\Sigma}$, the image of Σ in the cosphere bundle S^*X) then we have an invariant density $\omega'(d\tilde{\theta}')$ on $\tilde{\Sigma}$ defined by

$$\omega'(d\theta') = \frac{dr}{r}\, \tilde{\omega}'(d\theta').$$

Our main result is

<u>Theorem 0.1.</u> There are three cases:

1. If $md-n > 0$ then

$$N(\lambda) \sim \frac{1}{(2\pi)^n}\, \lambda^{n/m} \iint_{p(x,\xi) \leq 1} dxd\xi \qquad \lambda \to +\infty.$$

2. If $md-n = 0$ then

$$N(\lambda) \sim \frac{1}{(2\pi)^{n-d}}\, \frac{\lambda^{n/m} \log \lambda}{n(m-1)\Gamma(d)} \int_{\tilde{\Sigma}} \tilde{\omega}'(d\tilde{\theta}'), \qquad \lambda \to \infty$$

3. If $md-n < 0$ then

$$N(\lambda) \sim \frac{\lambda^{(n-d)/(m-1)}}{(2\pi)^{n-d}\Gamma(\frac{n-d}{m-1}+1)} \int_{\Sigma} e^{-(\frac{1}{2}\widetilde{tr}F_{\rho} + S_P(\rho))} \prod_{1}^{n} \frac{\mu_j}{1-e^{-\mu_j}} \omega'(d\rho)$$

$$\lambda \to \infty .$$

where we set $\mu/(1-e^{-\mu}) = 1$ if $\mu = 0$.

Our proof will be to study the singularity of $tr(e^{-tP})$ as $t \downarrow 0$ so as to be able to apply Karamata's Tauberian Theorem to the formula

$$tr(e^{-tP}) = \int e^{-\lambda t} dN(\lambda).$$

We shall construct the heat kernel as a Fourier integral operator with complex phase function. Because P is not of order 1, the phase and amplitude functions will not be homogeneous in the usual sense but rather quasi-homogeneous in way involving a change of time. This necessitates a careful study of the behavior of the characteristic and transport equations as $t \to \infty$. When Σ is non-involutive we found in [5] that there is a very nice exponential stability. In the present situation we have only polynomial bounds rather than exponentially fast convergence in the characteristic equation as $t \to \infty$. Thanks to condition (0.7), this is more than made up for by the exponential decay of the solutions of the transport equations. The method we use is more direct than in [5] where we used formal complex changes of coordinates.

We make considerable use of the technical parts of our previous paper and we will feel free to refer to [5] when proofs are similar. References to related works will also be found there.

1. FAMILIES OF QUADRATIC FORMS

The first step in our construction of $\exp(-tP)$ is to solve at least approximately the characteristic equation

$$(1.1)\qquad \begin{cases} \phi_t' = ip(x,\phi_x') \\ \phi|_{t=0} = \langle x,\eta\rangle. \end{cases}$$

As a preliminary, we consider in this section the case where $p(x,\xi)$ is a non-negative quadratic form in (x,ξ). The solution of (1.1) will be the generating function of the canonical relation $C_t = \{(\rho,\mu): \rho,\mu \in \widetilde{T^*X\, 0},\ \rho = \exp(-it\mathcal{H}_p)\mu\}$, the graph of the exponential of the Hamilton field of p, and vice-versa. We study C_t first, and then draw conclusions about ϕ.

We begin by recalling some facts about quadratic forms on symplectic vector spaces. Let M be a real 2n dimensional vector space, and σ a fixed symplectic form on M, i.e., a non-degenerate skew symmetric bilinear form on M. Let $p(t)$ be a non-negative quadratic form on M and $p(t,s)$ its corresponding symmetric bilinear form. The Fundamental map of p is the σ-skew symmetric linear map on M determined by the equation

$$(1.2)\qquad 2p(t,s) = \sigma(t,Fs).$$

Note that the Hamilton field of p at the point t is given by $F(t)$. We can extend σ, p and F to $\tilde{M}$, the complexification of M. We denote by Σ the set of real zeros of p. Then, $\tilde{\Sigma} = \{t \in M: p(t,\bar{t}) = 0\} = \ker F$. The eigenvalues of F are purely imaginary and come in conjugate pairs. Let $V_\lambda = \{t \in \tilde{M}: Ft = \lambda t\}$ denote the eigenspace belonging to the eigenvalue λ. We have $V_\lambda \perp V_\mu$ if $\lambda + \mu \neq 0$ (orthogonality being with respect to σ). Set $\Lambda_\pm = \bigoplus_{\pm\lambda>0} V_{i\lambda}$, then Λ_+ (Λ_-) are positive (negative) planes in $\tilde{M}$ in the sense that

$$\frac{1}{i}\,\sigma(t,\bar{t}) > 0\,, \qquad 0 \neq t \in \Lambda_+$$

$$\left(\frac{1}{i}\,\sigma(t,\bar{t}) < 0, \qquad 0 \neq t \in \Lambda_-\right)$$

Of course $\dim \Lambda_+ = \dim \Lambda_-$. The range of F may be decomposed as

$$\text{Im } F = \tilde{\Sigma} = (\tilde{\Sigma} \cap \tilde{\Sigma}^{\perp}) \oplus \Lambda_+ \oplus \Lambda_- . \tag{1.3}$$

For Σ of fixed even dimension, the dimension of $\Lambda_{\pm}$ is maximum when $\Sigma \cap \Sigma = 0$, i.e., when Σ is non-involutive. It is always possible to choose coordinates $(x,\xi) = (x_1,\dots,x_n, \xi_1,\dots,\xi_n)$ in M so that

$$\sigma((x,\xi), (y,\eta)) = \langle y,\xi\rangle - \langle x,\eta\rangle \tag{1.4}$$

and

$$p(x,\xi) = \sum_{1\le j\le r'} x_j^2 + \sum_{r'< j\le r} \mu_j(\xi_j^2 + x_j^2). \tag{1.5}$$

(See [2] or [5]).

The representation of p in (1.5) is not suitable for our purposes because it is not stable when we vary p, in particular if the rank of σ on Σ changes as p varies. We now find a representation of p which is stable as p varies as long as the dimension of Σ is held fixed. (It is similar to the form of the principal part to the Kohn-Laplacian on the boundary of a domain in $\mathbb{C}^n$.) We have to consider separately the cases that the dimension of Σ is even or odd, but we shall leave out the details of the later case.

First we suppose that the dimension of Σ is even and that co-dim $\Sigma = 2d$. Let J be a complex subspace of $\tilde{M}$ of codimension d having the following properties:

$$\tilde{\Sigma} \subset J \subset p^{-1}(0), \qquad J^{\perp} \subset J \tag{1.6}$$

$$\frac{1}{i}\,\sigma(u,\bar{u}) \ge 0 \qquad u \in J^{\perp}. \tag{1.7}$$

That such a subspace exists is easy to see starting from the special coordinates of

(1.5) for instance. Since $2p(u,v) = p(u+v) - p(u-v) = 0$ for $u,v \in J$ it follows from (1.2) and (1.6) that

$$\sigma(Fu,v) = 0 \qquad u,v \in J. \tag{1.8}$$

From (1.6) we have that $\tilde{\Sigma} \subset J \cap \overline{J} = \widetilde{(J_{\mathbb{R}})} \subset \tilde{\Sigma}$ which tells us that $J \cap J = \tilde{\Sigma}$. Since $\operatorname{codim} J + \operatorname{codim} \overline{J} = 2d = \operatorname{codim} \tilde{\Sigma}$ we also have $J + \overline{J} = \tilde{M}$ i.e., J and $\overline{J}$ intersect each other transversally. It follows that $J^{\perp} + \overline{J}^{\perp} = \tilde{\Sigma}^{\perp}$ and $J^{\perp} \cap \overline{J}^{\perp} = M = 0$. Combining these observations with (1.3) gives us

$$J^{\perp} \oplus \overline{J}^{\perp} = \tilde{\Sigma} = \operatorname{Im} F = (\tilde{\Sigma} \cap \tilde{\Sigma}^{\perp}) \oplus \Lambda_+ \oplus \Lambda_- . \tag{1.9}$$

Property (1.8) says that $F: J \to J^{\perp}$ which implies that $J^{\perp}$ is F invariant. The non-negativity of $J^{\perp}$ ((1.7)) tells us that $(\frac{1}{i} F)|_{J^{\perp}}$ can only have non-negative eigenvalues and consequently that $(\frac{1}{i} F)|_{\overline{J}^{\perp}}$ can only have non-positive eigenvalues. Taking (1.9) into consideration shows that $\Lambda_+ \subset J^{\perp}$ and $\Lambda_- \subset \overline{J}^{\perp}$. Since $\overline{\Lambda_+} = \Lambda_-$, it is clear that $J^{\perp} = \Lambda_+ \oplus K$, $\overline{J}^{\perp} = \Lambda_- \oplus \overline{K}$ and $K \cap \overline{K} = 0$ for some subspace $K \subset \tilde{\Sigma} \cap \tilde{\Sigma}^{\perp}$.

We can find d independent (over $\mathbb{C}$) linear equations so that J is given by $q_1 = \ldots = q_d = 0$. Set $\tilde{q}(s) = \overline{q(\overline{s})}$ so that $\overline{J}$ is given by $\tilde{q}_1 = \ldots = \tilde{q}_d = 0$. Since J and $\overline{J}$ intersect transversally $q_1,\ldots,q_d$, $\tilde{q}_1,\ldots,\tilde{q}_d$ are linearly independent. The vanishing of p on J and $\overline{J}$ tells us that

$$p = \sum a_{jk} q_j \tilde{q}_k \tag{1.10}$$

where (a_{jk}) is a positive definite Hermitian matrix. Replacing the q by a linear combination we can arrange to have

$$p = \sum q_j \tilde{q}_j \tag{1.11}$$

The vectors $\mathcal{H}_{q_1},\dots,\mathcal{H}_{q_d}$ form a basis for $J^{\perp}$. (This is easily seen from the identity $\sigma(t,\mathcal{H}_q) = \langle t,dq\rangle$.) The positivity of J, property (1.7), then tells us that

$$\sum \frac{1}{i}\sigma(\sum t_j \mathcal{H}_{q_j}, \sum \bar{t}_k \mathcal{H}_{\bar{q}_k}) = \sum \frac{1}{i}\{q_j,\bar{q}_k\} t_j\bar{t}_k \geq 0.$$

In other words, the Levi matrix $\mathcal{L} = (\frac{1}{i}\{q_j,q_k\})$ is positive semi-definite. Here $\{p,q\} = \sigma(\mathcal{H}_p,\mathcal{H}_q)$ is the Poisson bracket of p and q. In passing we may point out that if $iF|_{\tilde{\Sigma}^{\perp}}$ has the eigenvalue $\pm\lambda_1,\dots,\pm\lambda_d$, then $\mathcal{L}$ has the eigenvalues $\lambda_1,\dots,\lambda_d$. In fact, the matrix of $-\frac{1}{i}F|_{J^{\perp}}$ is $\mathcal{L}$ in the basis $\mathcal{H}_{\tilde{q}}$.

Let $\mathcal{F}$ be a family of N-tuples of vectors in a fixed Banach space V. We'll say that $\mathcal{F}$ is a set of uniformly linearly independent vectors if there is a constant $C > 0$ such that for any $\{\ell_1,\dots,\ell_N\} \in \mathcal{F}$ and $\lambda_1,\dots\lambda_N \in \mathbb{C}$

$$||\sum_1^N \lambda_j\ell_j|| \geq C \sum_1^N |\lambda_j|.$$

<u>Lemma 1.1</u> Let $\mathcal{P}$ be a compact family of quadratic forms $p = \sum_{j=1}^N \ell_j^2$ on a fixed finite dimensional real vector space V for fixed N such that the ℓ_j are linearly independent. Then, the ℓ_j are automatically uniformly independent.

<u>Proof:</u> Fix a real inner product $\langle\cdot,\cdot\rangle$ on V. For $p \in \mathcal{P}$ let $\Sigma = \{x: p(x) = 0\}$ and let Σ^0 be the $\langle\ ,\rangle$ orthogonal complement of Σ.

By compactness there is a constant C such that

$$p(x) \geq C\,||x||^2, \qquad x \in \Sigma^0 \tag{1.12}$$

uniformly for $p \in \mathcal{P}$. Since $\Sigma = \{\ell_1 = \cdots \ell_N = 0\}$ and $\ell_1,\dots,\ell_N$ are independent, it follows that the map $\Sigma^0 \ni x \to (\ell_1(x),\dots,\ell_N(x)) \in \mathbb{R}^N$ is bijective. For any $\lambda = (\lambda_1,\ldots,\lambda_N) \in \mathbb{R}^N$ we have

$$\sup_{x\in\Sigma^0} \frac{|\sum \lambda_j \ell_j(x)|^2}{\sum |\ell_j(x)|^2} = \sum \lambda_j^2 = ||\lambda||^2$$

since (x) ranges over all $\mathbb{R}^N$. Taking note of the form of p we can express the last inequality as

$$\sup_{x\in\Sigma^0} \frac{|\sum \lambda_j \ell_j(x)|^2}{p(x)} = ||\lambda||^2$$

Using (1.12) we have uniformly that

$$C\,||\lambda||^2 \leq \sup_{x\in\Sigma^0} \frac{|\sum \lambda_j \ell_j(x)|^2}{||x||^2} \quad ||\sum \lambda_j \ell_j||^2$$

which is the desired conclusion

Applying Lemma 1.1 to $\operatorname{Re} q_j$, $\operatorname{Im} q_j$, we can conclude

<u>Proposition 1.2.</u> Let $\mathcal{P}$ be a compact family of non-negative quadratic forms on a symplectic vector space M such that for all $p \in \mathcal{P}$, $\{p(t) = 0\}$ has co-dimension 2d. Then, there are linear forms $q_1,\ldots,q_d$ such that $q_1,\ldots,q_d$, $\overline{q}_1,\ldots,\overline{q}_d$ are uniformly independent and uniformly bounded, p has the form (1.11), the Levi matrix $\mathcal{L} = (\frac{1}{i}\{q_j,\overline{q}_k\})$ is positive semi-definite, and $\{q_j,q_k\} = 0$.

Let $\mathcal{P}$ be a compact family of non-negative quadratic forms as above. We next study the canonical relation generated by $\mathcal{H}_p$. Let $C_t = \{(\rho,\mu);\ \rho = \exp(-it\mathcal{H}_p)\mu\}$ denote the graph of $\exp(-it\mathcal{H}_p)$. We will examine the behavior of C_t as $t \to \infty$. After applying Proposition 1.2 let $q = (q_1,\ldots,q_d)$ be a row vector and $\mathcal{H}_q = {}^t(\mathcal{H}_{q_1},\ldots,\mathcal{H}_{q_d})$ be a column vector. From the form of p we have

(1.13) $$\mathcal{H}_p = q\mathcal{H}_{\tilde{q}} + \tilde{q}\mathcal{H}_q.$$

We see that $\frac{1}{i}\mathcal{H}_p(\tilde{q}) = \frac{1}{i}\tilde{q}\,\mathcal{H}_q(\tilde{q})$ or rather

(1.14) $$\frac{1}{i}\mathcal{H}_p\tilde{q} = \tilde{q}\mathcal{L}$$

Similarly

(1.15) $$\frac{1}{i}\mathcal{H}_p q = -q\bar{\mathcal{L}}.$$

Setting $\mu_t = \exp(-it\mathcal{H}_p)\mu$ we find that

$$\frac{dq}{dt}(\mu_t) = -i\mathcal{H}_p\tilde{q} = \tilde{q}(\mu_t)\mathcal{L}$$

$$\frac{dq}{dt}(\mu_t) = -i\mathcal{H}_p q = -q(\mu_t)\bar{\mathcal{L}}$$

If $\rho = \exp(-it\mathcal{H}_p)\mu$ we have

(1.16) $$q(\rho) = q(\mu)e^{-t\bar{\mathcal{L}}} \qquad \tilde{q}(\rho) = \tilde{q}(\mu)e^{t\mathcal{L}}.$$

Using (1.16) in (1.13) tells us that

(1.17) $$\frac{1}{i}\mathcal{H}_p(\rho) = \frac{1}{i}q(\mu)e^{-t\bar{\mathcal{L}}}\mathcal{H}_{\tilde{q}} + \frac{1}{i}q\,(\mu)e^{t\mathcal{L}}\mathcal{H}_q.$$

The q being linear we note that $\mathcal{H}_q$, $\mathcal{H}_{\tilde{q}}$ are constant vectors. Integrating (1.17) we obtain

$$(1.18)\qquad \rho = \mu + \frac{1}{i}\, q(\mu) \int_0^t e^{-s\bar{\mathcal{L}}}\, ds\, \mathcal{H}_{\tilde{q}} + \frac{1}{i}\, \tilde{q}(\mu) \int_0^t e^{s\mathcal{L}}\, ds\, \mathcal{H}_q$$

$$= \mu + \frac{1}{i}\, q(\rho) \int_0^t e^{s\bar{\mathcal{L}}}\, ds\, \mathcal{H}_{\tilde{q}} + \frac{1}{i}\, q(\mu) \int e^{s\mathcal{L}}\, ds\, \mathcal{H}_q$$

Lemma 1.3. If A is a positive Hermitian matrix then for $t > 0$, $B(t) = \int_0^t e^{sA} ds$ is invertible and $||B^{-1}(t)|| \leq 1/t$.

Proof. Diagonalize A.

The C_t are 2n-dimensional (complex) linear subspaces of $\tilde{M} \times \tilde{M}$. Let $(\rho,\mu) \in C_t$ with $||\rho|| + ||\mu|| = 1$. (in some auxiliary norm on M). Then the uniform linear independence of the q and $\tilde{q}$ imply that $\mathcal{H}_q, \mathcal{H}_{\tilde{q}}$ are uniformly independent in $\tilde{M}$. Using the second part of (1.18), we have

$$||q(\rho) \int_0^t e^{s\bar{\mathcal{L}}}\, ds|| + ||\tilde{q}(\mu) \int_0^t e^{s\mathcal{L}}\, ds|| \leq C$$

uniformly. Both $\mathcal{L}$ and $\bar{\mathcal{L}}$ being positive Hermitian we may apply the last lemma to conclude that

$$||q(\rho)|| + ||\tilde{q}(\mu)|| \leq \frac{C}{t}$$

uniformly as p varies in $\mathcal{P}$. We have proved most of

Proposition 1.4. Let $\mathcal{P}$ be a compact family of quadratic forms as in Proposition 1.2. Then $C_t = \{(\rho,\mu)\colon \rho = \exp(-it\mathcal{H}_p)\mu\}$ converges uniformly at the rate $\mathcal{O}(1/t)$ to C_∞ where

$$(1.20)\qquad C_\infty = \{(\rho,\mu) \in J \times \bar{J}\colon\ \rho - \mu \in \tilde{\Sigma}^{\perp} = J^{\perp} \oplus \bar{J}^{\perp}\}.$$

Furthermore, C_∞ depends continuously on p.

By the rate of convergence being $\mathcal{O}(1/t)$, we mean that the distance between C_t and C_∞ in any metric on the Grassmanian of 2n subspace of $\tilde{M} \times \tilde{M}$ is $\mathcal{O}(1/t)$.

<u>Proof.</u> Let $e_1,\ldots,e_{2n-2d}$ be a uniformly linearly independent basis of Σ. Then since $\mathcal{H}_{q_j}, \mathcal{H}_{\tilde{q}_k}$ are uniformly independent, the vectors $(\mathcal{H}_{q_j},0)$, (e_j,e_j), $(0,\mathcal{H}_{\tilde{q}_j})$ are uniformly independent in $\tilde{M} \times \tilde{M}$. Thus C_∞ is given by the 2n uniformly independent equations

$$q_j(\rho) = 0 \qquad (\sigma(\rho, \mathcal{H}_{q_j}) = 0) \qquad 1 \le j \le d$$

$$\sigma(e_j,\rho-\mu) = 0, \qquad 1 \le j \le 2n-2d$$

$$\tilde{q}_j(\mu) = 0.$$

Since $\sigma(e_j,\rho-\mu) = 0$ for $(\rho,\mu) \in C_t$, the proposition follows from (1.19)

Alternatively we have

$$(1.21) \qquad C_\infty = \{(\rho' + \mu_+, \rho'' + \mu_-) : \rho',\rho'' \in \tilde{\Sigma},\ \rho' - \rho'' \in \tilde{\Sigma} \cap \tilde{\Sigma}^{\perp},\ \mu_\pm \in \Lambda_\pm\}$$

$$= \Delta(\tilde{\Sigma} \times \tilde{\Sigma}) \oplus (J^{\perp} \times \bar{J}^{\perp})$$

as may easily be checked.

We now examine the positivity of C_t as $t \to \infty$. Let $[a,b] = i^{-1}\sigma(a,b)$ define a Hermitian product on M. We know from [5] that there is a constant $B_t > 0$ such that

$$(1.22) \qquad [\rho,\rho] - [\mu,\mu] \ge B_t d((\rho,\mu), \Delta(\tilde{\Sigma} \times \tilde{\Sigma}))^2$$

for $(\rho,\mu) \in C_t$, and furthermore $B_t \ge Bt$ for $0 \le t \le 1$. We now ask how B_t

behaves as $t \to +\infty$.

Let $(\rho,\mu) \in C_t$ and write $\rho = u_t$, $\mu = u_0$, where

(1.23)
$$\frac{du_s}{ds} = \frac{1}{i}\mathcal{H}_p(u_s) , \qquad 0 \le s \le t.$$

Then a simple computation shows that

(1.24)
$$\frac{d}{ds}[u_s,u_s] = 2(||q(u_s))||^2 + ||\tilde{q}(u_s||^2)$$

and using (1.16) we get

(1.25)
$$\frac{d}{ds}[u_s,u_s] = 2(||q(\rho)e^{(t-s)\overline{\mathcal{L}}}||^2 + ||\tilde{q}(\mu)e^{s\mathcal{L}}||^2).$$

Integrating from 0 for t gives

(1.26)
$$[\rho,\rho] - [\mu,\mu] = 2\int_0^t ||q(\mu)e^{s\overline{\mathcal{L}}}||^2 ds + 2\int_0^t ||\tilde{q}(\mu)e^{s\mathcal{L}}||^2 ds$$

so from Cauchy-Schwartz inequality:

(1.27)
$$[\rho,\rho] - [\mu,\mu] \ge \frac{2}{t}(||q(\rho)\int_0^t e^{s\overline{\mathcal{L}}} ds||^2 + ||\tilde{q}(\mu)\int_0^t e^{s\mathcal{L}} ds||^2)$$

<u>Lemma 1.6.</u> We have

$$d((\rho,\mu), \Delta(\tilde{\Sigma}\times\tilde{\Sigma})) \le C(1+\frac{1}{t})(||\tilde{q}(\mu)\int_0^t e^{s\mathcal{L}} ds|| + ||q(\rho)\int_0^t e^{s\overline{\mathcal{L}}} ds||.)$$

uniformly for $t > 0$, $p \in \mathcal{P}$, $(\rho,\mu) \in C_t$.

<u>Proof.</u> The distance of (ρ,μ) to the diagonal of $\tilde{\Sigma}$ is of the same order of magnitude as

$$d(\rho,\tilde{\Sigma}) + d(\mu,\tilde{\Sigma}) + d(\rho,\mu). \tag{1.28}$$

In view of (1.18), if $(\rho,\mu) \in C_t$ then

$$d(\rho,\mu) \sim ||q(\rho)\int_0^t e^{s\bar{\mathcal{L}}}\, ds|| + ||\tilde{q}(\mu)\int_0^t e^{s\mathcal{L}}\, ds||$$

Now $d(\rho,\tilde{\Sigma}) \sim ||q(\rho)|| + ||\tilde{q}(\rho)||$ uniformly in $\mathcal{P}$ since $\tilde{\Sigma}$ is given by the uniformly independent equations $q = \tilde{q} = 0$. Combining these last observations we see that (1.28) is uniformly of the same order of magnitude as

$$\begin{aligned} &||q(\rho)|| + ||\tilde{q}(\rho)|| + ||q(\mu)|| + ||\tilde{q}(\mu)|| + \\ &||q(\rho)\int_0^t e^{s\bar{\mathcal{L}}}\, ds|| + ||\tilde{q}(\mu)\int_0^t e^{s\mathcal{L}}||. \end{aligned} \tag{1.29}$$

From Lemma 1.3 we have

$$||q(\rho)|| \leq \frac{1}{t}\, ||q(\rho)\int_0^t e^{s\bar{\mathcal{L}}}\, ds||$$

and

$$||\tilde{q}(\rho)|| \leq \frac{1}{t}\, ||\tilde{q}(\mu)\int_0^t e^{s\mathcal{L}}\, ds||$$

Also

$$\begin{aligned} ||\tilde{q}(\rho)|| &\leq C(||\tilde{q}(\mu)|| + d(\rho,\mu)) \\ &\leq C'(1 + \tfrac{1}{t})(||\tilde{q}(\mu)\int_0^t e^{s\mathcal{L}}\, ds|| + ||q(\rho)\int_0^t e^{s\bar{\mathcal{L}}}\, ds||) \end{aligned}$$

uniformly for $t > 0$. We can also bound $||q(\mu)||$ by the same expression and the lemma follows.

From (1.27) and the last lemma, we conclude

Proposition 1.7. Let $\mathcal{P}$ be a family of quadratic form as in Proposition 1.2. Then there is a constant $C > 0$ such that

$$\frac{1}{i}\sigma(\rho,\bar{\rho}) - \frac{1}{i}\sigma(\mu,\bar{\mu}) \geq \frac{t}{C(1+t^2)} d((\rho,\mu), \Delta(\tilde{\Sigma} \times \tilde{\Sigma}))^2$$

for $t \geq 0$ and $(\rho,\mu) \in C_t$, $p \in \mathcal{P}$.

Propositions 1.4 and 1.7 also hold when $\dim \Sigma$ is odd. If $\operatorname{co}\dim \Sigma = 2d + 1$ then there will be $d+1$ linear functions $q_0, q_1, \ldots d_d$ such that q_0 is real, $q_0, q_1, \ldots d_d, \tilde{q}_1, \ldots, \tilde{q}_d$ are uniformly independent,

$$p = q_0^2 + \sum_{j=1}^{d} q_j \tilde{q}_j.$$

$$\{q_0, q_j\} = \{q_0, \tilde{q}_j\} = \{q_j, q_n\} = \{\tilde{q}_j, \tilde{q}_n\} = 0,$$

and $\mathcal{L} = (\frac{1}{i}\{q_j, \tilde{q}_n\})$ is positive. The proofs are modifications of the above arguments and we shall omit them.

We now construct the solution of equation (1.1) where $p(x,\xi)$ is a quadratic form in (x,ξ) as in Proposition 1.2. We will suppose that the projection $C_\infty \ni (x,\xi,y,\eta) \longmapsto (x,\eta) \in \mathbb{C}^{2n}$ is bijective. This remains true even if we perturb p in $\mathcal{P}$, and put C_t in palce of C_∞. The 2-form $\sigma = d\xi\, dx - d\eta\, dy$ vanishes on $T(C_t)$. Since C_t is a linear space, if $(x,\xi,y,\eta) \in C_t$ it is also a tangent, so

$$\sigma((x,\xi,y,\eta),(t_x,t_\xi,t_y,t_\eta)) = \langle\xi,t_x\rangle - \langle x,t_\xi\rangle - \langle\eta,t_y\rangle + \langle y,t_\eta\rangle = 0$$

for $(t_x,t_\xi,t_y,t_\eta) \in T(C_t)$. This says that

$$(1.30) \qquad \xi dx - x d\xi - \eta dy + y d\eta = 0 \quad \text{on} \quad C_t.$$

Let $\phi(t,x,\eta) = \frac{1}{2}(\langle x,\xi\rangle + \langle y,\eta\rangle)$ for $(x,\xi,y,\eta) \in C_t$ which is well defined for a neighborhood of $p \in \mathcal{P}$. Taking the differential of ϕ and using (1.30) we see that

$$d\phi = \frac{1}{2}(\xi dx + x d\xi + \eta dy + y d\eta) = \xi dx + y d\eta.$$

In other words, $\phi'_x = \xi$, $\phi'_\eta = y$ and we have

$$(1.31) \qquad C_t = \{(x, \phi'_x(t,x,\eta), \phi'_\eta(t,x,\eta), \eta) : (x,\eta) \in \mathbb{C}^{2n}\},$$

which says that ϕ is a generating function for C_t. It is also easy to prove that ϕ solves (1.1). From Proposition 1.4 it follows that $\phi(t,x,\eta) = \phi(\infty,x,\eta) + \mathcal{O}(t^{-1})$, $t \to \infty$ uniformly near p. From Proposition 1.7 and the arguments of [4, section 3] there is a constant $C > 0$ such that

$$(1.32) \qquad \operatorname{Im} \phi(t,x,\eta) \geq \frac{t}{C(1+t^2)} d((x,\eta), \Sigma)^2, \quad (x,\eta) \in \mathbb{R}^{2n}$$

uniformly for $t \geq 0$, and in a neighborhood of p.

2. THE CHARACTERISTIC EQUATION

We shall construct suitable approximate solutions $\phi(t,x,\eta)$ of the characteristic equation

$$(2.1) \qquad \begin{cases} \dfrac{\partial \phi}{\partial t} + \dfrac{1}{t} p(x,\phi'_x) = 0 \\ \phi(0,x,\eta) = \langle x,\eta\rangle \end{cases}$$

where x are local coordinates in X (i.e., we shall here consider X as an open set in $\mathbb{R}^n$). From [5, Proposition 1.1] we know that there is a solution for (2.1) if we admit an error, which is $\mathcal{O}((\mathrm{Im}\,\phi)^N$ for all N on $K \times [0,T]$ for all $K \subset\subset X \times \mathbb{R}^n$, $T \in \mathbb{R}_+$. This solution is unique modulo $\mathcal{O}((\mathrm{Im}\,\phi)^N)$ for all N on $K \times [0,T]$ and we have

$$(2.2) \qquad \mathrm{Im}\,\phi(t,x,\eta) \geq C_{K,T}t \; d((x,\eta),\Sigma)^2,$$

$$(t,x,\eta) \in [0,T] \times K.$$

where $C_{K,T} > 0$. As in [5] the main problem is to study the behavior as $t \to \infty$. We also know from (2.2) that for $t > 0$, ϕ is essentially determined by its Taylor expansion on Σ. Working from now on only with Taylor expansions on Σ, we recall from [5] (or "notice that")

$$(2.3) \qquad C_t = \{(x,\phi'_x,\phi'_\eta,\eta);\ (x,\eta) \in \mathbb{C}^{2n}\}$$

is the graph of $\exp(-it\mathcal{H}_p)$. For $\rho \in \Sigma$ the tangent space of C_t at (ρ,ρ) is $T_{\rho,\rho}(C_t) = \text{graph} \exp -it\mathcal{H}_{p_2}$, where $p_2(\nu) = \frac{1}{2}\langle p''(\rho)\nu,\nu\rangle$ is the leading homogeneous term in the Taylor expansion of p at ν. The results of the last section tell us that $C^2_{t,\rho} = T_{\rho,\rho}(C_t)$ converges uniformly like $\mathcal{O}(\frac{1}{t})$ to a limiting linear canonical relation $C^2_{\infty,\rho}$ which depends continuously on ρ.

Now fix a point $\rho_0 = (x_0,\eta_0) \in \Sigma$ and suppose that the projection C^2_{∞,ρ_0} $(t_x,t_\xi,t_y,t_\eta) \longmapsto (t_x,t_\eta)$ is bijective. (This assumption will be removed later.) We seek the solution of (2.1) of the form

$$(2.4) \qquad \phi(t,x,\eta) \sim \sum_{\nu=0}^{\infty} \phi_\nu(t,x,\eta),$$

where ϕ_ν is a homogeneous polynomial in $(x-x_0,\eta-\eta_0)$ of degree ν, and let the formal Taylor expansion of p at ρ_0 be

$$p(x,\xi) \sim \sum_{j=2}^{\infty} p_j(x,\xi), \tag{2.5}$$

where p_j is homogeneous of degree j in $(x-x_0,\xi-\xi_0)$. Since ρ_0 is a stationary point for C_t it is clear that ϕ_0,ϕ_1 are independent of t: $\phi_0(t,x,\eta) = \langle x_0,\eta_0\rangle$, $\phi_1(t,x,\eta) = \langle x-x_0,\eta_0\rangle + \langle x_0,\eta-\eta_0\rangle$. Inserting the formal power series (2.4), (2.5) in (2.1) we get

$$\sum_{\nu \geq 2} \frac{\partial\phi_\nu}{\partial t} + \frac{1}{i} \sum_{j \geq 2} p_j(x,\eta_0 + \phi'_{2,x} + \phi'_{3,x} + \cdots) \sim 0. \tag{2.6}$$

Collecting terms of the same degree of homogeneity, we get

$$\frac{\partial\phi_2}{\partial t} + \frac{1}{i} p_2(x,\eta_0 + \phi'_{2,x}) = 0, \quad \phi_2(0,x,\eta) = \langle x-x_0,\eta-\eta_0\rangle, \tag{2.7}$$

and for $\nu \geq 3$

$$\frac{\partial\phi_\nu}{\partial t} + \frac{1}{i} \sum_{|\alpha|=1} (\partial_\xi^\alpha p_2)(x,\eta_0 + \phi'_{2,x})(\phi'_{\nu,x})^\alpha + F_\nu(t,x,\eta), \quad \phi_\nu(0,x,\eta) = 0 \tag{2.8}$$

where F_ν is a polynomial in $\phi'_2,\ldots,\phi'_{\nu-1}$.

Working in the coordinates $(x-x_0,\eta-\eta_0)$, we recognize in (2.7) the same type of equation as was studied in section 1. (Also ϕ_2 is the generating function of $T_{\rho,\rho}(C_t)$) and we therefore know that there is a limiting quadratic form in $(x-x_0,\eta-\eta_0)$; $\phi_2(\infty,x,\eta)$ such that

$$(2.9)\qquad ||\phi_2(t,\cdot,\cdot\cdot) - \phi_2(\infty,\cdot,\cdot\cdot)|| \leq \frac{C}{(1+t)},$$

$$(2.10)\qquad \operatorname{Im} \phi_2(t,x,\eta) \geq \frac{t}{C(1+t^2)}\, d((x,\eta)\Sigma)^2 ,$$

where the constant C is independent of the parameter $\rho_0 \in \Sigma$, if ρ_0 is allowed to vary in a small neighborhood of a fixed point. Here and below $||\cdot||$ is a norm in the suitable space of homogeneous polynomials in $(x-x_0,\eta-\eta_0)$.

As for the higher derivative, we have

<u>Lemma 2.1.</u> For any $k \geq 0$, $\nu \geq 2$ there are constants $C_{k,\nu}$, $a(k,\nu) \geq 0$, such that

$$||\partial_t^{\,k}\phi_\nu(t,\cdot,\cdot\cdot)|| \leq C_{k,\nu}(1+t)^{a(k,\nu)}$$

Here $C_{k,\nu}$, $a(k,\nu)$ are independe of ρ_0, when ρ_0 is allowed to vary in a small neighborhood of a fixed point.

<u>Proof.</u> For simplicity we introduce $(x-x_0,\eta-\eta_0)$ as new coordinates, so that (x_0,η_0) now becomes $(0,0)$ and we shall proceed by induction on ν. For $\nu = 2$ and $k = 0$ the statement follows from (2.9) and for larger k we obtain the result from (2.7) after differentiating a suitable number of times with respect to t. Suppose now the Lemma has been proved for $\phi_2,\dots,\phi_{\nu-1}$. Then (2.8) gives (in the new coordinates)

$$\frac{\partial\phi}{\partial t} + \frac{1}{i}\sum_j p_2^{(j)}(x,\phi_2'(t,x,\ ))\frac{\partial\phi}{\partial x_j} = \mathcal{O}(t^\alpha)$$

in the vector space of homogeneous polynomial of fixed degree. We can rewrite this as

$$\frac{\partial\phi}{\partial t} + A\phi = B\phi + \mathcal{O}(t^\alpha)$$

where

$$A = \frac{1}{i} \sum p_2^{(j)}(x, \phi_2'(\infty, x, \eta))\frac{\partial}{\partial x_j}$$

and $||B|| = \mathcal{O}((1+t)^{-1})$. If we consider ϕ as a homogeneous polynomial on C_∞^2 (under the map $C_\infty^2 \ni (x,\xi,y,\eta) \to (x,\eta)$) then

$$A\phi(x,\eta) = \frac{1}{i}\mathcal{H}_{p_2(x,\xi)}\phi.$$

We'll now suppose that $\sum$ has even codimension (the odd case is similar). Recall that $C_\infty^2 = \Delta(\tilde{\Sigma} \times \tilde{\Sigma}) + (J \otimes \bar{J}^{\perp})$. For $(\lambda,\mu) \in C_\infty^2$ we have

$$\mathcal{H}_{p_2(x,\xi)}\big|_{C_\infty^2} = \tilde{q}(\lambda)\mathcal{H}_q,$$

where $q_2,\dots,q_d$ are chosen as in section 1. Let $t_1,\dots,t_{2n} \in C^2$ be uniformly linearly independent such that

$$t_j = \mathcal{H}_{q_d(x,\xi)} \qquad 1 \leq j \leq d$$

$$t_{d+1},\dots,t_{2n-d} \quad \text{is a basis in } \Delta(T_\rho\tilde{\Sigma} \times T_\rho\tilde{\Sigma})$$

$$t_{2n-d+j} = \mathcal{H}_{q_d(y,\eta)}$$

and let $z_1,\dots,z_{2n}$ be dual coordinates to t_j so that $t_j = \partial/\partial z_j$. We now have

$$\frac{1}{i}\mathcal{H}_{p_2(x,\xi)}\Big|_{C_\infty^2} = \sum_1^d \frac{1}{i}\tilde{q}_j(x,\xi)\frac{\partial}{\partial z_j}$$

and

$$\frac{\partial q_k(x,\xi)}{\partial z_j} = \begin{cases} \{q_j,\tilde{q}_k\} & \text{if } j \leq d \\ 0 & \text{if } j > d. \end{cases}$$

so that

$$\frac{1}{i}\tilde{q}_k(x,\xi) = \frac{1}{i}\tilde{q}_k\,(\sum z_j t_j) = \sum_{j=1}^{d} \frac{1}{i}\{q_j,\tilde{q}_k\}\, z_j$$

and

$$A = \langle \bar{\mathcal{L}} z, \frac{\partial}{\partial z} \rangle$$

where $\mathcal{L}$, the Levi matrix, is positive. We know (from [5] Lemma 1.6) that $\exp(-tA)\phi = \phi\circ \exp(-t\bar{\mathcal{L}})$ and is consequently bounded independently of t. The proof of the lemma is completed by applying

Lemma 2.2. Let V be a Banach space, A an operator such that $||\exp(-tA)||$ is bounded, $B(t)$ an operator such that $||B(t)||$ is bounded, $f(t)\colon R^+ \to V$ such that $||f(t)|| = \mathcal{O}(t^\alpha)$. Then there is a $\gamma > 0$ such that if v is a solution of

$$\begin{cases} \dfrac{dv}{dt} + Av = \dfrac{B(t)}{1+t}\, v + f(t) \\ v(0) = 0 \end{cases}$$

Then $||v(t)|| \leq \mathcal{O}(t^\gamma)$ as $t \to \infty$.

We omit the elementary proof.

Next we show that the assumption that $C^2_{\infty,\rho_0} \ni (x,\xi,y,\eta) \mapsto (x,\eta)$ is bijective may be removed. First we observe that this projection can be made bijective by making a canonical change of coordinates in T^*X.

Lemma 2.3. Let M be a real symplectic vector space and $\Lambda \subset \tilde{M} \times \tilde{M}$ a Lagrangean plane. Then, there are real symplectic coordinates (x,ξ) in M so that the projection $\Lambda \ni (x,\xi,y,\eta) \to (x,\eta)$ is bijective.

Proof. We start with any system of real symplectic coordinates (x,ξ) on M. We may permute either x_j and ξ_j or y_j and η_j so that $\Lambda \ni (x,\xi,y,\eta) \mapsto (x,\eta)$ is bijective. This process does not damage the product structure of M. In other words we can find two Lagrangean planes Λ_1, Λ_2 M such that Λ is transverse to $\tilde{\Lambda}_1 \times \tilde{\Lambda}_2$. Moving Λ_2 slightly we can also have Λ_1 and Λ_2 transverse. Choose coordinates in M so that $\Lambda_1 = \{x = 0\}$ and $\Lambda_2 = \{\xi = 0\}$ then gives the desired coordinates.

The last lemma tells us the the conclusion of Lemma 2.1 holds for some system of canonical coordinates in T^*X. We now show that if Lemma 2.1 holds in one system, then it must hold for any other choice of real symplectic coordinates. We can consider C_t, the graph of $\exp(-it\mathcal{H}_p)$ as a formal power series object. We can go backwards and forwards between conclusions about C_t and $\phi(t,x,\eta)$. Namely, let C_t be given by the equation $(y,\xi) = f_t(x,\eta)$, then the codimension of Lemma 2.1 is equivalent to

$$|\partial^{\alpha}_{t,x,\eta} f_t(x,\eta)| = \mathcal{O}(t^{a(\alpha)}). \tag{2.11}$$

We will show that if (2.11) holds for one system of coordinates it will hold for any other. Since C_t is a positive relation for $t < \infty$, the projection $C_t^2 \ni (t_x,t_\xi,t_y,t_\eta) \mapsto (t_x,t_\eta)$ is always bijective. In fact, by Proposition 1.8 there is a constant $C > 0$ such that

$$\frac{1}{i}(\sigma(\lambda,\lambda) - \sigma(\mu,\mu)) \geq \frac{t}{C(1+t^2)} \operatorname{dist}((\lambda,\mu), \Delta(T\tilde{\Sigma} \times T\tilde{\Sigma}))^2 \tag{2.12}$$

for $t \geq 0$, $(\lambda,\mu) \in C^2_{t,\rho}$. There is then a linear map $L_{t,\rho}$ such that

$$(t_x,t_\xi,t_y,t_\eta) \in C^2_{t,} \iff (t_\xi,t_y) = L_t(t_x,t_\eta)$$

Lemma 2.4. $||L_{t,\rho}||, ||L_{t,\rho}^{-1}|| \leq \mathcal{O}(t)$ as $t \to \infty$ uniformly for ρ near ρ_0.

Proof. Noting that

$$\frac{1}{i}\sigma((t_x,t_\xi),(\bar{t}_x,\bar{t}_\xi)) = 2 \text{ Im } <\bar{t}_x,t_\xi>$$

we have for $(t_x,t_\xi,t_y,t_\eta) \in C_t^2$ that

$$\frac{1}{i}\sigma((t_x,t_\xi), (\bar{t}_x,\bar{t}_\xi) - \frac{1}{i}\sigma((t_y,t_\eta), (\bar{t}_y,\bar{t}_\eta))$$

$$= 2 \text{ Im}(< \bar{t}_x-\bar{t}_y,t_\xi - <\bar{t}_y,t_\eta - t_\xi>)$$

$$\leq 2 (||t_x - t_y|| + ||t_\eta - t_\xi||) (||t_\xi|| + ||t_y||).$$

We also have $d(t_x,t_\xi,t_y,t_\eta), \Delta(T_\rho \tilde{\Sigma} \times T_\rho \tilde{\Sigma})) \approx ||t_x-t_y|| + ||t_\xi- t_\eta|| +$ $d((t_x,t_h), T_\rho \tilde{\Sigma}))$. Set $X = (t_x,t_\eta)$ and $Y = (t_y,t_\xi)$. Using (2.12) we get

$$||X-Y|| \ ||Y|| \geq \frac{t}{C(1+t^2)} ||X - Y||^2 \qquad \text{when } Y = L_tX.$$

from which we deduce that for $t \geq 1$,

$$Ct \ ||Y|| \geq ||X||.$$

The same holds with X and Y reversed which proves the lemma.

Let $(\tilde{x},\tilde{\xi})$ be another set of real symplectic coordinates in T^*X. If $(x,\eta) = \alpha_t(\tilde{x},\tilde{\eta})$ is the change of local coordinates on C_t then the last lemma tells us that the Jacobian of α_t and its inverse are bounded by $\mathcal{O}(1+t)$ on Σ. It follows that if (2.11) holds for the (x,ξ) coordinate system, it holds for the $(\tilde{x},\tilde{\xi})$ one also.

Proposition 2.5. There is a smooth function $\phi \in C^\infty(R_t \times X \times \mathbb{R}^n)$ such that for every compact set $K \subset X \times \mathbb{R}^n$ and $(x,\xi) \in K$, $t \geq 0$:

(2.13) $$\operatorname{Im} \phi(t,x,\eta) \geq \frac{t}{C_K(1+t^2)} d((x,\eta),\Sigma)^2, \qquad C_K > 0,$$

(2.14) $$|\partial_t^k \partial_{x,\eta}^\alpha \phi(t,x,\eta)| \leq C_{K,k,\alpha}(1+t)^{a(K,k,\alpha)}$$

(2.15) $$\phi(0,x,\eta) = \langle x,\eta\rangle \quad \text{and}$$

$$|\partial_t{}^k \partial_{x,\eta}^{\alpha}(\frac{\partial\phi}{\partial t} + \frac{1}{i}\, p(x,\phi_x'))|$$

$$\leq C_{K,k,\alpha,N}(1+t)^{a(K,k,\alpha,N)}\ (\operatorname{Im}\phi)^N$$

for every $N \geq 0$.

Here p also denotes a homogeneous almost analytic extension (see [4]).

Proof. We have noticed earlier that there are no problems for $t \leq 1$, so we shall only sketch the construction of ϕ for $t > 1$. Choose local coordinates $z = (z',z'')$ in $X \times \dot{\mathbb{R}}^n$ such that Σ is given by $z'' = 0$. We know the Taylor expansion of ϕ at each point of Σ, so we are looking for a function with the Taylor series

$$\phi(t,x) \sim \Sigma\, \psi_j(t,x),$$

where ψ_j is a certain homogeneous polynomial of degree j in z'' satisfying

$$\partial_t^k \partial_z^\alpha \psi_j(t,x) = \mathcal{O}((1+t)^{a(k,\alpha,j)}|z''|^{\max(j-|\alpha|,0)}) .$$

Also, any function ϕ with this Taylor expansion will have the property that $\frac{\partial\phi}{\partial t} + \frac{1}{i}\, p(x,\phi_x')$ vanishes for infinite order on Σ. Let $\chi(z'') \in C_0^\infty$ be equal to 1 near the origin and put

$$\phi(t,z) = \psi_0(t,z) + \psi_1(t,z) + \psi_z(t,z) + \sum_{j\geq 3} \psi_j(t,z)\chi(\lambda_j(1+t)^{b(j)}z")$$

Then it is easy to verify that if $\lambda_j, b(j)$ are suitable (fast increasing) sequences of positive numbers, then the series above converges in C^∞ and (2.13) and (2.14) hold. Moreover $\frac{\partial\phi}{\partial t} + \frac{1}{i} p(x,\phi_x')$ satisfies the same estimates as in (2.14) and since it vanishes for infinite order on Σ, (2.15) follows from (2.13). So far the construction was only local in $X \times \dot{\mathbb{R}}^n$, but it is easy to globalize by a partition of unity.

Remark 2.6. With a more refined analytis we can show that $a(K,k,\alpha)$ and $a(K,k,\alpha,N)$ are $\mathcal{O}(|\alpha|)$ for fixed K,k,N.

3. A COMPOSITION FORMULA

Before proceeding further in our construction of $\exp(-tP)$, we pause for some technical results about Fourier integral operators with phase functions like the ϕ just constructed. The main result of this section is a composition formula for such operators. The proof will be a refinement of the one for Theorem 2.3 in [4], the explicit change of variables below, goes back to Kuranishi and has been used by Boutet de Monvel in the proof of the standard asymptotic formula for $P(ae^{i\lambda\phi})$ when ϕ is real.

Let $\Omega \subset \mathbb{R}^n$ be open and let $m > 1$ be fixed. We say that a function $a \in C^\infty(\bar{\mathbb{R}}_+ \times \Omega \times \dot{\mathbb{R}}^n)$ is quasi-homogeneous (q.h.) of degree $k \in \mathbb{R}$ if $a(t,x,\lambda\eta) = \lambda^k a(\lambda^{m-1}t,x,\eta)$ for all $\lambda > 0$. Similarly we say that a subset $M \subset \bar{\mathbb{R}}_+ \times \Omega \times \dot{\mathbb{R}}^n$ is quasi-conic if $(\lambda^{-(m-1)}t,x,\lambda\eta) \in M$ for all $\lambda \geq 1$ whenever $(t,x,\eta) \in M$, $|\eta| \geq 1$. We define $\tilde{S}^k(\bar{\mathbb{R}}_+ \times \Omega \times \dot{\mathbb{R}}^n) \subset C^\infty(\bar{\mathbb{R}}_+ \times \Omega \times \dot{\mathbb{R}}^n)$ to be the space of symbols a, such that for all $\alpha,\beta,\gamma,N > 0$ and $K \subset\subset \Omega$ we have

$$\left(\frac{1}{|\eta|^{m-1}} D_t\right)^{\gamma} D_x^{\alpha} D_\eta^{\beta} a(t,x,\eta) \leq \mathcal{O}(1)\, |\eta|^{k-|\beta|} (1+t|\eta|^{m-1})^{-N}$$

$$x \in K, \ |\eta| \geq 1 \quad t > 0.$$

The usual result about the existence of asymptotic sums in $\tilde{S}^k$ holds and such sums are unique modulo $\tilde{S}^{-\infty} \stackrel{\text{def.}}{=} \bigcap_{k \in \mathbb{R}} \tilde{S}^k$. We denote by $\tilde{S}^k_c$ the symbols in $\tilde{S}^k$ which are of the form $a \sim \sum_{-\infty}^{k} a_j$, where $a_j \in \tilde{S}^j$ is quasi-homogeneous of degree j.

Let $\phi(t,x,\eta) \in C^\infty(\bar{\mathbb{R}}_+ \times \Omega \times \mathbb{R}^n)$ be quasi-homogeneous of degree 1. We will assume that ϕ has the following properties:

(3.2) For all $K \subset\subset \Omega$ and all indices α, there are constants $C_{K,\alpha}$ and a_α such that

$$|D^\alpha_{t,x,\eta} \phi(t,x,\eta)| \leq C_{K,\alpha} (1+t)^{a_\alpha}$$

for $t \geq 0$, $x \in K$, $|\eta| = 1$.

(3.3) $$\operatorname{Im} \phi \geq 0.$$

(3.4) $$\operatorname{Im} \phi(t,x,\eta) = 0 \Longrightarrow \phi'_x = \eta, \quad \phi'_\eta = x$$

Moreover, there is a constant $N > 0$ such that

$$|\phi'_x - \eta| + |\phi'_\eta - x| \leq \text{Const.} \ (1+t)^{-N}, \quad \text{for } |\eta| = 1$$

uniformly for x in compact subsets of Ω. Let $a \in S^k$. We shall study the operator

(3.5) $$(Au)(t,x) = \int e^{i\phi(t,x,\eta)} a(t,x,\eta) \hat{u}(\eta) d\eta.$$

We see immediately that

(3.6) $$A:\ C_0^\infty(\Omega) \to C^\infty(\overline{\mathbb{R}}_+ \times \Omega).$$

The distribution kernel

$$K(t,x,y) = \int e^{i\phi(t,x,\eta)-\langle y,\eta\rangle} a(t,x,\eta)d\eta$$

makes sense as an oscillatory integral (as integrating by parts in y readily shows) and

$$K \in C^\infty(\overline{\mathbb{R}}_+ \times \Omega_x\ ;\ \mathcal{D}'(\Omega_y)).$$

By (3.4), if $|\phi_x'-\eta| + |\phi_\eta'-x| \geq \varepsilon > 0$ when $|\eta| = 1$, $x \in K \subset\subset \Omega$ we have $0 \leq t \leq t_{\varepsilon,K} < \infty$ and $\operatorname{Im}\phi \geq c_{\varepsilon,K} > 0$. After introducing a cutoff function (q.h. of degree 0) which does not change the singularities of K we can reduce ourselves to the case that

(3.7) $$\frac{1}{|\eta|}|\phi_x'-\eta| + |\phi_\eta'-x| \leq \varepsilon$$

where ε can be chosen arbirarily small. Using integration by parts in x (using the operator $|\phi_x'(t,x,\eta)|^{-2} \sum \overline{\frac{\partial\phi}{\partial x_j}} \frac{\partial}{\partial x_j}$) we can easily show that

(3.8) $$K \in C^\infty(\mathbb{R}_+ \times \Omega_y;\ \mathcal{D}'(\Omega_x))$$

so that

(3.9) $$A:\ \mathcal{E}'(\Omega) \to C^\infty(\mathbb{R}_+;\ \mathcal{D}'(\Omega))$$

Finally, by using integration by parts in η together with (3.7), we show that

(3.10) $$K|_{x\neq y} \in C^\infty(\bar{\mathbb{R}}_+ \times (\Omega \times \Omega \setminus \Delta(\Omega \times \Omega))).$$

(We use the operator $|\phi'_\eta - y|^{-2} \sum (\overline{\frac{\partial\phi}{\partial\eta_j}} - y_j) \frac{\partial}{\partial\eta_j} \cdot$)

Now that we have given a sense to the operator A, we study left composition with pseudo-differential operators. The main result is

Theorem 3.1. Suppose that $a \in \tilde{S}^k$ and ϕ satisfies (3.2) - (3.3). Let $P \in L^m(\Omega)$ and assume that either P is properly supported or that a has uniformly compact support in x. Then,

$$P(ae^{i\phi}) = be^{i\phi} + b_{-\infty}$$

where $b_{-\infty} \in \tilde{S}^{-\infty}$ and $b \in \tilde{S}^{m+k}$ is given by

(3.11) $$b(t,x,\eta) \sim \sum \frac{1}{\alpha!} D_y^\alpha(P^{(\alpha)}(x,\psi(t,x,y,\eta))a(t,y,\eta))|_{y=x},$$

Here $\psi(t,x,y,\eta) = \int_0^1 \phi'_x(t,y+s(x-y),\eta)ds$ so that

$$\phi(t,x,\eta)-\phi(t,y,\eta) = \langle x-y,\psi\rangle$$

and P denotes some almost analytic extension in S^m.

Proof. We assume that a has uniformly compact support in x. Put $\lambda = |\eta|$ and $s = \lambda^{m-1}t$. Considering $\eta/|\eta|$ as an uninteresting parameter, we'll write

$$\phi(t,x,\eta) = \lambda\Phi(s,x) \qquad (\Phi(s,x) = \phi(s,x,\eta/|\eta|)$$

$$a(t,x,\eta) = A(s,x,\lambda)$$

so that we have

$$D_s^\gamma D_x^\alpha \Phi = \mathcal{O}(\,(1+s)^{a_{\alpha,\gamma}})$$

$$D_s^\gamma D_x^\alpha A = \mathcal{O}(\,(1+s)^{-N}\lambda^k), \qquad \text{for all } N .$$

We have

$$(3.12) \qquad P(x,D_x)(ae^{i\phi}) = P(x,D_x)(A(s,x,\ )e^{i\lambda\Phi(s,x)})$$

$$= \iint P(x,\xi)A(x,y,\lambda)e^{i(<x-y,\xi>+\lambda\Phi(s,y))}\frac{dyd\xi}{(2\pi)^n}$$

Let $\chi(\xi) \in C_0^\infty(\mathbb{R}^n\backslash 0)$ be equal to 1 in a neighborhood of S^{n-1}. Since Φ'_y is close to the unit sphere, we have in the support of $(1-\chi(\xi/\lambda))A(x,y,\lambda)$:

$$(3.13) \qquad ||d_y(<x-y,\xi> + \lambda\Phi(s,y))|| \geq C\,(|\lambda| + |\xi|)$$

where $C > 0$. Using the operator

$$||\lambda\Phi'_y-\xi||^{-2}\sum(\lambda\overline{\Phi}'_{y_j}-\xi_j)\frac{\partial}{\partial y_j}$$

repeatedly to integrate by parts we can show that for all N

$$(3.14) \qquad \iint (1-\chi(\xi/\lambda))P(x,\xi)A(s,y,\lambda)e^{i(<x-y,\xi>+\lambda\Phi(s,y))}dyd\xi$$

$$\sim \mathcal{O}(1)\lambda^{-N}(1+s)^{-N}, \qquad \lambda \to \infty$$

(uniformly for $(x,\eta) \in K \times \mathbb{R}^n$, $K \subset\subset \Omega$).

We are reduced to studying

$$(3.15)\qquad \iint \chi(\xi/\lambda)P(x,\xi)A(s,y,\lambda)e^{i(<x-y,\xi>+\lambda\Phi(x,y))}\,\frac{dyd\xi}{(2\pi)^n}$$

$$=\left(\frac{\lambda}{\pi}\right)^n \iint \chi(\xi)A(s,y,\lambda)P(x,\lambda\xi)e^{i\lambda(<x-y,\xi>+\Phi(x,y))}dyd\xi.$$

Now setting

$$(3.16)\qquad B(s,x,y,\xi,\lambda) = \left(\frac{\lambda}{2\pi}\right)^n \chi(\xi)A(x,y,\lambda)P(x,\lambda\xi)$$

we notice that

$$(3.17)\qquad \partial^\alpha_{s,x,y,\xi}B = \mathcal{O}((1+s)^{-N}\lambda^{n+k+m}) \qquad \forall N > 0.$$

(3.18) B has (uniformly) compact support in (y,ξ).

We have to study

$$(3.19)\qquad I(s,x,\lambda) = \iint B(s,x,y,\xi,\lambda)e^{i\lambda(<x-y,\xi>+\Phi(x,y))}dyd\xi.$$

For $(s,x,y,\xi) \in \operatorname{supp} B$ we write

$$(3.20)\qquad \Phi(s,x) - \Phi(x,y) = <x-y,\ \Psi(s;x,y)>$$

where

$$(3.21)\qquad \Psi(s,x,y) = \int_0^1 \Phi'_x(s,y+\tau(x-y))d\tau$$

By introducing a cutoff function $\chi(\frac{x-y}{\delta})$ and using integration by parts, we can reduce ourselves to the case when $|x-y|$ is small and (3.21) then always makes sense.) From the polynomial bound on the second derivatives we get for some N_0

$$(3.22)\qquad ||\Psi(s,x,y)|| \le C(||x-y||\cdot(1+s)^{N_0} + ||\Phi'_x(s,x)||)$$

Taking into account the non-negativity of $\operatorname{Im}\Phi$ and the polynomial bound for the second derivatives, we get (with some new constant N_0):

$$(3.23)\qquad ||\operatorname{Im}\Psi(s,x,y)|| \leq C(1+s)^{N_0}(||x-y|| + (\operatorname{Im}\Phi(s,x))^{1/2})$$

$$||\operatorname{Im}\Psi(s,x,y)|| \leq C'(1+s)^{N_0}(||x-y|| + (\operatorname{Im}\Phi(s,y))^{1/2}).$$

For $t \in [0,1]$, let $\Gamma_t = \Gamma_t(s,x,y)$ be the cycle

$$\mathbb{R}^n \ni \xi \longmapsto \xi - t\Psi(s,x,y) \in \mathbb{C}^n.$$

Since $\langle x-y,\xi\rangle + \Phi(x,y) = \Phi(s,x) + \langle x-y,\xi - \Psi(s,x,y)\rangle$ we can write

$$(3.24)\qquad I(s,x,\lambda) = e^{i\lambda\Phi(s,x)}\int_{\mathbb{R}^n}\left(\int_{\Gamma_1} e^{i\lambda\langle x-y,\xi\rangle}B(s,x,y,\zeta+\Psi,\lambda)d\zeta\right)dy$$

We would like to replace Γ_1 by Γ_0. For $\zeta = \xi - t\,\Psi \in \Gamma_t$, $0 \leq t \leq 1$, we have

$$(3.25)\qquad \operatorname{Im}(\Phi(s,x) + \langle x-y,\zeta\rangle) = \operatorname{Im}(\Phi(s,x) - t\langle x-y,\Psi\rangle)$$

$$= (1-t)\operatorname{Im}\Phi(s,x) + t\operatorname{Im}\Phi(s,y)$$

We take an almost analytic extension of $B(s,x,y,\xi,\lambda)$ with respect to ξ such that

$$(3.26)\qquad \bar\partial_\zeta B(s,x,y,\zeta,\lambda) = \mathcal{O}(1)\ (1+s)^{-M}|\operatorname{Im}\zeta|^N\lambda^{m+n+k}$$

for all N, and $M > 0$. (The appropriate extension of P and χ will automatically give this extension of B. (See [4], Theorem 1.3.) For $\zeta = \xi - t\,\Psi \in \Gamma_t$ we have $\operatorname{Im}(\zeta+\Psi) = \operatorname{Im}(1-t)\Psi$ so that

$$(3.27)\qquad (\bar\partial_\zeta B)(s,x,y,\zeta+\Psi,\lambda) = \mathcal{O}(1)(1+s)^{-M}|\operatorname{Im}\Psi|^N\lambda^{n+k+m}$$

for all N, $M \geq 0$, when $\zeta \in \Gamma_t$. If we define $J(s,x,\lambda)$ by (3.24) with Γ_1 replaced by Γ_0 we have

(3.28) $$J(s,x,\lambda) - I(s,x,\lambda) =$$

$$i \int_0^1 \int_{\mathbb{R}^n} \int_{\Gamma_t} e^{i\lambda(\Phi(s,x)+<x-y,\xi>)} <\bar{\partial}_\zeta B(s,x,y,\zeta+\Psi,\lambda), \operatorname{Im} \Psi> d\xi dy dt.$$

We can write this even more explicitly as

(3.29) $$J(s,x,\lambda) - I(s,x,\lambda) =$$

$$i \int_0^1 \int_{\mathbb{R}^n} \int_{\mathbb{R}^n} e^{i\lambda(\Phi(s,x)+<x-y,\xi-t\Psi>)} <\bar{\partial}_\zeta B(s,x,y,\xi+(1-t)\Psi,\lambda), \operatorname{Im} \Psi> d\xi dy dt$$

In the region where $|x-y| \geq \lambda^{-1/2}$, partial integrations with respect to ξ show that the corresponding contribution to J-I is $\mathcal{O}((1+s)^{-M}\lambda^{-N})$, for all $N \geq 0$. In the region where $|x-y| \leq \lambda^{-1/2}$ we have

$$||\operatorname{Im} \Psi(s,x,y)|| \leq C(1+s)^{N_0}(\lambda^{-1/2} + ((1-t)\operatorname{Im} \Phi(s,x) + t \operatorname{Im} \Phi(s,y))^{1/2})$$

in view of (3.23). Combining this with (3.26) we see that

(3.30) $$| <\bar{\partial}_\zeta B(s,x,y,\xi + (1-t)\Psi,\lambda), \Psi>|$$

$$= \mathcal{O}(1)(1+s)^{-M}(\lambda^{-N} + ((1-t)\operatorname{Im} \Phi(s,x) + t \operatorname{Im} \Phi(s,y))^N)$$

for all N, M. Noting (3.25) we get

(3.31) $$J(s,x,\lambda) - I(s,x,\lambda) = \mathcal{O}((1+s)^{-M}\lambda^{-N}), \quad \lambda \geq 1.$$

(uniformly for $x \in K \subset\subset \Omega$ and $\eta/|\eta| \in S^{n-1}$. This proves that

$$(3.32) \qquad I(s,x,\lambda) = e^{i\lambda\Phi(s,x)} \iint_{\mathbb{R}^{2n}} e^{i\lambda\langle x-y,\xi\rangle} B(s,x,y,\xi + \Phi(s,x,y),\lambda)\,dy\,d\xi$$

$$+ \mathcal{O}((1+s)^{-M}\lambda^{-N}) \qquad \forall\, N,M.$$

We recall that for $b \in C_0(\mathbb{R}^{2n})$ with support in some fixed compact set, we have for every $N > 0$

$$(3.33) \qquad \iint e^{i\lambda\langle x-y,\xi\rangle} b(y,\xi)\,dy \frac{d\xi}{(2\pi)^n}$$

$$= \sum_{|\alpha|<N} \frac{i^{-|\alpha|}}{\alpha!} b^{(\alpha)}_{(\alpha)}(x,0)\lambda^{-(|\alpha|+n)} + R_N(x,\lambda)$$

where

$$R_N(x,\lambda) = \mathcal{O}(1)\lambda^{-(N+n)} \sup_{\mathbb{R}^{2n},\,|\beta|\le 2N+n+1} |\partial^{\beta}_{y,\xi} b(y,\xi)|$$

Applying this to (3.32) we conclude

$$(3.34) \qquad I(s,x,\lambda) = (2\pi)^n \sum_{|\alpha|<N} e^{i\lambda\Phi(s,x)} \frac{1}{\alpha!\, i^{|\alpha|}} \partial^{\alpha}_y \partial^{\alpha}_{\xi}(B(s,x,y,\xi+\Phi,\lambda))\Big|_{\substack{y=x\\ \xi=0}} \lambda^{-|\alpha|+n}$$

$$+ \mathcal{O}((1+s)^{-M}\lambda^{m+k-N} \qquad \forall\, N,M.$$

Letting $b \in \tilde{S}^{m+k}$ be given by (3.11), we have proved that

$$P(ae^{i\phi}) = be^{i\phi} + b_{-\infty}$$

where $b_{-\infty} = \mathcal{O}((1+s)^{-M}|\eta|^{-N})$ for all N, M when $|\eta| \geq 1$). To estimate the derivatives of $b_{-\infty}$ we simply note $\partial^{\alpha}_{t,x,\eta} ae^{i\phi} \in e^{i\phi}\tilde{S}^{a}_{\alpha}$ so that for all M

$$\partial^{\alpha}_{t,x,\eta} P(ae^{i\phi}) = \mathcal{O}(|\eta|^{a'_{\alpha}}(1 + t|\eta|^{m-1})^{-M})$$

The same holds for $be^{i\phi}$ and so for $b_{-\infty}$ also. Then using simple "interpolation" inequalities for the derivatives we have $\partial^{\alpha}_{t,x,\eta} b_{-\infty} = \mathcal{O}((1+t|\eta|^{m-1})^{-M}|\eta|^{-N})$ for all N, M. This completes the proof.

4. THE HEAT EQUATION

Let $\Omega \subset \mathbb{R}^n$ be open and $P \in L^m_c(\Omega)$ be property support with p its principal symbol. We suppose that P satisfies all the conditions of the introduction (with X replaced by Ω). The results of section 2 apply to p. Since p is homogeneous of degree m, it is clear that the function $\phi(t,x,\eta)$ constructed in Proposition 2.5 can be made quasi-homogeneous of degree 1.

The goal of this section is to construct an approximate solution of the heat equation

$$(4.1) \qquad \begin{cases} (D_t + \frac{1}{i} P(x,D_x))v = 0 & \text{in } \mathbb{R}_+ \times \Omega \\ v|_{t=0} = u & \text{in } \Omega \end{cases}$$

where u is a given function in either $C^{\infty}_0(\Omega)$ or $\mathcal{E}'(\Omega)$. We'll seek a solution of the form

$$(4.2) \qquad v(x,t) = A_t u(x) = \int e^{i\phi(t,x,\eta)} a(t,x,\eta)\hat{u}(\eta) \frac{d\eta}{(2\pi)^n}$$

where

$$a(t,x,\eta) \sim \sum_0^\infty a_j(t,x,\eta) \tag{4.3}$$

and a_j is q.h. of degree $-j$. We put (4.2) in (4.1), pass $D_t + \frac{1}{i}P$ through the integral sign, and then introduce the formal asymptotic expansion of $P(x,D)\{e^{i\phi}a\}$ as if a and ϕ were a symbol and phase function in the usual sense. We set $(D_t + iP)\{e^{i\phi}a\} \sim 0$ and group terms according to their degree of quasi-homogeneity to obtain the transport equations

$$(T_0) \qquad L(t,x,\eta,D_t,D_x)\, a_0 = 0$$

$$(T_j) \qquad L(t,x,\eta,D_t,D_x)a_j + \ell_j(t,x,a_0,\dots,a_{j-1}) = 0$$

We have

$$L(t,x,\eta,D_t,D_x) = D_t + \frac{1}{i}\sum_{\nu=1}^n p^{(\nu)}(x,\phi'_x)D_\nu - i\, q_{m-1}(t,x,\eta)$$

where

$$q_{m-1}(t,x,\eta) = P_{m-1}(x,\phi'_x) + \frac{1}{2i}\sum_{j,k}\frac{\partial^2 p(x,\phi'_x)}{\partial\xi_j\partial\xi_k}\;\frac{\partial^2\phi(t,x,\eta)}{\partial x_j\partial x_k}$$

and ℓ_j is a linear differential operator acting on $a_0,\dots,a_{j-1}$. The initial condition in (3.1) tells us that

$$(I_0) \qquad a_0(0,x,\eta) = 1$$

$$(I_j) \qquad a_j(0,x,\eta) = 0 \qquad j \geq 1.$$

As with the characteristic equation, it is easy to see that the system (T) (I) can be solved uniquely modulo $\mathcal{O}(|\mathrm{Im}\,\phi|^N)$, $\forall N$, locally uniformly in (x,η) and on bounded t intervals. The Taylor expansion of a_j is then uniquely determined on Σ. As in section 2, we construct solutions of the transport equation by first getting estimates for a and its derivatives on Σ and then using a Borel type construction.

Let $C_t = \{(x,\phi',\phi'_\eta,\eta)\}$ and $E = \{(t,\rho): \rho \in C_t\}$. Then $\nu = \partial/\partial t + \frac{1}{i}\mathcal{H}p(x,\xi)$ is a vector field on E. Using (t,x,η) as coordinates on E

$$(4.4) \qquad \nu = \frac{\partial}{\partial t} + \frac{1}{i}\sum_1^n p^{(j)}(x,\phi'_x)\frac{\partial}{\partial x_j}$$

and

$$(4.5) \qquad \mathrm{div}(\nu) = \frac{1}{i}\Big(\sum \frac{\partial^2 p(x,\xi)}{\partial x_j \xi_j} + \sum\sum \frac{\partial^2 p}{\partial \xi_j \partial \xi_k}\frac{\partial^2 \phi}{\partial x_j \partial x_k}\Big)$$

For a smooth function $a(t,x,\eta)$ we introduce the formal 1/2 density

$$\alpha = a(t,x,\eta)\sqrt{dtdxd\eta}$$

on E which is defined up to some factor i^μ.

The Lie derivative of α along ν is

$$(4.6) \qquad \mathcal{L}_\nu(\alpha) = (\nu(a) + \tfrac{1}{2}(\mathrm{div}\ \nu)a)\sqrt{dtdxd\mu}$$

so that we have

$$(4.7) \qquad La\sqrt{dtdxd\eta} = \frac{1}{i}(\mathcal{L}_\nu + S_p|_E)(a\sqrt{dtdxd\eta})$$

where S_p is the subprincipal symbol of P. This makes it easy to change coordinates. If $(\tilde{x},\tilde{\xi})$ is another system of homogeneous symplectic coordinates in $T^*\Omega$ define $\tilde{a}(t,\tilde{x},\tilde{\eta})$ by

$$(4.8) \qquad \begin{cases} \tilde{a}(0,\tilde{x},\tilde{\eta}) = a(0,x,\eta) \\ \tilde{a}(t,\tilde{x},\tilde{\eta})\sqrt{dtd\tilde{x}d\tilde{\eta}} = a(t,x,\eta)\sqrt{dtdxd\eta} \end{cases}$$

(The last equation is ambiguous by a factor of i^{μ} which we determine from the first.) From (4.7) we see that

$$(La)\sqrt{dtdxd\eta} = (\tilde{L}\,\tilde{a})\sqrt{dtd\tilde{x}d\tilde{\eta}}$$

where the formula for $\tilde{L}$ is the exact analogue of the one for L, obtained after conjugating P by an elliptic Fourier integral operator.

Proposition 4.1. For any continuous function $\lambda(x,\eta) > 0$ such that $\lambda < S_P + \frac{1}{2}\,\mathrm{t\tilde{r}}\,F$ on Σ, we can find $a_0, a_1, \ldots$, q.h. of degree $0,-1,\ldots$ such that

(4.9) $$\partial^{\alpha}_{t,x,\eta}\, a_j(t,x,\eta) = \mathcal{O}(e^{-\lambda(x,\eta)t})$$

uniformly for (x,η) in compact subsets of $T^*\Omega\setminus 0$ and $t \in \overline{\mathbb{R}_+}$.

(4.10) Let $\Delta_j(t,x,\eta)$ be the error with which (T_j) holds, then

$$\partial^{\alpha}_{t,x,\eta}\,\Delta_j(t,x,\eta) = \mathcal{O}(e^{-\lambda(x,\eta)t}d_{\Sigma}^{N}) \qquad \forall N$$

where $d_{\Sigma}(x,\eta)$ is the distance of (x,η) to Σ.

Proof. First we note that if we prove the estimates (4.9), (4.10) in one coordinate system they will be valid in any other. This is because half densities in different coordinates differ by the factor

$$\left(\frac{\partial(t,x,\eta)}{\partial(t,\tilde{x},\tilde{\eta})}\right)^{1/2}$$

and we have seen in section 2 that all derivatives of the Jacobian are of the order of magnitude of a power of t. In view of Lemma 2.3 we may then assume that at $\rho_0 \in \Sigma$, the projection $T_{\rho,\rho}(C_\infty) \ni (t_x,t_\xi,t_y,t_\eta) \mapsto (t_x,t_\eta)$ is bijective. We use induction on j and then on the degree of the terms in the Taylor expansion of a_j. As we have done for the characteristic equation, we rewrite (T_j) in coordinates on $T_{\rho_0,\rho_0}(C_\infty)$. In general we get an equation in a space of homogeneous polynomials of the form

$$\frac{\partial a}{\partial t} + (A + S_p(\rho_0) + \frac{1}{2} \operatorname{tr} F_{\rho_0})a = \frac{B(t)}{1+t}a + \mathcal{O}(e^{-\lambda(\rho_0)t})$$

where A and B are as in section 2. Using a variant of Lemma 2.2 it is easy to see that

$$||\partial_t^k a|| = \mathcal{O}(e^{-\lambda' t})$$

for $0 < \lambda' < \lambda$. Gluing the Taylor series of a_j together via a Borel type constuction we obtain a function which satisfies (T_j) to infinite order of Σ and the estimate (4.9). The bounds of (4.10) then follow by applying Taylor's theorem.

We next check that our formal solution is a legitimate solution of (4.1) modulo smoothing operators. In particular, we must verify that it's permissible to apply the results of section 3. The phase function ϕ clearly satisfies conditions (3.2) and (3.3). The first part of (3.4) holds since C_0 is the graph of the identity and $(C_t)_{\mathbb{R}} = \Delta(\Sigma\times\Sigma)$ for $t > 0$. As for the second part, $\phi'_x = \eta$, $\phi'_\eta = x$, when $t = 0$ or when $(x,\eta) \in \Sigma$. In view of the bounds on the second derivative, (3.4) will hold for $|\eta| = 1$ in a set of the form $d((x,\eta),\Sigma) < \frac{\delta}{(1+t)^{N_0}}$, for $\delta > 0$ small and N_0 large. Outside this set (for $|\eta| = 1$) we can easily modify ϕ to satisfy (3.4), so that Proposition 2.5 remains valid. From Theorem 3.1 we get

<u>Proposition 4.2.</u> Let P, ϕ and a be as above, then the operator A_t defined by (4.2) is a continuous map $C_0^\infty(\Omega) \to C^\infty(\overline{\mathbb{R}}_+;\ C^\infty(\Omega))$ and extends continuously to a map $\mathcal{E}'(\Omega) \to C^\infty(\overline{\mathbb{R}}_+;\ \mathcal{D}'(\Omega))$. Moreover, the kernel $A_t(x,y)$ is C^∞ for $x \neq y$ or $t \neq 0$, $A_0 = I$ and the kernel of $(D_t + \frac{1}{2} P(x,D_x)) \circ A_t$ is in

$C^{\infty}(\mathbb{R}_+ \times \Omega \times \Omega)$.

To make the last statement clear, notice that

$$(D_t + \frac{1}{i} P(x,D_x))a(t,x,\eta)e^{i\phi(t,x,\eta)} \equiv b(t,x,\eta)e^{i\phi(t,x,\eta)} \bmod \tilde{S}^{-\infty}$$

where for all N we have an estimate of the form

$$|b(t,x,\eta)| \leq C(|\eta|^{-N} + (1 + |\eta|^{m-1}t)^N d_\Sigma^N |\eta|^m)e^{-\mathrm{Im}\phi - \lambda(x,\eta)t|\eta|^{m-1}}$$

Since

$$\mathrm{Im}\ \phi \geq \frac{c\ t|\eta|^{m-1}}{(1+(t|\eta|)^{m-1})^2} d_\Sigma^2(x,\eta)|\eta|$$

it is easy to see that b is rapidly decreasing in $|\eta|$.

We now return to the original situation in which P is given on a compact manifold X. Taking a finite covering of X by coordinate patches for which the conclusions of Propositon 4.2 hold we can construct, using a partition of unity, a global solution B_t satisfying the conclusion of Proposition 4.2 with Ω replaced by X. With some further arguments we see that if $E_t = \exp(-tP)$ in the L^2-sense then we have

$$\mathrm{tr}\ E_t - \mathrm{tr}\ B_t \in C^{\infty}(\overline{\mathbb{R}}_+).$$

5. THE SINGULARITY OF THE TRACE

Our final task is to study $\mathrm{tr}\ \chi' A_t \chi$ where A_t is the operator constructed in section 4, and $\chi, \chi' \in C_0^{\infty}(\Omega)$ with $\chi'\chi = \chi$. Retaining the notation of the last section, we have

$$(5.1)\qquad I_\chi(t) = \operatorname{tr} \chi' A_t \chi = \iint e^{i\Phi(t,x,\eta)} a(t,x,\eta)\chi(x) \frac{dxd\eta}{(2\pi)^n} ,$$

where $\Phi(t,x,\eta) = \phi(t,x,\eta) - \langle x,\eta\rangle$. The inequality (2.13) implies (taking into account the quasihomogeneity)

$$(5.2)\qquad \operatorname{Im} \Phi(t,x,\eta) \geq \frac{tp(x,\eta)}{C(1+t|\eta|^{m-1})^2}$$

Also $a(t,x,\eta) \leq Ce^{-\frac{1}{C} t|\eta|^{m-1}}$, and if we use the inequality

$$(5.3)\qquad \frac{tp(x,\eta)}{(1+t|\eta|^{m-1})^2} + t|\eta|^{m-1} \geq \frac{1}{C} (tp+t|\eta|^{m-1})^{1/3} - C$$

we can estimate the integral (5.1) by

$$\iint F(tp(x,\eta) + t|\eta|^{m-1})\, dxd\eta ,$$

where F is rapidly decreasing near $+\infty$. If now

$$V(\tau) = \iint_{p(x,\xi)+|\xi|^{m-1} \leq \tau} dxd\xi$$

we have $V(a\tau) \leq a^{\frac{n}{m-1}} V(\tau)$ for $a \geq 1$, and it is easy to see that the integral above:

$$\int F(t\tau)dV(\tau)$$

can be estimated from above by $C\cdot V(\frac{1}{t})$. Noting that $A_t \equiv A_{\frac{t}{2}} A^*_{\frac{t}{2}}$ by the self-adjointness, and the semigroup-property, one can also estimate the integral (5.1) from below.

For $\varepsilon > 0$ small we can split $I_\chi(t)$ in two parts $I_\chi^1(t)$, $I_\chi^2(t)$ where the integration is restricted to $t|\eta|^{m-1} \leq |\eta|^{-\varepsilon}$ and $t|\eta|^{m-1} > |\eta|^{-\varepsilon}$ respectively. It is easy to see that

$$I_\chi^1(t) \sim \int_{t|\eta|^{m-1} \leq |\eta|^{-\varepsilon}} e^{-tp(x,\eta)}\chi(x)\, dx \frac{d\xi}{(2\pi)^n}$$

Using similar estimates as above we can show that modulo an error $o(1)E(t)$ where

$$E(t) = \iint_{p(x,\xi)+|\xi|^{m-1} \leq \frac{1}{t}} dx d\xi$$

we have

$$I_\chi^2(t) \equiv \iint_{t|\eta|^{m-1} \geq |\eta|^{-\varepsilon}} \frac{a_0(t,\theta',0)\chi(\theta',0)}{(\det \frac{1}{i}\Phi''_{\theta''\theta''}(t,\theta',0))^{1/2}} \frac{d\theta'}{(2\pi)^{n-d}}$$

Here θ, d are as defined in the introduction and we simply write $a_0(t,\theta',0)$, $\chi(\theta',0)$ etc. for the restrictions of a_0, χ to Σ. Now $I_\chi(t)$ can be analyzed further precisely as in [5] and one gets for <u>md-n > 0:</u>

$$(5.6) \qquad I_\chi(t) \equiv \frac{1}{(2\pi)^n} \iint e^{-tp(x,\eta)}\chi(x,\eta)\, dx d\eta$$

For <u>md-n = 0:</u>

$$(5.7) \qquad I_\chi(t) \equiv \frac{t^{-\frac{n}{m}} \log t^{-1}}{(m-1)(2\pi)^{n-d}} \int_{\tilde{\Sigma}} \chi\, \tilde{\omega}' \quad .$$

In the third case $\underline{md-n < 0}$, we get

(5.8) $$I_\chi(t) \equiv \iint \frac{a_0(t,\theta',0)}{(\det \frac{1}{i} \Phi''_{\theta''\theta''}(t,\theta',0))^{1/2}} \chi(\theta',0) \frac{d\theta'}{(2\pi)^{n-d}}$$

Applying Karamata's Tauberian theorem as in [5], we deduce Theorem 0.1 from (5.6), (5.7) in the first two cases. In the third case, the analysis of [5], is not immediately applicable and we have to study the density

(5.9) $$\frac{a_0(t,\theta',0)}{(\det \frac{1}{i} \Phi''_{\theta''\theta''}(t,\theta',0))^{1/2}} \quad d\theta'$$

more closely.

Put $\omega''_\rho(d\theta'') = (\det p''_{\theta''\theta''})^{1/2} d\theta''$, $\rho \in \Sigma$. This is an invariant density on the normal space $N_\rho(\Sigma) = T_\rho(T^*X)/T_\rho(\Sigma)$, for if p_2 is the quadratic form on $N_\rho(\Sigma)$ induced by p, then $\int_{p_2(\theta'')\leq 1} \omega''_\rho(d\theta'') = \int_{\theta''^2 \leq 1} d\theta'' \cdot 2^{\frac{d}{2}}$.

Then $\omega'_\rho(d\theta')$ is invariably defined on $T_\rho(\Sigma)$ by the property that $\omega'_\rho \wedge \omega''_\rho = dxd\xi$.

We can express C_t, using Φ as $\{(x,\eta + \Phi'(x,\eta), x + \Phi'_\eta(x,\eta),\eta)\}$. Since $\Delta(\Sigma\times\Sigma) = C_t \cap \Delta(T^*X \times T^*X)$, Φ vanishes to exactly second order on Σ. The tangent space $T_{\rho,\rho}(C_t)$ for $\rho \in \Sigma$ is

$$\{(t_x, t_\eta + \Phi''_{xx} t_x + \Phi''_{x\eta} t_\eta, t_x + \Phi''_{\eta x} t_x + \Phi''_{\eta\eta} t_\eta, t_\eta)\}$$

Let $t_X = (t_x, t_\xi)$ and $t_Y = (t_y, t_\eta)$, then for $(t_X, t_Y) \in T_{\rho,\rho}(C_t)$ we have

(5.10) $$t_Y - t_X = F_\Phi(t_x, t_y)$$

where

$$F_{\Phi} = \begin{pmatrix} \Phi_{\eta x} & \Phi_{\eta\eta} \\ -\Phi_{xx} & -\Phi_{x\eta} \end{pmatrix}$$

is the Fundamental matrix of Φ in (x,η) coordinates. The kernel of F_{Φ} is $T_{\rho}(\Sigma)$ and its range $T_{\rho}(\Sigma)$. Consequently F_{Φ}: $N_{\rho}(\Sigma) \to T_{\rho}(\Sigma)$ is a bijection, and similarly for F_p. Consider the map

(5.11) $$\kappa : T_{\rho,\rho}(C_t) \to T_{\rho}(\Sigma) \xrightarrow{F_p^{-1}} N_{\rho}(\Sigma)$$

$$(t_X, t_Y) \longmapsto t_Y - t_X \mapsto F_p^{-1}(t_Y - t_X).$$

The form $\kappa^*(\omega''_{\rho})$ is the invariantly defined on $T_{\rho,\rho}(C_t)$. Identifying C_t with T^*X via the map C_t $(x,\xi,y,\eta) \mapsto (x,\eta)$, we get

$$\kappa^*(\omega''_{\rho}) = \det (F_p^{-1} F_{\Phi})\Big|_{N_{\rho}} \omega''_{\rho}(x,\eta)$$

Indeed, this follows at once from (5.10). To compute $\det F_p^{-1} F_{\Phi}|_{N_{\rho}}$ we take $\delta\theta''$ and $\mathcal{H}_{\theta''}$ as basis in N_{ρ} and $T_{\rho}(\Sigma)^{\perp}$ respectively. The matrices of F_{Φ} and F_p are then $\Phi''_{\theta''\theta''}$ and $p''_{\theta''\theta''}$ respectively. We see this by observing that $\delta\theta''$ and $\mathcal{H}_{\theta''}$ are dual bases with respect to σ and that

$$\sigma(\delta\theta_k, F_{\Phi}\,\delta\theta_j) = \Phi''_{\theta_j\theta_k}$$

so that the $\mathcal{H}_{\theta_k}$ component of $F_{\Phi}(\delta\theta_j)$ is $\Phi''_{\theta_j\theta_k}$. We get that

$$\det(F_p^{-1}F_{\Phi})\Big|_{N_{\rho}(\Sigma)} = \det\,(p''_{\theta''\theta''})^{-1}(\Phi''_{\theta''\theta''}).$$

This leads to the conlcusion that

(5.12) the form $\det((p''_{\theta''\theta''})^{-1}\, \Phi''_{\theta''\theta''})\, \omega''_\rho(x,\eta)$ on $T_{\rho,\rho}(C_t)$ depends only on p and C_t, but not on the choice of symplectic coordinates nor on the corresponding generating function ϕ.

Every linear factor in this form vanishes on $\Delta(T_\rho(\Sigma) \times T_\rho(\Sigma))$ so that if we multiply it by $\omega'_\rho(x,\eta)$, which is invariantly defined on this subspace we get

Lemma 5.1. The form

$$\beta = (\det \Phi''_{\theta''\theta''}) \cdot (\det p''_{\theta''\theta''})^{-1}\, dxd\eta$$

on $T_{\rho,\rho}(C_t)$ depends only on C_t and p but not on the choice of symplectic coordinates.

Assume now that we are given an invariant half density α on $T_{\rho,\rho}(C_t)$ which we can express as $\alpha = a\sqrt{dxd\eta}$ for any choice of symplectic coordinates.

Lemma 5.2 The form

$$\omega = \frac{a}{(\det \frac{1}{i}\,\Phi''_{\theta''\theta''})^{1/2}}\, d\theta' \qquad \text{on } T_\rho(\Sigma)$$

depends only on C_t, but not on the choice of (x,ξ) or θ.

Proof.

$$\omega = a\sqrt{dxd\eta} \cdot \left(\frac{\det \frac{1}{i}\,\Phi''_{\theta''\theta''}}{\det \frac{1}{2}\, p''_{\theta''\theta''}}\, dxd\eta\right)^{-1/2} (\det \tfrac{1}{2}\, p''_{\theta''\theta''})^{-1/2}\, d\theta'$$

$$= \alpha \cdot \beta^{-1/2}\, \omega'(d\rho),$$

the product of three invariants.

In order to compute the density (5.9) at a point $\rho \in \Sigma$, we choose a complex canonical transformation which takes p, modulo terms vanishing to the third order at ρ, into

$$\tilde{p}(x,\xi) = \sum_{1}^{r'} i\mu_j x_j \xi_j + \sum_{r'+1}^{r} \frac{1}{2} \xi_j^2 .$$

Put

$$P(x,D) = p(x,D) + K$$

where $K = S_p + \frac{1}{2} \tilde{tr}$. Then if we consider $\tilde{p}$ as the principal symbol and the constant K as the lower order symbol (corresponding to p_{m-1}), we see that $S_{\tilde{p}} = S_p$. We saw in section 6 how the first transport equation is interpreted invariantly by introducing the "principal symbol" $a_0(t,x,\eta)\sqrt{dxd\eta}$ and, we will get the principal symbol by solving the characteristic-, and transport-equations for $\tilde{P}$. Hence by Lemma 5.2 we only have to compute the phase and the amplitude corresponding to $\tilde{P}$

The characteristic equation has the solution

$$\tilde{\phi}(t,x,\eta) = \sum_{1}^{r'} e^{-\mu_j t} x_j \eta_j + \langle x'',\eta''\rangle + i\frac{t}{2}\eta''^2 + \langle x''',\eta'''\rangle .$$

The first transport equation gives

$$\tilde{a}_o(t) = e^{-tK} = e^{-t(S_p + \frac{1}{2}\tilde{tr}\,F)}$$

and apart from factors $i^{k/2}$ we get

$$\frac{\tilde{a}_0(t)}{(\det \frac{1}{i}\tilde{\Phi}''_{\tilde{\theta}''\tilde{\theta}''})^{1/2}}\, d\tilde{\theta}' = \frac{e^{-t(S_p + \frac{1}{2}\mathrm{tr}\,F)}}{\prod_{1}^{r'}(1 - e^{-\mu_j t})\; t^{\frac{1}{2}\dim x''}}\, dx'' dx''' d\eta''' .$$

Here we have put $\tilde{\theta}' = (x'',x''',\eta''')$, $\tilde{\theta}'' = (x',\eta',\eta'')$. Since $\omega'(d\tilde{\theta}') = \frac{dx''dx'''d\,'''}{\mu_1 \ldots \mu_{r'}}$, we conclude

Lemma 5.3. The density (5.9) equals

$$t^{-d} e^{-t(S_p + \frac{1}{2}\tilde{tr}F)} \prod \frac{\mu_j t}{(1-e^{-\mu_j t})} \qquad \omega'(d\theta').$$

where the product is taken over all the positive eigenvalues of $\frac{1}{i} F$.

Thus in the case $md-n < 0$, we have

$$\operatorname{tr} e^{-tP} \sim \frac{t^{-d}}{(2\pi)^{n-d}} \int_\Sigma e^{-t(S_p + \frac{1}{2}\operatorname{tr} F)} \prod \frac{\mu_j t}{(1-e^{-\mu_j t})} \quad \omega'(d\theta').$$

(Here we use also that $e^{-tP} - B_t$ has C^∞ kernel.) Choosing $\theta' = (\omega,r)$ where r is homogeneous of degree 1 and ω,θ'' are homogeneous of degree 0 and making $r \to t^{-\frac{1}{m-1}} r$, we get

$$\operatorname{tr} e^{-tP} \sim \frac{t^{-\frac{(n-d)}{m-1}}}{(2\pi)^{n-d}} \int_\Sigma e^{-(S_p + \frac{1}{2}\tilde{tr}F)} \prod \frac{\mu_j}{(1-e^{-\mu_j})} \quad \omega'(d\theta') \;.$$

Again Karamata's Tauberian theorem applies to give Theorem 0.1 in this case.

REFERENCES

1. Boutet de Monvel, L., Grigis, A., Helffer, B., Paramétrixes d'operateur pseudo-differential à caractéristiques multiples, Astérisque, 1976.

2. Hörmander, L., A class of hypoelliptic pseudodifferential operators with double characteristics, Math. Ann. 217(1975), 165-188.

3. Melin, A., Lower bounds for pseudo-differential operators, Arkiv för Math. 9 (1971), 117-140.

4. Melin, A. and Sjöstrand, J., Fourier integral operators with complex-valued phase function, Lecture Notes in Math. 459, pp. 120-233, Berlin: Springer 1975.

5. Menikoff, A. and Sjöstrand, J., On the eigenvalues of a class of hypoelliptic operators, Math. Ann. 235(1978), 55-85.

On the smoothness of the time t-map of the KdV equation and the bifurcation of the eigenvalues of Hill's operator

Tudor Ratiu

§1. Introduction

The remarkable properties of the Korteweg-de Vries equation, such as existence of solutions for all time and its complete integrability as a Hamiltonian system, make the question of smoothness of the time t-map natural. Two methods for attacking the problem are outlined in §2. The first one, using Kato's methods on quasi-linear equations, only gives differentiability as a function $H^{s+1} \to H^{s}$. The second uses the inverse scattering method. It turns out in 4, after a bifurcation result on the eigenvalues of Hill's operator is obtained, that this approach also fails to give the result. In proving the bifurcation result, the density of simple potentials is needed which is proved here using infinite dimensional manifolds; this result was already known and proved using analytic function theory by McKean and Trubowitz [1976] and by Simon [1976] using operator theory.

The trouble in proving the smoothness of the time t-map comes from the term uu_x. But such difficulties have been overcome by Ebin and Marsden [1970] when working with the Euler equations of a perfect fluid. These methods strongly use the co-adjoint orbit theory and properties of Hamiltonian systems on Lie groups. §5 briefly reviews the non-periodic Toda lattice, a finite dimensional completely integrable Hamiltonian system behaving formally like the KdV equation, obtaining all classical coordinate expressions and the abstract formula for the integrals in involutions; from there onwards the methods described by Adler in this conference or by Kostant [1978] are more natural in dealing with this system. §6 attempts to apply the group theoretic techniques to the KdV equation using results of Ebin and Marsden [1970] and to accomplish a program similar to the one in §5 but yielding a smooth time t-map. Formally, this has been done by Adler who recovers from group theory the complete integrability of the KdV equation, but does not eliminate the troublesome term uu_x. Regarding the KdV equation as an expression in space coordinates of a certain "vector field" on the group $\mathcal{D}^s$ of H^s diffeomorphisms on the circle, it is shown that its time t—map is also an expression in space coordinates of a smooth flow on $\mathcal{D}^s$. As a sample corollary of this approach, one can prove that for short time the solutions of $u_t + uu_x + \varepsilon u_{xxx} = 0$ converge as $\varepsilon \to 0$ to the solutions of the equation $u_t + uu_x = 0$. One can also prove the convergence of product formulas obtained by iterating $u_t + uu_x = 0$ and $u_t + u_{xxx} = 0$, using Chorin , etal. [1978]. It is also shown that the symplectic structure on a certain co-adjoint orbit of $\mathcal{D}^s$ coincides with the classical one of Fadeev and Zakharov [1971] and Adler [1978]. The trouble with this approach is that the KdV equation is *not* Hamiltonian on $\mathcal{D}^s$. It seems that it will be on the group of invertible Fourier integral operators.

I want to thank J. Marsden and P. van Moerbeke for their support and help in writing this paper. Especially, §6 is an outgrowth of discussions with J. Marsden who pointed out the smoothness result using Trotter's product formula. Conversations with M. Adler, J. Duistermaat and B. Kostant are also gratefully acknowledged.

§2. Two approaches towards the smoothness of the time t-map

Consider the (periodic or non-periodic) Korteweg-de Vries equation

$$(KdV)\, u_t + uu_x + u_{xxx} = 0.$$

The following result of Kato [1974] gives existence and uniqueness of the solutions of this equation as well as continuity of the time t-map.

Let W *be an open bounded ball in* H^s *,* $s \geq 3$ *centered at the origin of radius* R*, and* $T \geq 0$*. If* $f \in W$*,* KdV *has a unique solution*

$$u \in C^0([0,T],W) \cap C^1([0,T],H^{s-3}), \quad u(0) = f$$

Moreover, letting $S: [0,T] \times W \to H^s$, $S_t f = u(t) =$ *the unique solution of* KdV *with* $u(0) = f$, $S_t: W \to H^s$ *is continuous in the* H^s *-norm, uniformly in* $t \in [0,T]$.

It is also clear that if $v(s) = u(-s)$, v satisfies

$$v_s - vv_x - v_{xxx} = 0$$

so that again by the above theorem this equation has a unique solution. The conclusion is that in the above theorem we can replace everywhere the interval $[0,T]$ by $[-T,T]$. Since $T \geq 0$ is arbitrary, solutions exist for all time. This results from the local theory plus à priori estimates provided by the well-known integrals for the KdV equation (convenient references are McKean's article in this volume or Abraham and Marsden [1978]). The question we pose is: is the map $S_t: H^s \to H^s$ smooth? For example, Burger's equation $u_t + uu_x + u_{xx} = 0$ has a smooth time t-map*. However, $u_t + uu_x = 0$ has non-smooth time t-map, so the question above cannot be settled easily. Recall that the solution of $u_t + uu_x = 0$ with $u(0,x) = u_0(x)$, u_0 a C^1 bounded map, is given by $u(t,x) = u_0(y)$ for $x = y + tu_0(y)$; for short t (e.g. $t < 1/\|\partial u/\partial x\|_{L^\infty}$) the correspondence $x \mapsto y$ is invertible; Kato has shown in [1975] section 5.3 that even for short t, its evolution operator $U_t: H^s \to H^s$, $s \geq 2$ is continuous, but is not Hölder continuous for any exponent α, $0 < \alpha < 1$. One might hope for smoothness of S_t; the idea is that, by analogy with

---

*This is readily proved directly or using the techniques of Marsden and McCracken [1976], or by methods below.

the *dissipative* term u_{xx} in Burger's equation, the *dispersive* term u_{xxx} smoothes out the shock-waves. That's exactly what happens when one proves long-time solutions of KdV; also the fact that KdV is a completely integrable Hamiltonian system makes one believe this.

A first approach towards the smoothness of S_t is via Kato's methods of quasi-linear evolution equations; see Kato [1974], [1977] and Marsden and McCracken [1976]. We outline here the strategy of computing the derivative DS_t using this method and how far can one go. Denote by $T: H^s \to H^{s-3}$, $u \mapsto -u_{xxx} - uu_x$, the KdV operator which is C^∞ since $\frac{\partial^3}{\partial x^3}: H^s \to H^{s-3}$ is linear continuous and $(u,v) \mapsto uv_x: H^{s-1} \times H^s \to H^{s-1}$ is bilinear continuous. Its derivative is $(DT)(u) = -\frac{\partial^3}{\partial x^3} - u\frac{\partial}{\partial x} - u_x I \in L(H^s, H^{s-3})$. Rewrite the KdV equation as

$$\frac{d(S_t f)}{dt} = T(S_t f), \quad S_0 f = f, \; f \in H^s.$$

If S_t where differentiable, we could take the differentials in the above equation and obtain the *linear* evolution equation in the unknown $(DS_t)(f)$:

$$\frac{d}{dt}(DS_t)(f) = (DT)(S_t f)\circ(DS_t)f, \; (DS_0)f = I, \quad f \in H^s .$$

Kato's theorems on the existence of solutions for linear evolution equations apply for the operator $(DT)(S_t f)$ and hence the above equation has a unique solution $(DS_t)f$ with $(DS_0)f = I$. This is a tedious but straightforward verification ; one checks, for example, all conditions of Theorem 1.9 in Kato [1977]. Thus, if the derivative exists, it is the solution of the above linear equation. At this moment we have a candidate for the derivative and it is here that the trouble begins. One should use Theorem 7 of Kato [1974] to prove existence of DS_t. This theorem states that under a multitude of hypotheses, a sequence of quasi-linear evolution equations converging to a fixed equation has solutions converging in H^s to the solution of the limit equation. But when trying to apply this theorem, one estimate is not obtainable and all one can show is that $S_t: H^{s+1} \to H^s$ is differentiable. Moreover, the differential $(DS_t)(f)$, $f \in H^{s+1}$ extedns to a linear continuous operator $H^s \to H^s$. These statements are again proved by a series of tedious calculations using the Lemmas in Kato [1974] or [1977]. That's how far these methods can take us.*

A second approach is using the inverse scattering method, and here the special nature of the KdV equation is fully exploited; the previous methods held for quasi-linear hyperbolic evolution equations in general. From here onwards we deal only with the periodic KdV equation. Denote by $H^s_{per}[0,1]$ the periodic H^s-maps on the interval $[0,1]$. Let $Q = -\frac{d^2}{dx^2} + q(x)$ be the Hill operator with $q \in H^s_{per}[0,1]$, $s \geq 3$. The following material can be found in Magnus and Winkler [1966] or in McKean's series of talks in this volume.

If $y_1(x;\lambda,q)$, $y_2(x;\lambda,q)$ denotes a fundamental set of solutions of $Qy = \lambda y$ with $y_1(0) = y_2'(0) = 1$, $y_1'(0) = y_2(0) = 0$, then $\Delta(\lambda,q) = y_1(1;\lambda,q) + y_2'(1;\lambda,q)$ is called the *discriminant* of Q. The roots of $\Delta(\lambda) = 2$ comprise the eigenvalues of Q with periodic boundary conditions and the roots of $\Delta(\lambda) = -2$ are the eigenvalues of Q with anti-periodic boundary conditions. A typical graph of $\Delta(\lambda)$ is shown below. The intervals $(-\infty,\lambda_0)$, (λ_1,λ_2), (λ_3,λ_4),...

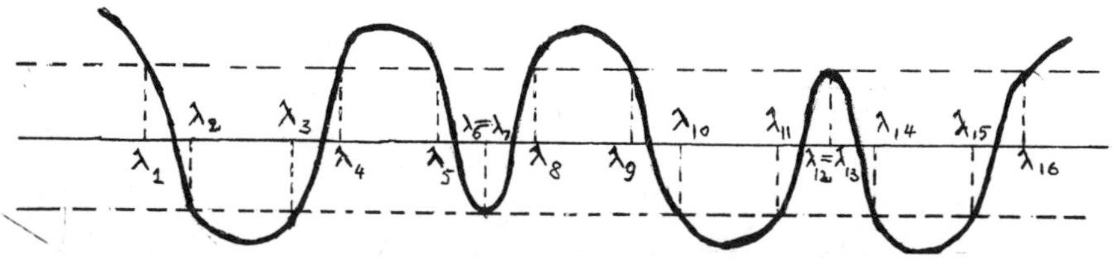

are called *forbidden bands* since no solution of $Qy = \lambda y$ is bounded on the real line if λ is in one of these intervals. Except for the

---

*Abstract results of R. Graff and J. Dorroh (preprint) also apply here. They refer to these properties of S_t by saying that S_t is "β-differentiable."

last forbidden band, all others may collapse to a point. q is called a *finite band potential* if all but a finite number of bands are collapsed. q is called a *simple potential,* if no band is collapsed.

N. Meimann [1977] proved that the set of finite band potentials is C^0-dense in $C^0_{per}[0,1]$. His proof can be modified to obtain H^s-density in $H^s_{per}[0,1]$. Meimann's proof relies on complex function theory and uses properties of conformal maps of the upper half-plane derived from Δ. McKean and Trubowitz [1976], proved, using analytic function theory, and B. Simon via operator theory, that the set of simple potentials is dense in $H^s_{per}[0,1]$. Another proof using infinite dimensional manifolds is given in §3 and these techniques lead then to the bifurcation result in §4.

In what follows the derivative with respect to λ is denoted by $\cdot$ and with respect to x by $'$. The following formula holds

$$\dot{\Delta}(\lambda,q) = (y_1(1)-y_2'(1))\int_0^1 y_1(x)y_2(x)dx + y_1'(1)\int_0^1 y_2^2(x)dx - y_2(1)\int_0^1 y^2(x)dx.$$

The eigenvalues of $Qy = \lambda y$ with the boundary conditions $y(0) = y(1) = 0$ (respectively $y'(0) = y'(1) = 0$) are called the *tied* (respectively *reflecting*) *spectrum*. If μ_i denotes the tied and ν_i the reflecting spectrum, then $\mu_i, \nu_i \in [\lambda_{2i-1}, \lambda_{2i}]$ and they are all simple roots of $y_2(1) = 0$ (respectively $y_1'(1) = 0$). One has

$$\int_0^1 y_2^2(x;\mu_i)dx = y_2'(1;\mu_i)\dot{y}_2(1;\mu_i)$$

and

$$-\int_0^1 y_1^2(x;\nu_i)dx = y_1(1;\nu_i)\dot{y}_1'(1;\nu_i).$$

It is known that the eigenvalues λ_i are preserved by the flow of the KdV equation and the behavior of μ_i under S_t has been completely described by Trubowitz [1977]. In the same paper, starting with a formula of McKean and van Moerbeke [1975], Trubowitz expresses the potential q in terms of $\{\lambda_i, \mu_i(t)\}$, namely $q(t) = \lambda_0 + \sum_{i=1}^{\infty} (\lambda_{2i-1} + \lambda_{2i} - 2\mu_i(t))$. The inverse scattering method consists of the following commutative square:

$$\begin{array}{ccc} q & \overset{①}{\longmapsto} & \{\lambda_i, \mu_i\} \\ S_t q \; \Big\downarrow ④ & & \Big\downarrow ② \\ q(t) & \overset{③}{\longleftarrow\!\!\!\!\dashv} & \{\lambda_i, \mu_i(t)\} \end{array}$$

Arrow ① is the association of $\{\lambda_i, \mu_i\}$ to the Hill operator $-\frac{d^2}{dx^2} + q$, $q \in H^s_{per}[0,1]$ being given. Arrow ② is explained in Trubowitz [1977] and describes the variation of μ_i when q changes according to the KdV equation. Arrow ③ is the formula $q(t) = \lambda_0 + \sum_{i=1}^{\infty} (\lambda_{2i-1} + \lambda_{2i} - 2\mu_i(t))$ and arrow ④ is the time t-map of the KdV equation. In view of Trubowitz' work is seems plausible that arrows ② and ③ represent smooth functions. The strategy of proving that S_t is smooth consists, therefore, of showing that the three arrows ①-③ are each smooth. Now is is clear that $q \mapsto \mu_i$ is a smooth function; this follows from the implicit function theorem and the fact that μ_i are *simple* roots of $y_2(1;\lambda,q) = 0$. Thus we have to examine $q \mapsto \lambda_i$ which is the subject of the next two sections. The bifurcation result of §4 will show that this is *not* a smooth function and thus the inverse scattering method seems not to yield the desired result on the smoothness of S_t, and in fact suggests that the time t-map may not be smooth.

There is one more way to attempt to prove smoothness, using recent results of Deift and Trubowitz [1978]. This method hasn't been tried out yet as far as I know.

§3. Density of the simple potentials

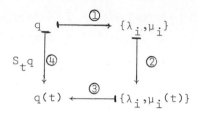

The aim of this section is to prove that the set S of simple potentials $q \in H^s_{per}[0,1]$, $s \geqslant 3$ is dense in $H^s_{per}[0,1]$. This will be done by exhibiting S as the set of regular points of a certain smooth map defined on a submanifold of $\mathbb{R} \times H^s_{per}[0,1]$.

Lemma 3.1. *The following equality holds*

$$\{(\lambda,q) \in \mathbb{R} \times H^s_{per}[0,1] \mid \Delta(\lambda,q) \mp 2 = 0, \quad \dot{\Delta}(\lambda,q) = 0\}$$

$$= \{(\lambda,q) \in \mathbb{R} \times H^s_{per}[0,1] \mid y_2(1;\lambda,q) = 0,$$

$$y_1'(1;\lambda,q) = 0, \quad y_1(1;\lambda,q) = \pm 1\}.$$

Proof. Call the two sets D_1 and D_2. If $(\lambda^0,q^0) \in D_1$, then λ^0 is a double eigenvalue for $-\frac{d^2}{dx^2} + q^0(x)$, so it coincides with the corresponding μ^0 from the tied and ν^0 from the reflecting spectrum, i.e.

$$y_2(1;\lambda^0,q^0) = 0, \quad \text{since } \lambda^0 = \mu^0$$

$$y_1'(1;\lambda^0,q^0) = 0, \quad \text{since } \lambda^0 = \nu^0 .$$

From $y_1y_2' - y_1'y_2 = 1$ (the Wronskian relation) computed at $x = 1$ and (λ^0,q^0), we get

$$y_1(1;\lambda^0,q^0)y_2'(1;\lambda^0,q^0) = 1$$

which together with

$$\Delta(\lambda^0,q^0) = y_1(1;\lambda^0,q^0) + y_2'(1;\lambda^0,q^0) = \pm 2$$

yields

$$y_1(1;\lambda^0,q^0) = y_2'(1;\lambda^0,q^0) = \pm 1$$

and hence $(\lambda^0,q^0) \in D_2$.

Conversely, if $(\lambda^0, q^0) \in D_2$, the Wronskian relation at $(1;\lambda^0,q^0)$ and $y_2(1;\lambda^0,q^0) = 0$, $y_1'(1;\lambda^0,q^0) = 0$ gives

$$y_1(1;\lambda^0,q^0)y_2'(1;\lambda^0,q^0) = 1.$$

Since $y_1(1;\lambda^0,q^0) = \pm 1$, this implies that $y_2'(1;\lambda^0,q^0) = \pm 1$ too, so that

$$\Delta(\lambda^0,q^0) = \pm 2.$$

The known formula

$$\dot{\Delta}(\lambda,q) = (y_1(1)-y_2'(1))\int_0^1 y_1(x)y_2(x)dx + y_1'(1)\int_0^1 y_2^2(x)dx - y_2(1)\int_0^1 y_1^2(x)dx$$

computed at (λ^0,q^0) together with the previous result yields $\dot{\Delta}(\lambda^0,q^0) = 0$ and hence $(\lambda^0,q^0) \in D_1$. ▼

In what follows, this set will be called D.

For computational purposes which will become immediately clear we state the following, whose proof is a direct verification.

Lemma 3.2. *The solution of the initial value problem*

$$(\lambda - Q)\phi = f, \ \phi(0) = \phi'(0) = 0 \quad \textit{is}$$

$$\phi(x) = \int_0^x [y_2(x;\lambda)y_1(\eta;\lambda) - y_1(x;\lambda)y_2(\eta;\lambda)]f(\eta)d\eta .$$

The following two lemmas can be found in McKean-Trubowitz [1976], p. 156.

<u>Lemma 3.3</u>. *If* $v \in H^s_{per}[0,1]$, *and* D_2 *denotes the partial derivative with respect to* q

$$D_2y_1(x;\cdot,\cdot)(\lambda,q)\cdot v =$$

$$= \int_0^x [y_2(x;\lambda,q)y_1(\xi;\lambda,q) - y_1(x;\lambda,q)y_2(\xi;\lambda,q)]y_1(\xi)v(\xi)d\xi$$

$$D_2y_2(x;\cdot,\cdot)(\lambda,q)\cdot v =$$

$$= \int_0^x [y_2(x;\lambda,q)y_1(\xi;\lambda,q) - y_1(x;\lambda,q)y_2(\xi;\lambda,q)]y_2(\xi)v(\xi)d\xi$$

$$D_2\Delta(\lambda,q)\cdot v = (y_2'(1;\lambda,q) - y_1(1;\lambda,q))\int_0^1 y_1(\xi)y_2(\xi)v(\xi)d\xi -$$

$$- y_1'(1;\lambda,q)\int_0^1 y_2^2(\xi)v(\xi)d\xi + y_2(1;\lambda,q)\int_0^1 y_1^2(\xi)v(\xi)d\xi \ .$$

<u>Proof</u>. Take the partial derivative with respect to q evaluated at (λ,q) and applied to $v \in H^s_{per}[0,1]$ of the relation $Qy_i = \lambda y_i$, $i = 1,2$ and get $vy_i(\cdot;\lambda,q) + QD_2y_2(\cdot;\lambda,q)\cdot v = \lambda D_2y_2(\cdot;\lambda,q)$, or

$$(\lambda-Q)D_2y_i(\cdot;\lambda,q)\cdot v = vy_i(\cdot;\lambda,q).$$

Now since $y_i(0)$ = constant and $y_i'(0)$ = constant, we conclude that $D_2y_i(0;\lambda,q)\cdot v = 0$ and $D_2y_i'(0;\lambda,q)\cdot v = 0$. Now apply Lemma 3.2.

The last relation follows from

$$D_2\Delta(\lambda,q)\cdot v = D_2y_1(1;\lambda,q)\cdot v + D_2y_2'(1;\lambda,q). \blacktriangledown$$

<u>Lemma 3.4</u>. *Consider the skew-symmetric operator*

$$L = q\frac{d}{dx} + \frac{d}{dx}q - \frac{1}{2}\frac{d^3}{dx^3} \ .$$

The product $\phi_1\phi_2$ *of any two solutions of* $Qy = \lambda y$ *satisfies* $L(\phi_1\phi_2) = 2\lambda(\phi_1\phi_2)'$. *The three functions* y_1^2, y_1y_2, y_2^2 *are independent and hence form a basis of the nullspace of* $L - 2\lambda\frac{d}{dx}$ *which has dimension equal to* $\deg L = 3$.

The proof is a direct verification.

Lemma 3.5. *If* λ *is a double eigenvalue for* $-\frac{d^2}{dx^2} + q(x)$, *then*

$$\dot{y}_1(1;\lambda,q) = \pm\int_0^1 y_1(x)y_2(x)dx = -\dot{y}_2'(1;\lambda,q)$$

$$\dot{y}_2(1;\lambda,q) = \pm\int_0^1 y_2^2(x)dx$$

$$\dot{y}_1'(1;\pm,q) = \mp\int_0^1 y_1^2(x)dx .$$

Proof. The following known relations have been mentioned in §1.

$$\int_0^1 y_2^2(x;\mu_i)dx = y_2'(1;\mu_i)\dot{y}_2(1,\mu_i)$$

$$-\int_0^1 y_1^2(x,\nu_i)dx = y_1(1;\nu_i)\dot{y}_1'(1,\nu_i)$$

where μ, ν denote the tied respectively reflecting spectrum. Since λ is a double eigenvalue, $\lambda = \mu = \nu$, $y_1(1) = y_2'(1) = \pm 1$ and the last two relations are proved.

To compute $\dot{y}_1(1;\lambda,q)$, start with $Qy_1 = \lambda y_1$, differentiate with respect to λ and get $Q\dot{y}_1 = y_1 + \lambda\dot{y}_1$, or $(\lambda-Q)\dot{y}_1(x) = -y_1(x)$. Since $y_1(0;\lambda,q) = 1$ and $y_1'(0;\lambda,q) = 0$, we have $\dot{y}_1(0;\lambda,q) = 0$, $\dot{y}_1'(0,\lambda,q) = 0$. Now apply Lemma 3.2 and the fact that $y_2(1;\lambda,q) = 0$ and $y_1(1;\lambda,q) = \pm 1$. Do the same thing for $\dot{y}_2(x)$ using Lemma 3.2 (thus recovering in particular the formula for $\dot{y}_2(1)$); then take the derivative of this expression with respect to x at $x = 1$ and use the fact that $y_2(1) = y_1'(1) = 0$ to find the given expression for $\dot{y}_2'(1;\lambda,q)$. ▼

Theorem 3.1. *D is a closed codimension 3 submanifold of* $\mathbb{R} \times H^s_{per}[0,1]$. *If* $(\lambda,q) \in D$

$$T_{(\lambda,q)}D = [\{0\} \times (\mathrm{Ker}(L - 2\lambda \tfrac{d}{dx}))^\perp] \oplus \mathrm{Span}\,(1,1)$$

and an orthogonal complement to $T_{(\lambda,q)}D$ *in* $\mathbb{R} \times H^s_{per}[0,1]$ *is*

$$\{0\} \times \mathrm{Ker}(L - 2\lambda \tfrac{d}{dx}) .$$

(⊥ refers to the L^2 *-orthogonal complement).*

Proof. We shall exhibit D as the inverse image of a regular point of a smooth map. Define

$$F: \mathbb{R} \times H^s_{per}[0,1] \to \mathbb{R}^3$$

by $F(\lambda,q) = (y_1(1;\lambda,q) \mp 1,\ y_2(1;\lambda,q),\ y_1'(1;\lambda,q))$. Then $D = F^{-1}(0,0,0)$, so that if we prove that 0 is a regular value for F, D will be a codimension 3 submanifold of $\mathbb{R} \times H^s_{per}[0,1]$.

Strictly speaking, we are dealing here with two maps F_- and F_+ (for - and + in the definition of F) and D is the union of $F_-^{-1}(0,0,0)$ with $F_+^{-1}(0,0,0)$. What we shall show in one computation is that each of them is a manifold. Since they are disjoint D will be a manifold. F_- picks up all the pairs (eigenvalue, eigenfunction) which are double and periodic, whereas F_+ those which are anti-periodic.

Let $(\lambda,q) \in F^{-1}(0,0,0)$. From Lemma 3.1 we conclude

$$y_1(1;\lambda,q) = y_2'(1;\lambda,q) = \pm 1$$

$$y_2(1;\lambda,q) = y_1'(1;\lambda,q) = 0.$$

For $(t,v) \in \mathbb{R} \times H^s_{per}[0,1]$, we have

$$D[y_1(1;\cdot,\cdot) \mp 1](\lambda,q)\cdot(t,v) =$$

$$= D_1[y_1(1;\cdot,\cdot)](\lambda,q)\cdot t + D_2[y_1(1;\cdot,\cdot)](\lambda,q)\cdot v$$

$$\overset{\text{(Lemma 3.3)}}{=} t\dot{y}_1(1;\lambda,q) + \int_0^1 [y_2(1)y_1(\xi) - y_1(1)y_2(\xi)]y_1(\xi)v(\xi)d\xi$$

$$\overset{(3.5)}{=} \pm t \int_0^1 y_1(x)y_2(x)dx \mp \int_0^1 y_1(x)y_2(x)v(x)dx$$

$$D[y_2(1;\cdot,\cdot)](\lambda,q)\ (t,v) =$$

$$= D_1(y_2(1;\cdot,\cdot))(\lambda,q)\cdot t + D_2[y_2(1;\cdot,\cdot)](\lambda,q)\cdot v$$

$$\overset{(3.3)}{=} t\dot{y}_2(1;\lambda,q) + \int_0^1 [y_2(1)y_1(\xi) - y_1(1)y_2(\xi)]y_2(\xi)v(\xi)d\xi$$

$$\overset{(3.5)}{=} \pm\ t\int_0^1 y_2^2(x)dx \mp \int_0^1 y_2^2(x)v(x)dx.$$

$$D[y_1'(1;\cdot,\cdot)](\lambda,q)\cdot(t,v) =$$

$$= D_1[y_1'(1;\cdot,\cdot)](\lambda,q)\cdot t + D_2[y_1'(1;\cdot,\cdot)](\lambda,q)\cdot v$$

$$\overset{(3.3)}{=} t\dot{y}_1'(1;\lambda,q) + \int_0^1 [y_2'(1)y_1(\xi) - y_1'(1)y_2(\xi)]y_1(\xi)v(\xi)d\xi$$

$$\overset{(3.5)}{=} \mp\ t\int_0^1 y_1^2(x)dx \pm \int_0^1 y_1^2(x)v(x)dx\ .$$

Hence

$$DF(\lambda,q)\cdot(t,v) = \begin{bmatrix} \pm t \int_0^1 y_1(x)y_2(x)dx \mp \int_0^1 y_1(x)y_2(x)v(x)dx \\ \pm t \int_0^1 y_2^2(x)dx \mp \int_0^1 y_2^2(x)v(x)dx \\ \mp t \int_0^1 y_1^2(x)dx \pm \int_0^1 y_1^2(x)v(x)dx \end{bmatrix}.$$

We claim that

$$\text{Ker } DF(\lambda,q) = (\{0\} \times \text{Ker}(L - 2\lambda\tfrac{d}{dx})^{\perp}) \oplus \text{Span}(1,1).$$

If this is shown, since $\text{Ker}(L - 2\lambda\frac{d}{dx})$ is 3-dimensional, $\{0\} \times \text{Ker}(L - 2\lambda\frac{d}{dx})$ has codimension 4 in $\mathbb{R} \times H^s_{per}[0,1]$ and hence $\text{Ker } DF(\lambda,q)$ has codimension 3 in $\mathbb{R} \times H^s_{per}[0,1]$. Thus the dimension of the range of $DF(\lambda,q)$ is 3-dimensional and hence $DF(\lambda,q)$ is surjective for any $(\lambda,q) \in F^{-1}(0,0,0)$, i.e. $(0,0,0) \in \mathbb{R}^3$ is a regular value for F. Then D is a codimension 3 submanifold of $\mathbb{R} \times H^s_{per}[0,1]$ and $T_{(\lambda,q)}D = \text{Ker } DF(\lambda,q)$.

We now prove the above claim. If $v \in \text{Ker}(L - 2\lambda\frac{d}{dx})^{\perp}$ since y_1^2, y_1y_2, y_2^2 is a basis for $\text{Ker}(L - 2\lambda\frac{d}{dx})$ (Lemma 3.4), we have $\int_0^1 y_1^2 v = \int_0^1 y_1y_2v = \int_0^1 y_2^2 v = 0$. Hence $(0,v) \in \text{Ker } DF(\lambda,q)$. Clearly $(1,1) \in \text{Ker } DF(\lambda,q)$ and $\supseteq$ is proved. Conversely, if $(t,v) \in \text{Ker } DF(\lambda,q)$, then either $t = 0$ or $t \neq 0$. If $t = 0$, then $(0,v) \in \text{Ker } DF(\lambda,q)$ actually says that $\int_0^1 y_1^2 v = \int_0^1 y_1y_2v = \int_0^1 y_2^2 v = 0$, i.e. $v \in \text{Ker}(L - 2\lambda\frac{d}{dx})^{\perp}$ (by Lemma 3.4). If $t \neq 0$, then by rescaling, we can assume $t = 1$, i.e. $(1,v) \in \text{Ker } DF(\lambda,q)$. Since $(1,1) \in \text{Ker } DF(\lambda,q)$, $(0,v-1) \in \text{Ker } DF(\lambda,q)$ and hence $v - 1 \in \text{Ker }(L - 2\lambda\frac{d}{dx})^{\perp}$ by what we just proved above. But then $(1,v) = (1,1) + (0,v-1) \in (\{0\} \times \text{Ker}(L - 2\lambda\frac{d}{dx})^{\perp}) \oplus \text{Span}(1,1)$ proving $\subseteq$.

We now exhibit an explicit orthogonal complement to $T_{(\lambda,q)}D$. In $\mathbb{R} \times H^s_{per}[0,1]$ with the inner product inherited from $\mathbb{R} \times L^2_{per}[0,1]$

$\langle (t,u),(s,v)\rangle = ts + \langle u,v\rangle_{L^2}$, note that $\{0\} \times \mathrm{Ker}(L - 2\lambda\frac{d}{dx})$ is orthogonal to $T_{(\lambda,q)}D$. This subspace is 3-dimensional with basis $(0,y_1^2)$, $(0,y_1y_2)$, $(0,y_2^2)$. Since $T_{(\lambda,q)}D$ has codimension 3, $\{0\} \times \mathrm{Ker}(L - 2\lambda\frac{d}{dx})$ is the orthogonal complement to $T_{(\lambda,q)}D$. ∎

Theorem 3.2. $S = \{q \in H^s_{per}[0,1] \mid Q = -\frac{d^2}{dx^2} + q(x)$ *has only simple eigenvalues*$\}$ *is dense in* $H^s_{per}[0,1]$.

Proof. Define $\Pi: D \to H^s_{per}[0,1]$, by $\Pi(\lambda,q) = q$. We have

$$T_{(\lambda,q)}\Pi : T_{(\lambda,q)}D \to H^s_{per}[0,1]$$

$$T_{(\lambda,q)}\Pi(a(0,v) + b(1,1)) = av + b, \quad a, b \in \mathbb{R}$$

$$v \in (\mathrm{Ker}(L - 2\tfrac{d}{dx}))^{\perp}$$

Thus $a(0,v) + b(1,1) \in \mathrm{Ker}\, T_{(\lambda,q)}\Pi$ iff $av + b = 0$ iff $a = 0$ (in which case $b = 0$ too) or $v = -\frac{b}{a}$ iff $a = b = 0$ or $v = 0$ and $b = 0$ (since the constant functions are not in $\mathrm{Ker}(L - 2\lambda\frac{d}{dx})^{\perp}$). Hence $\mathrm{Ker}\, T_{(\lambda,q)}\Pi = \{0\}$.

The range of $T_{(\lambda,q)}\Pi$ is clearly $\mathrm{Ker}(L - 2\lambda\frac{d}{dx})^{\perp} \oplus \mathrm{Span}(1)$ which is a codimension 2 subspace of $H^s_{per}[0,1]$.

The conclusion is that Π is a Fredholm map of index -2. The argument above also shows that $T_{(\lambda,q)}\Pi$ is *never* surjective, its range having a 2-dimensional complement. Thus $\Pi(D)$ consists only of critical values of Π. By the Sard-Smale theorem, the set of regular values of Π, i.e. $H^s_{per}[0,1] \setminus \Pi(D) = S$ is residual and hence dense in $H^s_{per}[0,1]$. ∎

This last result has been obtained already by McKean-Trubowitz [1976] using analytic function theory, by B. Simon [1976] using operator theory, and by Bob Carlson [1978] using Kato's perturbation theory. Our method of proof, however, enables us to obtain a bifurcation result on the eigenvalues of Hill's operator in the next section.

§4. A degenerate bifurcation result on the eigenvalues

Our interest is to study the bifurcation of the eigenvalues λ of Q at double eigenvalues as function of thepotential q. By the previous section, D gives the manifold where bifurcations do not occur; in other words, we deal with a problem with degenerate directions of bifurcation. Another preculiarity of the problem is that the parameter space is infinite dimensional.

Thus, we are compelled to consider the case of a smooth map $g: H \to \mathbb{R}$, H an infinite dimensional Hilbert space, $g(x_0) = 0$, $Dg(x_0) = 0$ with $D = \{x \in H | g(x) = 0, \quad Dg(x) = 0\}$ a submanifold of H. Assume that for each $x \in D$, $T_x D$ is the maximal degenerate subspace of $D^2 g(x)$ and that on the orthogonal complement E_x of $T_x D$ in H, the bilinear form $D^2 g(x)$ is strongly non-degenerate (i.e. for each $w \in E$ the map $v \mapsto D^2 g(x)(v,w)$ is an isomorphism of E). We then call D a non-degenerate critical manifold of g.

Theorem 4.1. *Under the above hypothesis, the zero set of* g *is locally the manifold* D *together with a family of cones each lying in* E_x, *varying smoothly in* $x \in D$. *("Cone" means the zero set of a quadratic form.) The cone at* x *is given modulo a diffeomorphism by* $D^2 g$ *restricted to* E_x.

The proof of this statement is straightforward. Locally, since D is a submanifold, $H = E_1 \oplus E_2$, where $E_1 \cong T_{x_0} D$, $E_2 \cong E_{x_0}$. Hence, in this chart with the submanifold property $g(x_1, 0) = 0$, $Dg(x_1, 0) = 0$. The non-degeneracy condition on the second derivative on E_2 applied at each $(x_1, 0)$, gives by the implicit function theorem, locally a *unique* solution $x_2 = x_2(x_1)$ of $D_2 g(x_1, x_2) = 0$. Hence $x_2(x_1) \equiv 0$. In other words, locally, in a tubular neighborhood of E_1, the only critical points of g are given by E_1. In each fiber apply the Morse-Palais lemma to obtain $g(x_1,x_2) = \frac{1}{2}D^2g_{(x_1,0)}(x_2,x_2)$. Thus $g(x_1,x_2) = 0$ iff either $x_2 = 0$ (i.e. we obtain E_1, that is D) or $D^2g_{(x_1,0)}(x_2,x_2) = 0$ (i.e. we obtain a cone in the orthogonal to T_xD).

We shall apply this result to study the bifurcation of the eigenvalues λ at a double eigenvalue. The verification of the hypotheses is contained in the following Lemmas.

<u>Lemma 4.1</u>. *Let* λ *be a double eigenvalue for* $-\frac{d^2}{dx^2} + q(x) = Q$. *Then*

$$\begin{aligned} D^2\Delta(\lambda,q)\cdot((t,v),(s,w)) &= \pm 2\left\{\left(\int_0^1 y_1y_2\right)^2 - \int_0^1 y_1^2\int_0^1 y_2^2\right\}ts \\ &\mp \left\{2\int_0^1 y_1y_2\int_0^1 y_1y_2w - \int_0^1 y_1^2\int_0^1 y_2^2w - \int_0^1 y_2^2\int_0^1 y_1^2w\right\}t \\ &\mp \left\{2\int_0^1 y_1y_2\int_0^1 y_1y_2v - \int_0^1 y_1^2\int_0^1 y_2^2v - \int_0^1 y_2^2\int_0^1 y_1^2v\right\}s \\ &\pm \left\{2\int_0^1 y_1y_2v\int_0^1 y_1y_2w - \int_0^1 y_1^2v\int_0^1 y_2^2w - \int_0^1 y_1^2w\int_0^1 y_2^2v\right\} \end{aligned}$$

$$= ts\ddot{\Delta}(\lambda,q) + D_1D_2\Delta(\lambda,q)(t,w) + D_1D_2\Delta(\lambda,q)(s,v) + D_2^2\Delta(\lambda,q)(v,w).$$

(The terms in the above equality correspond.)

<u>Proof</u>. Take the derivative with respect to λ of the formula

$$\dot{\Delta}(\lambda,q) = (y_1(1) - y_2'(1))\int_0^1 y_1y_2 + y_1'(1)\int_0^1 y_2^2 - y_2(1)\int_0^1 y_1^2$$

and recall that for λ a double eigenvalue

$$y_1(1) = y_2'(1) = 1, \quad y_1'(1) = y_2(1) = 0,$$

so that

$$\ddot{\Delta}(\lambda,q) = (\dot{y}_1(1) - \dot{y}_2'(1)\int_0^1 y_1y_2 + \dot{y}_1'(1)\int_0^1 y_2^2 - \dot{y}_2(1)\int_0^1 y_1^2$$

Lemma 3.5 then will yield

$$\ddot{\Delta}(\lambda,q) = \pm 2\left\{\left(\int_0^1 y_1y_2\right)^2 - \int_0^1 y_1^2\int_0^1 y_2^2\right\}. \tag{1}$$

From Lemma 3.3 we have

$$D_2\Delta(\lambda,q)\cdot w = (y_2'(1) - y_1(1))\int_0^1 y_1y_2w - y_1'(1)\int_0^1 y_2^2w + y_2(1)\int_0^1 y_1^2w$$

so that taking the derivative with respect to λ at a double eigenvalue we get

$$D_1D_2\Delta(\lambda,q)\cdot(t,w) = t(\dot{y}_2'(1) - \dot{y}_1(1))\int_0^1 y_1y_2w - t\dot{y}_1'(1)\int_0^1 y_2^2w + t\dot{y}_2(1)\int_0^1 y_1^2w$$

Lemma 3.5 yields

$$D_1D_2\Delta(\lambda,q)\cdot(t,w) = \mp t\left\{2\int_0^1 y_1y_2\int_0^1 y_1y_2w - \int_0^1 y_1^2\int_0^1 y_2^2w - \int_0^1 y_2^2\int_0^1 y_1^2w\right\} \tag{2}$$

Finally, we have to compute

$$D_2^2\Delta(\lambda,q)\cdot(v,w) = D_2^2[y_1(1;\cdot,\cdot)](\lambda,q)\cdot(v,w) + D_2^2[y_2'(1;\cdot,\cdot)](\lambda,q)\cdot(v,w).$$

In the course of the proof of Lemma 3.3 we obtained

$$(\lambda-Q)(D_2y_i)(\lambda,q)\cdot v = vy_i \ .$$

Take the derivative of this relation with respect to q applied to w and get

$$-w(D_2y_i)(\lambda,q)\cdot v + (\lambda-Q)(D_2^2y_i)(\lambda,q)\cdot(v,w) = v(D_2y_i)(\lambda,q)\cdot w$$

or

$$(\lambda-Q)(D_2^2y_i)(\lambda,q)\cdot(v,w) = v(D_2y_i)(\lambda,q)\cdot w + w(D_2y_i)(\lambda,q)\cdot v \equiv \phi_i \ .$$

By Lemma 3.3 (since $y_i(0)$ = constant implies $D_2y_i(\lambda,q)\cdot(v,w) = 0$)

$$(D_2y_i)(\lambda,q)\cdot(v,w) = \int_0^x [y_2(x)y_1(\eta) - y_1(x)y_2(\eta)]\phi_i(\eta)d\eta$$

If $i = 1$, compute this at $x = 1$ and keep in mind that $y_2(1;\lambda,q) = 0$, $y_1(1;\lambda,q) = \pm 1$ to obtain

$$D_2^2y_1(1;\lambda,q)\cdot(v,w) = \mp\int_0^1 y_2(\eta)\phi_1(\eta)d\eta \tag{3}$$

If $i = 2$, take the derivative with respect to x at $x = 1$ of this relation and get (using $y_2'(1) = \pm 1$, $y_1'(1) = 0$)

$$\begin{aligned} D_2^2y_2'(1;\lambda,q)\cdot(v,w) &= \int_0^1 (y_2'(1)y_1(\eta) - y_1'(1)y_2(\eta)]\phi_2(\eta)d\eta \\ &= \pm\int_0^1 y_1(\eta)\phi_2(\eta)d\eta \end{aligned} \tag{4}$$

Relations (3) and (4) give

$$D_2^2\Delta(\lambda,q)\cdot(v,w) = \pm\int_0^1 [y_1(\eta)\phi_2(\eta) - y_2(\eta)\phi_1(\eta)]d\eta \ .$$

A straightforward direct computation of the expression inside the integral gives

$$y_1\phi_2 - y_2\phi_1 = \frac{d}{d\eta}\left\{2\int_0^\eta y_1y_2v \int_0^\eta y_1y_2w - \int_0^\eta y_1^2v\int_0^\eta y_2^2w - \int_0^\eta y_1^2w\int_0^\eta y_2^2v\right\}$$

so that finally

$$D_2^2\Delta(\lambda,q)\cdot(v,w) = \pm\left\{2\int_0^1 y_1y_2v\int_0^1 y_1y_2w - \int_0^1 y_1^2v\int_0^1 y_2^2w - \int_0^1 y_1^2w\int_0^1 y_2^2v\right\} \qquad (5)$$

Relations (1), (2), (5) give the result. ▼

Lemma 4.2. *The tangent space* $T_{(\lambda,q)}D$ *of the manifold* D *is the maximal degenerate subspace of* $D^2\Delta(\lambda,q)$.

Proof. We have to show that the $D^2\Delta(\lambda,q)$-orthogonal of $\mathbb{R}\times H^s_{per}[0,1]$ is $T_{(\lambda,q)}D = \{\{0\}\times \mathrm{Ker}(L-2\frac{d}{dx})^\perp\} \oplus \mathrm{Span}(1,1)$. (t,v) is in the $D^2\Delta(\lambda,q)$-orthogonal if and only if $D^2\Delta(\lambda,q)\cdot((t,v),(s,w)) = 0$ for all $(s,w)\in \mathbb{R}\times H^s_{per}[0,1]$.

Notice now that by the formula of the preceding lemma, $D^2\Delta(\lambda,q)\cdot((1,1),(s,w)) = 0$ for all $(s,w)\in\mathbb{R}\times H^s_{per}[0,1]$, $D^2\Delta(\lambda,q)\cdot((0,v),(s,w)) = 0$ for all $(s,w)\in\mathbb{R}\times H^s_{per}[0,1]$ if $v\in(\mathrm{Ker}(L-2\lambda\frac{d}{dx}))^\perp$ and hence $[\{0\}\times\mathrm{Ker}(L-2\lambda\frac{d}{dx})^\perp]\oplus\mathrm{Span}(1,1)\subset$ $D^2\Delta(\lambda,q)$-orthogonal of $\mathbb{R}\times H^s_{per}[0,1]$.

Conversely, if $(t,v)\in D^2\Delta(\lambda,q)$-orthogonal of $\mathbb{R}\times H^s_{per}[0,1]$, there are two possibilities: $t = 0$ and $t\neq 0$. If $t = 0$, then the formula for $D^2\Delta(\lambda,q)$ at $s = 0$ yields

$$2\int_0^1 y_1y_2v\int_0^1 y_1y_2w - \int_0^1 y_1^2v\int_0^1 y_2^2w - \int_0^1 y_1^2w\int_0^1 y_2^2v = 0 \quad \text{for all} \quad w\in H^s_{per}[0,1].$$

We shall prove that this implies that $\int_0^1 y_1y_2v = \int_0^1 y_1^2v = \int_0^1 y_2^2v = 0$ whence $v\in\mathrm{Ker}(L-2\lambda\frac{d}{dx})^\perp$. If $t\neq 0$, then since $\mathrm{Span}(1,1)\subset D^2\Delta(\lambda,q)$-orthogonal of $\mathbb{R}\times H^s_{per}[0,1]$, $(0,v-t) = (t,v)-(t,t)\in D^2\Delta(\lambda,q)$-orthogonal of $\mathbb{R}\times H^s_{per}[0,1]$ and by the claim above $v-t\in\mathrm{Ker}(L-2\lambda\frac{d}{dx})^\perp$ and hence $(t,v) = (0,v-t)+(t,t)\in$ $[\{0\}\times\mathrm{Ker}(L-2\lambda\frac{d}{dx})^\perp]\oplus\mathrm{Span}(1,1)$, i.e. $D^2\Delta(\lambda,q)$-orthogonal of $\mathbb{R}\times H^s_{per}[0,1]\subset[\{0\}\times\mathrm{Ker}(L-2\lambda\frac{d}{dx})^\perp]\oplus\mathrm{Span}(1,1)$ and the lemma is proved.

Thus we are left to show that

$$2\int_0^1 y_1y_2v\int_0^1 y_1y_2w - \int_0^1 y_1^2v\int_0^1 y_2^2w - \int_0^1 y_1^2w\int_0^1 y_2^2v = 0 \quad \text{for all} \quad w \in H^s_{per}[0,1]$$

implies $v \in \operatorname{Ker}(L-2\lambda\frac{d}{dx})^{\perp}$. Denoting $A = \int_0^1 y_1^2 v$, $B = \int_0^1 y_1y_2v$, $C = \int_0^1 y_2^2v$, this amounts to showing that $A = B = C = 0$. In these notations the above relation becomes

$$\int_0^1 (2\,By_1y_2 - Ay_2^2 - Cy_1^2)w = 0 \quad \text{for all} \quad w \in H^s_{per}[0,1]$$

and hence $Ay_2^2 - 2By_1y_2 + Cy_1^2 \equiv 0$ on $[0,1]$. But y_2^2, y_1y_2, y_1^2 are linearly independent and so $A = B = C = 0$. ▼

Lemma 4.3. *The manifold* D *is a non-degenerate critical manifold of* $\Delta(\lambda,q) \mp 2$ *on which* $D_2\Delta \equiv 0$, $\dot{\Delta} \equiv 0$.

Proof. If $(\lambda,q) \in D$, $\Delta(\lambda,q) \mp 2 = 0$, $\dot{\Delta}(\lambda,q) = 0$. According to Lemma 3.3, $D_2\Delta(\lambda,q) = 0$ and the previous lemma shows that $T_{(\lambda,q)}D$ is the maximal degenerate subspace of $D^2\Delta(\lambda,q)$. ▼

Theorem 4.2. *The zero set of* $\Delta(\lambda,q) \mp 2$ *consists locally of the manifold* D *together with a family of two-dimensional cones lying in the orthogonal complement of* $T_{(\lambda,q)}D$, *varying smoothly in* $(\lambda,q) \in D$.

Proof. It is clear that $D = \{(\lambda,q) \in \mathbb{R} \times H^s_{per}[0,1] \mid \Delta(\lambda,q) \mp 2 = 0,\ D\Delta(\lambda,q) = 0\}$. All hypotheses of theorem 4.1 have been verified in the previous lemmas. The only thing left to show is that over each $(\lambda^0,q^0) \in D$, we actually do have a 2-dimensional cone, not shrunk to a point. A straightforward computation shows that $D^2\Delta(\lambda^0,q^0)(v,v) = 0$ if and only if

$$a^2[\|y_1^2\|^2\|y_1y_2\|^2 - (\int_0^1 y_1^3y_2)^2] + b^2[\int_0^1 y_1^3y_2\int_0^1 y_1y_2^3 - \|y_1y_2\|^4] +$$

$$+ c^2[\|y_1y_2\|^2\|y_2^2\|^2 - (\int_0^1 y_1y_2^3)^2] + ab[\|y_1^2\|^2\int_0^1 y_1y_2^3 - \|y_1y_2\|^2\int_0^1 y_1^3y_2] +$$

$$+ bc[\|y_2^2\|^2\int_0^1 y_1^3y_2 - \|y_1y_2\|^2\int_0^1 y_1y_2^3] +$$

$$+ ac[\|y_1y_2\|^4 + \|y_1^2\|^2\|y_2^2\|^2 - 2\int_0^1 y_1^3y_2\int_0^1 y_1y_2^3] = 0$$

where $v \in \mathrm{Ker}(L-2\lambda^0\frac{d}{dx}) \cong \mathbb{R}^3$, the isomorphism being given by Lemma 3.4, namely

$$\mathbb{R}^3 \ni (a,b,c) \leftrightarrow ay_1^2 + by_1y_2 + cy_2^2 \in \mathrm{Ker}(L-2\lambda^0\frac{d}{dx}).$$

If in the canonical form of this quadratic form, all squares have positive sign at q^0, then by continuity, in a neighborhood of q^0 it still will have positive signs. In other words, for a neighborhood of q^0, the zeros of $\Delta(\lambda,q) \mp 2$ would lie on D only, i.e. λ must be a double eigenvalue for all q in this neighborhood. But this is impossible since S is dense in $H^s_{per}[0,1]$ by Theorem 3.2. ■

The following degenerate bifurcation result holds.

<u>Theorem 4.3</u>. *If* λ^0 *is a simple eigenvalue for* $Q = -\frac{d^2}{dx^2} + q^0(x)$, *then there is a neighborhood of* q^0 *in* $H^s_{per}[0,1]$ *such that* λ *is a smooth function of* q. *If* λ^0 *is a double eigenvalue, then at* q^0, $\lambda(q)$ *is not differentiable. However, the function* $\lambda_{2i-1} + \lambda_{2i}$ *which at* q^0 *has the value* $2\lambda^0$ *is a smooth function of* q *and its differential is given by*

$$D(\lambda_{2i-1} + \lambda_{2i})(q^0)\cdot v = \frac{1}{2}\int_0^1 (f_1^2 + f_2^2)v$$

where f_1, f_2 *are the eigenfunctions of* L^2*-norm* 1 *of* λ^0.

Proof. The first statement is a direct consequence of the implicit function theorem. The second statement is a corollary of the cone picture described in the previous theorem: In passing through the double eigenvalue, there is no consistent way of labelling the simple eigenvalues arising around λ^0 since as a function of (λ,q), the zeros must lie on a 2-dimensional cone. (If the bifurcation theorem above would have provided us with a one-dimensional cone in $\mathbb{R}^2$, i.e. with two smooth intersecting curves, such a labelling would have been possible.) The third statement is due P. Lax; for completeness we give here his proof (see also Abraham & Marsden [1978], section 5.3). Let $v \in H^s_{per}[0,1]$ and denote $Q(\varepsilon) = -\frac{d^2}{dx^2} + q + \varepsilon v$ and by f_1, f_2 the two eigenfunctions of λ^0 of L^2-norm one. Then

$$Q(\varepsilon)f_j(q^0 + \varepsilon v) = \lambda(q^0 + \varepsilon v)f_j(q^0 + \varepsilon v), \quad j = 1,2$$

where $f_j(q^0) = f_j$, $\lambda(q^0) = \lambda^0$. Take the derivative of this expression (formally) at $\varepsilon = 0$ and obtain

$$vf_j + Q \frac{d}{d\varepsilon}\Big|_{\varepsilon=0} f_j(q^0 + \varepsilon v) = \frac{d}{d\varepsilon}\Big|_{\varepsilon=0}\lambda(q^0 + \varepsilon v)f_j + \\ + \lambda^0 \frac{d}{d\varepsilon}\Big|_{\varepsilon=0} f_j(q^0 + \varepsilon v).$$

But since Q is symmetric we have

$$\int_0^1 Q \frac{d}{d\varepsilon}\Big|_{\varepsilon=0} f_j(q^0 + \varepsilon v)f_j(q^0)dx = \int_0^1 \frac{d}{d\varepsilon}\Big|_{\varepsilon=0} f_j(q^0 + \varepsilon v)Qf_j dx \\ = \int_0^1 \frac{d}{d\varepsilon}\Big|_{\varepsilon=0} f_j(q^0 + \varepsilon v)\lambda(q^0)f_j(q^0)dx.$$

Now take in the above relation the scalar product with $f_j(q^0)$ and add

$$\int_0^1 v(f_1^2 + f_2^2)dx = \frac{d}{d\varepsilon}\Big|_{\varepsilon=0}\lambda_1(q^0 + \varepsilon v)\int_0^1 f_1^2 + \frac{d}{d\varepsilon}\Big|_{\varepsilon=0}\lambda_2(q^0 + \varepsilon v)\int_0^1 f_2^2$$

$$= 2\frac{d}{d\varepsilon}\Big|_{\varepsilon=0}\lambda(q^0 + \varepsilon v) = 2d\lambda(q^0)\cdot v.$$

Hence

$$d\lambda(q^0)\cdot v = \frac{1}{2}\int_0^1 (f_1^2 + f_2^2)v\,dx \; . \quad \blacksquare$$

The final picture that emerges now is the following:

-- the finite band potentials are dense;

-- the simple potentials are dense;

-- there exists a codimension 3 submanifold in $\mathbb{R} \times H^s_{per}[0,1]$ covering the simple potentials, on which Q has at least a double eigenvalues;

-- at each point of this submanifold a bifurcation of the eigenvalues occurs that makes the eigenvalues non-smooth (and this occurs on a dense subset of $H^s_{per}[0,1]$).

From the inverse scattering diagram described at the end of §2 it seems that the KdV time t-map is not smooth. This is in accordance with remarks in McKean and Trubowitz [1976]. After this paper was written I found out that E. Schechter [1978] showed, by estimating the difference between two solitons, that for the KdV equation on $\mathbb{R}$, S_t for $t > 0$ is not even Lipschitz in $L^p(\mathbb{R})$, $1 \leq p \leq \infty$ (example 2.5). His idea of proof might work in the periodic case too. What we accomplished is that neither Kato's methods on quasi-linear equations, nor the inverse scattering method yield results in either direction; however, they point towards non-smoothness. However, the differentiability of the time t map from H^s to H^{s-3} is still a bit of a mystery from the point of view above.

Difficulties coming from the term uu_x have been overcome by Ebin and Marsden [1970] making a non-smooth change of coordinates for the Euler equations of a perfect fluid. We shall apply their methods in §6 to get a smooth time t-map for the KdV equation in Lagrangian coordinates. Another key result regarding the KdV equation is its complete integrability as a Hamiltonian system on a co-adjoint orbit. The next section describes a finite dimensional analogous problem.

§5. A brief review of the Toda lattice

In order to understand the problems raised in the next section and to get used to a certain technique, the Toda lattice -- a completely integrable Hamiltonian system -- is briefly reviewed here since it has strong formal connections with the KdV equation. The Toda lattice has been studied from the view point of an inverse scattering problem by van Moerbeke [1976] and from the Lie group point of view by Adler [1978] and Kostant [1978].

Let G be a Lie group, Ad_g and $Ad^*_{g^{-1}}$ the adjoint and co-adjoint actions of G on $\mathfrak{g}$ and $\mathfrak{g}^*$ respectively. The theorem of Kostant-Kirillov-Souriau states that for any $\mu \in \mathfrak{g}^*$, the co-adjoint orbit $G\cdot\mu \subseteq \mathfrak{g}^*$ of G through μ is a symplectic manifold with symplectic form ω_μ given by

$$\omega_\mu(Ad^*_{g^{-1}}\mu)((ad\ \xi_1)^*(Ad^*_{g^{-1}}\mu),(ad\ \xi_2)^*(Ad^*_{g^{-1}}\mu)) = (Ad^*_{g^{-1}}\mu)[\xi_2,\xi_1]$$

for $g \in G$, $\xi_1,\xi_2 \in \mathfrak{g}$. One of the facts used in this formula is the expression of a tangent vector to $G\cdot\mu$ at the point $Ad^*_{g^{-1}}(\mu)$ as $(ad\ \xi)^*(Ad^*_{g^{-1}}\mu)$ for some $\xi \in \mathfrak{g}$. A proof of this theorem can be found in the more general context of reduction in Abraham-Marsden [1978], Chapter 4.

Let G be the identity component of the group of invertible lower triangular real $n \times n$ matrices. Then its Lie algebra consists of all lower triangular matrices and its dual $\mathfrak{g}^*$ of all upper triangular matrices, the pairing of $\mu \in \mathfrak{g}^*$ and $\xi \in \mathfrak{g}$ being $\mu(\xi) = \mathrm{Trace}(\mu\xi)$ where $\mu\xi$ denotes matrix multiplication. The adjoint action of G on $\mathfrak{g}$ is given by $\mathrm{Ad}_g\xi = g\xi g^{-1}$. In order to compute the co-adjoint action, note that if A^+ denotes the upper triangular part of the matrix A and $\xi \in \mathfrak{g}$, then $\mathrm{Trace}(A\xi) = \mathrm{Trace}(A^+\xi)$. Thus

$$\begin{aligned}\mathrm{Ad}^*_{g^{-1}}(\mu)\cdot\xi &= \mu(\mathrm{Ad}_{g^{-1}}\xi) = \mu(g^{-1}\xi g) = \mathrm{Trace}(\mu g^{-1}\xi g)\\ &= \mathrm{Trace}\,(g\mu g^{-1}\xi) = \mathrm{Trace}\,((g\mu g^{-1})^+\xi)\\ &= (g\mu g^{-1})^+(\xi), \quad \text{i.e.}\end{aligned}$$

$$\mathrm{Ad}^*_{g^{-1}}\mu = (g\mu g^{-1})^+ .$$

Let μ be the following matrix:

$$\mu = \begin{bmatrix} c & \varepsilon_1 & 0 & 0 & \cdots & 0 & 0\\ 0 & c & \varepsilon_2 & 0 & \cdots & 0 & 0\\ - & - & - & - & --- & - & -\\ 0 & 0 & 0 & 0 & & c & \varepsilon_{n-1}\\ 0 & 0 & 0 & 0 & & 0 & c\end{bmatrix} = cI + \mu'$$

where c, ε_i are constants, $\varepsilon_i > 0$, I is the identity $n \times n$ matrix and μ' the matrix having all entires zero except the super-diagonal equal to $(\varepsilon_1,\ldots,\varepsilon_{n-1})$. We shall determine explicitly the co-adjoint orbit $G\cdot\mu$. First, a short computation shows that if $g = (g_{ij})_{i\geqslant j} \in G$ and $g^{-1} = (\bar{g}_{ij})_{i\geqslant j} \in G$ is its inverse, then $\bar{g}_{ii} = 1/g_{ii}$, $\bar{g}_{i+1,i} = -g_{i+1,i}/g_{ii}g_{i+1,i+1}$. Also, it is clear that $g\mu'$ is the matrix obtained from g by adding a zero column to the left of g and cutting off its last column, each non-zero column being multiplied by $\varepsilon_1,\ldots,\varepsilon_{n-1}$. Another direct computation shows that $(g\mu' g^{-1})^+$ has diagonal entries equal to $-\varepsilon_1 g_{21}/g_{22}$,

$(\varepsilon_1 g_{21}/g_{22}) - (\varepsilon_2 g_{32}/g_{33})$, ..., $(\varepsilon_{n-2} g_{n-1,n-2}/g_{n-1,n-1}) - (\varepsilon_{n-1} g_{n,n-1}/g_{nn})$, $\varepsilon_{n-1} g_{n,n-1}/g_{nn}$, superdiagonal entries equal to $\varepsilon_1 g_{11}/g_{22}, \ldots, \varepsilon_{n-1} g_{n-1,n-1}/g_{nn}$ and all other entries equal to zero. Thus all entries of the superdiagonal are strictly positive and $\mathrm{Trace}(g\mu' g^{-1})^+ = 0$, these being the only relations between the entries of $(g\mu g^{-1})^+$. Thus

$$G\cdot\mu = \left\{ \begin{bmatrix} b_1 & a_1 & 0 & \cdots & 0 & 0 \\ 0 & b_2 & a_2 & \cdots & 0 & 0 \\ - & - & - & & - & - \\ 0 & 0 & 0 & & b_{n-1} & a_{n-1} \\ 0 & 0 & 0 & & 0 & b_n \end{bmatrix} \middle| \, a_i, b_i \in \mathbb{R},\ a_i > 0,\ \sum_{i=1}^{n} b_i = nc \right\}.$$

The dimension of $G\cdot\mu$ is $2(n-1)$ and the symplectic form on it is given by

$$\omega_\mu(\beta)([\beta,\xi_1]^+,[\beta,\xi_2]^+) = \mathrm{Trace}(\beta\cdot[\xi_2,\xi_1])$$

for $\beta = \mathrm{Ad}^*_{g^{-1}}\mu$, a typical element of $G\cdot\mu$. This formula uses the fact that for $\beta \in G\cdot\mu$, the tangent space at β to $G\cdot\mu$ is given by

$$T_\beta(G\cdot\mu) = \{[\beta,\xi]^+ \mid \xi \in \mathfrak{g}\}$$

which in turn follows from the equality $(\mathrm{ad}\,\xi)^*\beta = [\beta,\xi]^+$.

Let $\beta \in G\cdot\mu$, $\beta = \begin{bmatrix} b_1 & a_1 & & 0 \\ & \ddots & \ddots & \\ & & \ddots & a_{n-1} \\ 0 & & & b_n \end{bmatrix}$. Around β, a chart is given by $(a_1,\ldots,a_{n-1}, b_1,\ldots,b_{n-1})$. Note that this is actually a global chart making $G\cdot\mu$ diffeomorphic to $\mathbb{R}^{2n-2}$. ω_μ will now be expressed in this chart. This involves a three-step computation. First, it is shown that

$$[\beta,\xi]^+ = \sum_{i=1}^{n-1}(\xi^{i+1,i}a_i - \xi^{i,i-1}a_{i-1})\frac{\partial}{\partial b_i} +$$

$$+ \sum_{i=1}^{n-1}(\xi^{i+1,i+1} - \xi^{ii})a_i \frac{\partial}{\partial a_i}$$

with the conventions $a_0 = \xi^{1,0} = 0$. In matrix notation this shows that $[\beta,\xi]^+$ is a trace zero matrix whose only non-zero entries are on the diagonal and superdiagonal. Second, it is shown that

$$\mathrm{Trace}(\beta[\xi_2,\xi_1]) = \sum_{i=1}^{n-1}[\xi_1^{i+1,i}(\xi_2^{i+1,i+1} - \xi_2^{ii}) - \xi_2^{i+1,i}(\xi_1^{i+1,i+1} - \xi_1^{ii})]a_i .$$

Third, it is shown that

$$\omega_\mu\left(\frac{\partial}{\partial a_i}, \frac{\partial}{\partial a_j}\right) = 0, \quad \omega_\mu\left(\frac{\partial}{\partial b_i}, \frac{\partial}{\partial b_j}\right) = 0$$

$$\omega_\mu\left(\frac{\partial}{\partial a_i}, \frac{\partial}{\partial b_j}\right) = \begin{cases} 0 & \text{for } i < j \\ -1/a_i, & \text{for } i \geqslant j \end{cases}$$

Finally, putting these three results together we get

$$\omega_\mu = -\sum_{j=1}^{n-1}\left(\sum_{i=j}^{n-1}\frac{da_i}{a_i}\right)\wedge db_j .$$

It follows then that the index lowering and raising actions of ω_μ in this chart are given by

$$\left(\frac{\partial}{\partial a_i}\right)^\flat = -\frac{1}{a_i}\sum_{j=1}^{i} db_j, \quad \left(\frac{\partial}{\partial b_i}\right)^\flat = \sum_{j=1}^{n-1}\frac{da_j}{a_j}, \quad i = 1,\dots,n-1$$

$$(da_i)^\# = a_i\left(\frac{\partial}{\partial b_i} - \frac{\partial}{\partial b_{i+1}}\right), \quad i = 1,\dots,n-1, \quad \text{with the convention } \frac{\partial}{\partial b_n} \equiv 0$$

$$(db_i)^\# = -a_i\frac{\partial}{\partial a_i} + a_{i-1}\frac{\partial}{\partial a_{i-1}}, \quad i = 1,\dots,n, \quad \text{with the conventions } a_0 = a_n = 0.$$

Thus, if $f: G\cdot\mu \to \mathbb{R}$ is smooth, the Hamiltonian vector field $X_f = (df)^{\#}$ is given by

$$X_f(\beta) = \sum_{i=1}^{n-1}\left[a_i\left(\frac{\partial f}{\partial b_{i+1}} - \frac{\partial f}{\partial b_i}\right)\frac{\partial}{\partial a_i} + \left(a_i\frac{\partial f}{\partial a_i} - a_{i-1}\frac{\partial f}{\partial a_{i-1}}\right)\frac{\partial}{\partial b_i}\right], \quad a_0 = 0$$

$$= [\beta,\xi]^+$$

where $\xi \in \mathfrak{g}$ has diagonal entries $\frac{\partial f}{\partial b_1},\ldots,\frac{\partial f}{\partial b_n}$ and subdiagonal entries $\frac{\partial f}{\partial a_1},\ldots,\frac{\partial f}{\partial a_{n-1}}$. In these coordinates the Poisson bracket of two functions f and g is given by

$$\{f,g\} = \sum_{i=1}^{n-1}\left[\frac{\partial f}{\partial b_i}\left(-a_{i-1}\frac{\partial g}{\partial a_{i-1}} + a_i\frac{\partial g}{\partial a_i}\right) - \right.$$

$$\left. - \frac{\partial g}{\partial b_i}\left(-a_{i-1}\frac{\partial f}{\partial a_{i-1}} + a_i\frac{\partial f}{\partial a_i}\right)\right], \quad a_0 = 0.$$

The symplectic form ω_μ in this chart is not canonical. However, it arises as the push-forward of the restriction of the canonical symplectic structure $\sum_{i=1}^{n} dq^i \wedge dp_i$ of $T^*\mathbb{R}^n$ to the submanifold $M = \{(q,p) \in T^*\mathbb{R}^n \mid \sum_{i=1}^{n} q^i = 0,\ \sum_{i=1}^{n} p_i = \text{constant}\}$. This push-forward is known in the literature as "Flaschka's transformation." We shall explain this below.

First, the restriction of $\sum_{i=1}^{n} dq^i \wedge dp_i$ to M is non-degenerate and hence symplectic since its matrix in the basis

$$\left(\frac{\partial}{\partial q^1} - \frac{\partial}{\partial q^n},\ldots,\frac{\partial}{\partial q^{n-1}} - \frac{\partial}{\partial q^n},\ \frac{\partial}{\partial p_1} - \frac{\partial}{\partial p_n},\ldots,\frac{\partial}{\partial p_{n-1}} - \frac{\partial}{\partial p_n}\right) \text{ is}$$

$\begin{pmatrix} 0 & A_{n-1} \\ -A_{n-1} & 0 \end{pmatrix}$ where $A_{n-1} = \begin{bmatrix} 2 & 1 & \cdots & 1 \\ 1 & 2 & \cdots & 1 \\ 1 & 1 & \cdots & 2 \end{bmatrix}$. Second, the diffeomorphism $\phi: M \to \{(a,b) \in \mathbb{R}^{2n} \mid a_i > 0,\ i = 1,\ldots,n-1,\ a_n = 0,\ \sum_{i=1}^{n} b_i = -\text{constant}\}$ (Flaschka's transformation) given by

$$a_i = \exp(q^i - q^{i+1}),\ b_i = -p_i,\ i = 1,\ldots,n$$

with the convention $q^{n+1} = +\infty$ transforms ω_μ to

$$\phi^*\omega_\mu = \sum_{i=1}^{n-1}(dq^i - dq^n) \wedge dp_i = \sum_{i=1}^{n} dq^i \wedge dp_i .$$

The non-periodic Toda lattice Hamiltonian $H: T^*\mathbb{R}^n \to \mathbb{R}$ is given by

$$H(q,p) = \frac{1}{2}\sum_{i=1}^{n} p_i^2 + \frac{1}{4}\sum_{i=1}^{n} \exp 2(q^i - q^{i+1})$$

where we impose the conditions $q^0 = -\infty$, $q^{n+1} = +\infty$, $\sum_{i=1}^{n} q^i = 0$, i.e. we fix the center of mass. Hamilton's equations become

$$\dot{q}^i = p_i,\ \dot{p}_i = -\frac{1}{2}\exp 2(q^i - q^{i+1}) + \frac{1}{2}\exp 2(q^{i-1} - q^i)$$

from which it follows that $\left(\sum_{i=1}^{n} p_i\right)^{\bullet} = 0$ and hence $\sum_{i=1}^{n} p_i = c =$ constant. Push forward via ϕ this Hamiltonian system on M to $G\cdot\mu$ where μ has diagonal entries all equal to $-c/n$, superdiagonal entries all equal to 1 and the rest of the entries zero. Since ϕ is a symplectic diffeomorphism, we see that the Hamiltonian system defined by H is actually a Hamiltonian system on the co-adjoint orbit $G\cdot\mu$. In the chart $(a_1,\ldots,b_{n-1})$ Hamilton's equations become

$$(*)\qquad \dot{b}_i = \frac{1}{2}(a_i^2 - a_{i-1}^2),\quad \dot{a}_i = a_i(b_{i+1} - b_i),\quad i = 1,\ldots,n$$

with the conventions $a_0 = a_n = 0$ and the Hamiltonian H transforms to $H_\mu(\beta) = \frac{1}{2}\sum_{i=1}^{n} b_i^2 + \frac{1}{4}\sum_{i=1}^{n-1} a_i^2 = \frac{1}{2}\operatorname{Trace} L^2$, where L is the tridiagonal symmetric matrix

$$L = \frac{1}{2}(\beta + \beta^*) = \frac{1}{2}\begin{bmatrix} 2b_1 & a_1 & 0 & \cdots & 0 & 0 \\ a_1 & 2b_2 & a_2 & \cdots & 0 & 0 \\ - & - & - & - - - & - & - \\ 0 & 0 & 0 & \cdots & 2b_{n-1} & a_{n-1} \\ 0 & 0 & 0 & & 0 & 2b_n \end{bmatrix}$$

Putting $B = \frac{1}{2}(\beta - \beta^*)$ a short computation shows that the Hamiltonian system (*) is equivalent to

$$\dot{L} = [B,L]$$

which is Lax' isospectral equation. Thus all eigenvalues of the Jacobi matrix L are preserved by the flow of the Hamiltonian vector field X_{H_μ}. Thus $F_k(\beta) = \frac{1}{k}\,\mathrm{Trace}(L^k) = \frac{1}{k}\sum_{i=1}^{n}\lambda_i^k$ will be preserved by this flow. Let $X_k = X_{F_k}$. Then the claim is that

$$X_k(\beta) = [\beta,(L^{k-1})^-]^+ \in T_\beta(G\cdot\mu)$$

where A^- denotes the lower triangular part of the matrix A. The easiest proof of this formula is the definition of X_k, i.e. we have to show that $i_{X_k}\omega_\mu = dF_k$. We have for any $[\beta,\eta]^+ \in T_\beta(G\cdot\mu)$, $dF_k(\beta)\cdot[\beta,\eta]^+ = \mathrm{Trace}(L^{k-1}[\beta,\eta]^+)$ and

$$i_{X_k}\omega_\mu(\beta)\cdot[\beta,\eta]^+ = \omega_\mu(X_k(\beta),\ [\beta,\eta]^+) = \omega_\mu([\beta,(L^{k-1})^-]^+,[\beta,\eta]^+) =$$

$$= \mathrm{Trace}(\beta[\eta,(L^{k-1})^-]) = \mathrm{Trace}\ ((L^{k-1})^-[\beta,\eta]) = \mathrm{Trace}((L^{k-1})^-[\beta,\eta]^+) =$$

$$= \mathrm{Trace}\ (L^{k-1}[\beta,\eta]^+) = dF_k(\beta)\cdot[\beta,\eta]^+ .$$

In particular $X_{H_\mu}(\beta) = X_2(\beta) = [\beta,L^-]^+$, a formula which can be also proved directly. It follows that $\{F_2,F_i\} = 0$ for all i.

At this moment we refer to the work of Adler ([1978] and his paper in this Proceedings), who has explained, using the Kostant-Symes integrability theorem, how this formula arises naturally from the structure of the Lie algebra $\mathfrak{g}$. The formula for $X_k(\beta)$ is the key to his beautiful abstract formulation and it represents a Lax isospectral equation. At this point the facts which will be stated below are proved easier in the abstract context. A direct proof using matrix entries can be found in van Moerbeke [1976]. At this abstract level, the Toda lattice and the KdV equation have the same behavior; see Adler [1978].

The flows defined by X_k are all isospectral for the matrix L; the folows of X_i and X_j commute, i.e. $\{F_i, F_j\} = 0$ for all i,j. Moreover $F_2,\ldots,F_n$ form an independent set of integrals. The conclusion is that the non-periodic Toda system is completely integrable, n-1 independent integrals in involution being $F_2,\ldots,F_n$. See Moser [1974] for a complete solution of the non-periodic Toda system and Kostant [1978] for generalizations.

The periodic Toda system is discussed in van Moerbeke [1976] from the inverse scattering method point of view. The group theory behind it is more complicated; one has to work with Moody Lie algebras, i.e. infinite dimensional Lie algebras of the form $\mathfrak{g} \otimes \mathbb{R}[t,\frac{1}{t}]$. Another striking difference between the periodic and non-periodic case is the fact that algebraic geometry plays a fundamental role in the periodic case; see Mumford-van Moerbeke [1978] for generalizations and Adler and van Moerbeke [1979] for a unitary set up of all periodic Toda-like systems on the simple Lie algebras. It should be noted that there are also other Hamiltonian systems which can be imbedded in a bigger system in a Lie group; the Calogero systems have been discussed in this way by Kazhdan, Kostant, Sternberg [1978].

This program is to be accomplished for the KdV equation. We want a group theoretic set-up that recovers the complete integrability of the KdV equation as well as the smoothness of the time t-map of its flow. A first step towards this program is described in the next section. It turns out that the tricky part is to obtain the smoothness of the time t-map; the Hamiltonian character has already been caught by Adler in [1978].

§6. The KdV equation as a "vector field" on an infinite dimensional "Lie group"

This section attempts to apply the group theoretic techniques of §5 to the periodic KdV equation. One way of doing that is to generalize directly the methods and Lie groups of the Toda lattice; this has been successfully done in a formal way by M. Adler (see his paper [1978] and his talk in this volume) who recovers the Hamiltonian character as well as the complete integrability of the KdV equation. However, his Hamiltonian equations at the group level still contain the term uu_x since he works on a co-adjoint orbit rather than on the group and thus the time t-map of this new Hamiltonian system seems to be again not smooth.

The trouble due to a convective term uu_x has been overcome by Ebin and Marsden [1970] who showed that the solutions of the Euler equations for a perfect fluid are geodesics in the group of volume preserving diffeomorphisms. Since the KdV equation is built of the shock-wave equation $u_t + uu_x = 0$ and the linear hyperbolic equation $u_t + u_{xxx} = 0$, we shall analyze both of them separately in the manner of Ebin & Marsden, using their results; for the proofs of the facts stated below see their paper [1970], Marsden [1974] and references therein.

If M is a compact n-dimensional boundaryless manifold, for $s > (n/2) + 2$, let $\mathcal{D}^s$ denote the diffeomorphisms of M of Sobolev class H^s, i.e. $\eta \in \mathcal{D}^s$ iff η is bijective and $\eta, \eta^{-1} : M \to M$ are of class H^s. Right multiplication $R_\eta : \mathcal{D}^s \to \mathcal{D}^s$, $\xi \mapsto \xi \circ \eta$ is C^∞ for each $\eta \in \mathcal{D}^s$ and if $\eta \in \mathcal{D}^{s+\ell}$, left multiplication by η $L_\eta : \mathcal{D}^s \to \mathcal{D}^s$, $\xi \mapsto \eta \circ \xi$ is of class C^ℓ. Thus $\mathcal{D}^s$ is not a Lie group: right multiplication is smooth whereas left multiplication is only continuous. Also $\eta \mapsto \eta^{-1}$ is only a continuous map in $\mathcal{D}^s$. The tangent maps of R_η and L_η are given by $TR_\eta : X \mapsto X \circ \eta$ and $TL_\eta : X \mapsto T_\eta \circ X$ where $X \in H^s(TM)$, the set of all H^s-vector fields on M. $H^s(TM)$ is the "Lie algebra" of $\mathcal{D}^s$ in the sense that

$H^s(TM) = T_e\mathcal{D}^s$, $\mathcal{D}^s$ being open in the Hilbert manifold of all H^s maps of M into itself. The usual bracket of vector fields is the Lie algebra bracket on $H^s(TM) = T_e\mathcal{D}^s$, i.e. if $X, Y \in H^{s+\ell}(TM)$, $\ell \geqslant 1$ and $\tilde{X}$, $\tilde{Y}$ denote the *right-invariant* vector fields corresponding to X and Y, then $[\tilde{X},\tilde{Y}]_e = [X,Y]$. Moreover, if X is an H^s-vector field on M with flow η_t, then $t \mapsto \eta_t$ is a C^1-one parameter subgroup of $\mathcal{D}^s$. Then one can define as in the classical case the exponential map $\mathrm{Exp}\colon T_e\mathcal{D}^s \to \mathcal{D}^s$, $X \mapsto \eta_1$. Exp is continuous but is not even C^1 since it does not cover a neighborhood of the identity. That's how far $\mathcal{D}^s$ is a "Lie group." We record here for later use that the tangent space at $\eta \in \mathcal{D}^s$ is given by $T_\eta\mathcal{D}^s = \{X \in H^s(M,TM) \mid \tau_M \circ X = \eta\}$, where $\tau_M\colon TM \to M$ is the canonical tangent bundle projection. Similarly, if $X \in T\mathcal{D}^s$, $T_X(T\mathcal{D}^s) = \{Y \in H^s(M,T^2M) \mid \tau_{TM}\circ Y = X\}$, where $\tau_{TM}\colon T^2M \to TM$ is the canonical bundle projection.

Assume now that the oriented compact boundaryless manifold M has a metric $\langle\, , \rangle$ and a volume form μ induced by the metric. Define then a bilinear form on $T_\eta\mathcal{D}^s$ by

$$(X,Y) = \int_M \langle X(m),Y(m)\rangle_{\eta(m)}\mu(m) .$$

It is proved then in Ebin and Marsden [1970] that $(\ ,\ )$ is a weak Riemannian metric on $\mathcal{D}^s$. Moreover $(\ ,\ )$ has an associated smooth affine connection. Also, if $\exp\colon TM \to M$ is the exponential map corresponding to the metric $\langle\ ,\ \rangle$ on M, then $E\colon T\mathcal{D}^s \to \mathcal{D}^s$, $E(X) = \exp\circ X$ is the exponential map of the weak metric $(\ ,\ )$ on $\mathcal{D}^s$; E is defined only on a neighborhood of the zero section of $T\mathcal{D}^s$ and is C^∞ onto a neighborhood of $e \in \mathcal{D}^s$. Note that the two statements regarding the connection and the exponential map have to be proved separately since $(\ ,\ )$ is only a *weak* metric. The C^∞-spray $\bar{Z}\colon T\mathcal{D}^s \to T^2\mathcal{D}^s$ associated to $(\ ,\ )$ is given by $\bar{Z}(X) = Z\circ X$, where

$Z: TM \to T^2M$ is the spray of $<\ ,\ >$. In what follows we shall denote by ∇ the connection of $<,>$.

We will express now the vector field $\bar{Z}$ in space coordinates. Recall that if G is a Lie group, its tangent bundle TG is in two ways isomorphic to the trivial bundle $G \times \mathfrak{g}$, where $\mathfrak{g}$ is the Lie algebra of G:

$$\lambda: TG \to G \times \mathfrak{g},\quad \lambda(v_g) = (g, T_e L_g^{-1}(v_g)),\ L_g = \text{left multiplication by } g$$

$$\rho: TG \to G \times \mathfrak{g},\quad \rho(v_g) = (g, T_e R_g^{-1}(v_g)),\ R_g = \text{right multiplication by } g.$$

Motivated by the classical rigid body problem, $\lambda(v_g)$ and $\rho(v_g)$ are said to define the *body*, respectively the *space coordinates* of the vector $v_g \in T_gG$. In our case $G = \mathcal{D}^s$, $\mathfrak{g} = H^s(TM)$, $\rho(v_\eta) = (\eta, T_e R_\eta^{-1}(v_\eta)) = (\eta, v_\eta \circ \eta^{-1})$ for $v_\eta \in T_\eta \mathcal{D}^s$. Thus we have to compute $\rho_*(\bar{Z})$.

In the computation of $\rho_*(\bar{Z})$ that follows, one more ingredient from Riemannian geometry is needed. If $v, w \in T_mM$ the *lift of* w *relative to* v is defined by

$$(w)_v^\ell = \frac{d}{dt}(v + tw)\Big|_{t=0} \in T_v(TM).$$

In a chart, $(w)_v^\ell = (m,v,0,w)$. Then it is an easy computation in coordinates, starting with the formula $Z(q^i,\dot{q}^i) = (\dot{q}^i, -\Gamma^j_{k\ell}\dot{q}^k\dot{q}^\ell)$, to see that for X a vector field on M,

$$Z\circ X = TX\circ X - (\nabla_X X)_X^\ell .$$

We are now ready to compute $(\rho_*\bar{Z})(\eta,X)$.

$$(\rho_*\bar{Z})(\eta,X) = (T_{X\circ\eta}\rho\circ\bar{Z})(\rho^{-1}(\eta,X))$$

$$= (T_{X\circ\eta}\rho\circ\bar{Z})(X\circ\eta)$$

$$= T_{X\circ\eta}\rho(Z\circ X\circ\eta)$$

$$= T_{X\circ\eta}\rho(TX\circ X\circ\eta - (\nabla_X X)^{\ell}_X\circ\eta)$$

$$= (X\circ\eta, X, -\nabla_X X) \in T_\eta\mathcal{D}^s \times H^s(TM) \times H^{s-1}(TM)$$

The last equality follows in the following way. Since $\rho^{-1}(\eta,X) = X\circ\eta$, an easy computation shows that $T_{(\eta,X)}\rho^{-1}(v_\eta,X,Y) = Y^{\ell}_X\circ\eta + TX\circ v_\eta \in T_{X\circ\eta}(T\mathcal{D}^s)$. Thus $T_{(\eta,X)}\rho^{-1}(X\circ\eta,X, -\nabla_X X) = -(\nabla_X X)^{\ell}_X \circ\eta + TX\circ X\circ\eta$ and the equality is proved. Note that we obtained in this way a new "vector field" $\bar{Y}: H^s(TM) \to H^{s-1}(TM)$ defined by

$$\bar{Y}(X) = -\nabla_X X.$$

The crucial remark is that $\bar{Y}$ looses derivatives and it brings to light the troublesome convective term $\nabla_X X$. If the metric would be right invariant (which happens in the case of volume preserving diffeomorphisms) $\bar{Y}$ would be the Euler vector field from classical mechanics (see Abraham-Marsden [1978], Chapter 4).

We apply these results to the shock-wave equation $u_t + uu_x = 0$ with non-smooth time t-map as we noted earlier in §2. In all that follows let $s \geqslant 3$. Consider $M = SO(2) = \left\{\begin{pmatrix}\cos\theta & -\sin\theta\\ \sin\theta & \cos\theta\end{pmatrix} \middle| \theta \in \mathbb{R}\right\}$. M is a Lie group itself in this case and its Lie algebra is $so(2) = \left\{\begin{pmatrix}0 & -\xi\\ \xi & 0\end{pmatrix} \middle| \xi \in \mathbb{R}\right\}$ and it has the metric

$$\left\langle\begin{pmatrix}0 & -\xi\\ \xi & 0\end{pmatrix}, \begin{pmatrix}0 & -\eta\\ \eta & 0\end{pmatrix}\right\rangle = \xi\eta$$

which of course is bi-invariant, SO(2) being commutative. The exponential map so(2) → SO(2) is given by

$$\begin{pmatrix} 0 & -\xi \\ \xi & 0 \end{pmatrix} \mapsto \begin{pmatrix} \cos\xi & -\sin\xi \\ \sin\xi & \cos\xi \end{pmatrix}$$

Given $\begin{pmatrix} \cos\theta & -\sin\theta \\ \sin\theta & \cos\theta \end{pmatrix} \in SO(2)$ an arbitrary tangent vector v_θ at this point is obtained by left translation of a certain vector $\begin{pmatrix} 0 & -\xi \\ \xi & 0 \end{pmatrix} \in so(2)$ and hence $T_{\begin{pmatrix} \cos\theta & -\sin\theta \\ \sin\theta & \cos\theta \end{pmatrix}} SO(2) =$

$= \left\{ v_\theta = \begin{pmatrix} 0 & -\xi \\ \xi & 0 \end{pmatrix}\begin{pmatrix} \cos\theta & -\sin\theta \\ \sin\theta & \cos\theta \end{pmatrix} \mid \xi \in \mathbb{R} \right\}$. Thus the metric on this tangent space, obtained by left translation of the metric in so(2) is given by

$$\langle v_\theta, w_\theta \rangle = \frac{1}{2} \operatorname{Trace} \xi \begin{pmatrix} -\sin\theta & -\cos\theta \\ \cos\theta & -\sin\theta \end{pmatrix} \eta \begin{pmatrix} -\sin\theta & -\cos\theta \\ \cos\theta & -\sin\theta \end{pmatrix}^t = \xi\eta$$

With respect to this metric, the geodesic which at $t = 0$ passes through $\begin{pmatrix} \cos\theta & -\sin\theta \\ \sin\theta & \cos\theta \end{pmatrix}$ and is tangent to $\begin{pmatrix} 0 & -\xi \\ \xi & 0 \end{pmatrix}\begin{pmatrix} \cos\theta & -\sin\theta \\ \sin\theta & \cos\theta \end{pmatrix}$ is given by

$$t \mapsto \begin{pmatrix} \cos t\xi & -\sin t\xi \\ \sin t\xi & \cos t\xi \end{pmatrix}\begin{pmatrix} \cos\theta & -\sin\theta \\ \sin\theta & \cos\theta \end{pmatrix} = \begin{pmatrix} \cos(\theta+t\xi) & -\sin(\theta+t\xi) \\ \sin(\theta+t\xi) & \cos(\theta+t\xi) \end{pmatrix},$$

the geodesic flow $F_t\colon TSO(2) \to TSO(2)$ by

$$F_t(v_\theta) = \begin{pmatrix} 0 & -\xi \\ \xi & 0 \end{pmatrix}\begin{pmatrix} \cos(\theta+t\xi) & -\sin(\theta+t\xi) \\ \sin(\theta+t\xi) & \cos(\theta+t\xi) \end{pmatrix} = v_\theta \begin{pmatrix} \cos t\xi & -\sin t\xi \\ \sin t\xi & \cos t\xi \end{pmatrix}$$

and the geodesic spray $Z\colon TSO(2) \to T^2SO(2)$ by

$$Z(v_\xi) = \begin{pmatrix} 0 & -\xi \\ \xi & 0 \end{pmatrix} v_\theta \in T_v\,(TSO(2)),$$

where $v_\theta = \begin{pmatrix} 0 & -\xi \\ \xi & 0 \end{pmatrix}\begin{pmatrix} \cos\theta & -\sin\theta \\ \sin\theta & \cos\theta \end{pmatrix}$. Finally, note that a vector field on SO(2) is uniquely determined by a 2π-periodic smooth function and thus can be written as $X\frac{\partial}{\partial\theta}$, where $X(\theta)\frac{\partial}{\partial\theta}|_\theta =$ $= \begin{pmatrix} 0 & -X(\theta) \\ X(\theta) & 0 \end{pmatrix}\begin{pmatrix} \cos\theta & -\sin\theta \\ \sin\theta & \cos\theta \end{pmatrix}$. Thus it is clear that the Levi-Cività connection ∇ determined by $<,>$ on SO(2) is given by

$$\nabla_{X\frac{\partial}{\partial\theta}}(Y\frac{\partial}{\partial\theta}) = X\frac{\partial Y}{\partial\theta}\frac{\partial}{\partial\theta}.$$

We pass now to the "Hilbert Lie group" $\mathcal{D}^s$ with Lie algebra $H^s(TSO(2))$. The weak Riemannian metric on $\mathcal{D}^s$ is given by

$$(v,w)_\eta = \int_{S^1} <v(\eta(\theta)),w(\eta(\theta))>\, d\theta \quad , \; v,w \in T_\eta\mathcal{D}^s$$

and its spray $\bar{Z}: T\mathcal{D}^s \to T^2\mathcal{D}^s$ by $\bar{Z}(v) = Z\circ v$; $\bar{Z}$ is C^∞. The expression of $\bar{Z}$ in space coordinates is $\rho_*(\bar{Z})(\eta,X) = (X\circ\eta, X, -X\frac{\partial X}{\partial\theta})$ as we saw before. Thus the "Euler vector field" $\bar{Y}: H^s \to H^{s-1}$ is given by $\bar{Y}(X) = -XX_0$ and the equations of the geodesics in $\mathcal{D}^s$ become in space coordinates

$$X_t + XX_\theta = 0$$

and so we have regained the first part of the KdV equation, namely $u_t + uu_x = 0$. In this way the equation $u_t + uu_x = 0$ comes from a geodesic spray on $\mathcal{D}^s$ and regarded as such has a smooth flow. The relation between the time t-map H_t of the "vector field" $\bar{Y}(X) = -XX_\theta$ and the flow F_t of $\bar{Z}$ is the following. Let ϕ_t denote the flow of $X \in H^s(TM)$ and denote by $\bar{X}: \mathcal{D}^s \to T\mathcal{D}^s$ the vector field $\bar{X}(\eta) = X\circ\eta$. Then the flow of $\bar{X}$ is clearly $\phi_t\circ\eta$. Since $\rho_*(\bar{Z}) = (\bar{X},\bar{Y})$, the flow of $\rho_*(\bar{Z})$ is $(\phi_t\circ\eta, H_t(X))$ on one hand and $\rho\circ F_t\circ\rho^{-1}$ on the other, so we must have

$$(\rho\circ F_t\circ\rho^{-1})(\eta,X) = (\phi_t\circ\eta, H_t(X)), \text{ i.e.}$$

$$F_t(v) = H_t(v\circ\eta^{-1})\circ\phi_t\circ\eta \quad \text{for} \quad v \in T_\eta \mathcal{D}^s$$

In particular, if $\eta = e \in \mathcal{D}^s$, then $F_t(X)\circ\phi_t^{-1} = H_t(X)$ a formula that gives H_t and also explains why H_t is *not* smooth even though F_t is: taking inverses in $\mathcal{D}^s$ is only C^0 as was remarked at the at the beginning.

We pass to the study of the linear equation $u_t + u_{xxx} = 0$. Consider the linear continuous operator $A = -\frac{\partial^3}{\partial\theta^3}: T_e\mathcal{D}^{s+3} \to T_e\mathcal{D}^s$ with $s \geqslant 3$; we have $\|A\| < 1$. A is clearly a skew adjoint operator (in the strong sense) so by Stone's theorem it generates a group of isometries $\tilde{G}_t: T_e\mathcal{D}^s \to T_e\mathcal{D}^s$. We extend this flow to $G_t: T\mathcal{D}^s \to T\mathcal{D}^s$ by

$$G_t(v) = (T_eR_\eta\circ\tilde{G}_t\circ T_eR_{\eta^{-1}})(v), \quad v \in T_\eta\mathcal{D}^s$$

and A to a "vector field" $\bar{A}: T\mathcal{D}^{s+3} \to T^2\mathcal{D}^s$ by

$$\bar{A}(v) = (TR_\eta\circ A\circ TR_{\eta^{-1}}(v))_0^\ell$$

The operation $(\cdot)_0^\ell$ of canonical vertical lift is clearly smooth and the map inside the parenthesis is smooth in η by Lemma 2 of Appendix A of Ebin-Marsden [1970]. Thus $\bar{A}$ is smooth and it is easy to see that G_t is its flow: for η fixed $\bar{A}_\eta = TR_\eta\circ A\circ TR_{\eta^{-1}}$ is generated by $T_eR_\eta\circ\tilde{G}_t\circ T_eR_{\eta^{-1}}$. Another result of Marsden [1972] guarantees then that G_t is a smooth map for each fixed $t \in \mathbb{R}$. Under these hypotheses the Trotter formula holds, that is

$$K_t(v) = \lim_{n\to\infty} (G_{t/n}\circ F_{t/n})^n(v)$$

converges locally uniformly in v and t and defines a local uniformly Lipschitz flow $K_t: T\mathcal{D}^s \to T\mathcal{D}^s$ which is generated by

$\bar{Z} + \bar{A} : T\mathcal{D}^{s+3} \to T^2\mathcal{D}^s$; this flow K_t is unique. Furthermore, note that the right hand side exists à priori only for small t whereas we know from classical results on the KdV equation that K_t exists for all $t \in \mathbb{R}$; the formula above thus holds for all $t \in \mathbb{R}$, not only for small t. (By the smoothness of both G and F once v is picked there exists a neighborhood of v so that for all initial conditions v' close to v, for big n, $(G_{t/n} \circ F_{t/n})(v')$ is still close to v. Moreover K_t is smooth for fixed t.) The proof of all these results can be found in Ebin-Marsden [1970], Marsden [1973] (the smoothness of K_t) and Chorin, Hughes, McCracken, Marsden [1978].

Finally, if we regard the "vector field" $\varepsilon\bar{A}$, it has flow $G_t^\varepsilon = G_{\varepsilon t}$. Thus the Trotter formula gives a flow K_t^ε and by the local uniform convergence, we get for short time (the time of existence of F_t)

$$\lim_{\varepsilon \to 0} K_t^\varepsilon(v) = F_t(v)$$

Note that this last limit holds only for short time, even though K_t exists for all $t \in \mathbb{R}$.

Let us express $\bar{A}$ in space coordinates. We claim that $(\rho_*\bar{A})(\eta, X) = (X \circ \eta, 0, A(X))$. Indeed

$$\begin{aligned}(T\rho \circ \bar{A} \circ \rho^{-1})(\eta, X) &= T_{X\circ\eta}\rho\bar{A}(X\circ\eta) \\ &= T_{X\circ\eta}\rho(TR_\eta \circ A \circ TR_{\eta^{-1}}(X\circ\eta))_0^\ell \\ &= T_{X\circ\eta}\rho(A(X)\circ\eta)_0^\ell = \\ &= T_{X\circ\eta}\rho(A(X)_0^\ell \circ\eta) \\ &= (X\circ\eta, 0, A(X))\end{aligned}$$

the last equality holding since by the formula for $T_{(\eta,X)}\rho^{-1}$ we get $T_{(\eta,X)}\rho^{-1}(X\circ\eta, 0, A(X)) = A(X)^{\ell}_{0}\circ\eta$. Thus the "vector field" $\bar{Z} + \varepsilon\bar{A}$ expressed in space coordinates has as last component the KdV operator:

$$\rho_{*}(\bar{Z} + \varepsilon\bar{A})(\eta,X) = (X\circ\eta,\ \dot{X}\ ,\ -XX_{\theta} - \varepsilon X_{\theta\theta\theta})$$

As we saw before, the relation between the time t map S_t of the KdV equation and the flow K_t is the following:

$$K_t(v) = S_t(v\circ\eta^{-1})\circ\phi_t\circ\eta,\ v \in T\mathcal{D}^s\ ,\quad \phi_t \text{ flow of } v\circ\eta^{-1},$$

and conversely,

$$S_t(X) = K_t(X)\circ\phi_t,\quad X \in H^s \text{ and } \phi_t \text{ flow of } X.$$

The limit result in ε is expressed in the following way:

The solutions of the equation $u_t + uu_x + \varepsilon u_{xxx} = 0$ *therefore converge for short time in* H^s *to the solutions of* $u_t + uu_x = 0$.

So far we have accomplished the following:

- the time t-map S_t of the KdV equation is the expression in space coordinates of a smooth flow on $T\mathcal{D}^s$;
- the Trotter formula made clear the vague statement that S_t is built from a shock wave and a dispersive term compensating each other to give existence for all $t \in \mathbb{R}$;
- for short time (smaller than the first shock time) the solutions of $u_t + uu_x + \varepsilon u_{xxx} = 0$ converge as $\varepsilon \to 0$ to $u_t + uu_x = 0$.

It may look like we have begun to accomplish the program outlined in §5 where we dealt with the Toda lattice, but there is a catch: $\bar{A}$ is *not* Hamiltonian as it is easy to see (for example, in $\mathbb{R}^{2n}$, if T is a skew symmetric operator on $\mathbb{R}^n$, the operator $\begin{pmatrix} 0 & 0 \\ 0 & T \end{pmatrix}$ is not Hamiltonian.) Thus in order to regain the Hamiltonian character of the KdV equation, as well as the smoothness of its time t-map after a change of coordinates, a bigger group is needed, namely a group that includes the diffeomorphism group $\mathcal{D}^s$ and Adler's group of pseudo-differential smoothing operators (see Adler [1978]). A candidate is the group of invertible Fourier integral operators with some conditions on their Lagrangian submanifolds; this suggestion is due to J. Duistermaat and J. Marsden. Hopefully, on this group results like those in Kazhdan, Kostant, Sternberg [1978] will be obtainable.

We finish this section with one more additional argument why the group $\mathcal{D}^s$ is good even from the symplectic view point. Namely we will exhibit on co-adjoint orbits of $\mathcal{D}^s$, the Fadeev-Zakharov symplectic structure [1971], the same one that Adler obtains for his co-adjoint orbits in the Lie algebra of pseudo-differential operators.

The group is $G = \mathcal{D}^s(SO(2)) = \mathcal{D}^s(S^1)$, S^1 = the unit circle, the "Lie algebra" $\mathfrak{g}$ of G is $H^s(TS^1)$ and hence its L^2-dual $\mathfrak{g}^*$ consists of the 1-form densities, i.e. $\mathfrak{g}^* = \Omega(S^1) \otimes \Omega(S^1)$, where $\Omega(S^1)$ denotes the H^s *one*-forms on S^1. Since $Ad_\eta = T_e(L_\eta \circ R_{\eta^{-1}})$ is the definition of the adjoint action, $Ad_\eta(X) = \eta_*(X)$ for $\eta \in G$, $X \in \mathfrak{g}$. To compute the co-adjoint action of G on $\mathfrak{g}^*$, we note that $\mathfrak{g}^*$ is generated by elements of the form $\alpha \otimes \mu \in \Omega(S^1) \otimes \Omega(S^1)$ where α stands for a 1-form and μ for the fixed Riemannian volume form. Thus $\mu = dx$ on S^1. We have

$$\mathrm{Ad}^*_{\eta^{-1}}(\alpha \otimes \mu)\cdot X = (\alpha \otimes \mu)(\mathrm{Ad}_{\eta^{-1}}(X)) = (\alpha \otimes \mu)(\eta^* X)$$

$$= \int_{S^1} \alpha(\eta^* X)\mu = \int_{S^1} \eta^*((\eta_*\alpha)X)\mu$$

$$= \int_{S^1} (\eta_*\alpha)(X)\eta_*\mu = \eta_*(\alpha \otimes \mu)\cdot X$$

so that $\mathrm{Ad}^*_{\eta^{-1}}(\alpha \otimes \mu) = \eta_*(\alpha \otimes \mu)$. As we saw at the beginning of this section $\mathrm{ad}\, X = L_X$, where L_X is the Lie derivative of H^s-vector fields with respect to X. Let now $\alpha \equiv \alpha dx$ and $X \equiv X \frac{\partial}{\partial x}$ be a one-form and a vector field on S^1 with α, X H^s-functions on S^1. We have

$$L_X\alpha = di_X\alpha + i_X d\alpha = d(\alpha(X)) = d(\alpha X) \quad \text{(as functions)}$$

so that denoting by α', X' the derivatives of the functions α and X

$$(\mathrm{ad}\, X)(\alpha \otimes \mu) = L_X(\alpha \otimes \mu) = L_X\alpha \otimes \mu + \alpha \otimes L_X\mu$$

$$= d(\alpha X) \otimes \mu + \alpha \otimes d(\mu(X))$$

$$= (\alpha' X + \alpha X')dx \otimes dx + \alpha dx \otimes dX$$

$$= (2\alpha X' + \alpha' X)dx \otimes dx.$$

From here we are able to compute $(\mathrm{ad}\, X)^*(\alpha \otimes \mu)$. We have for any $Y \in \mathfrak{g}$, $Y \equiv Y\frac{\partial}{\partial x}$, by integration by parts in the fourth equality:

$$(\mathrm{ad}\, X)^*(\alpha \otimes \mu)\cdot Y = (\alpha \otimes \mu)\cdot[X,Y] = \int_{S^1} \alpha\cdot[X,Y]\mu$$

$$= \int_{S^1} \alpha(X'Y - Y'X)dx$$

$$= \int_{S^1} \alpha X' Y dx + \int_{S^1} (\alpha X)' Y dx$$

$$= \int_{S^1} (2\alpha X' + \alpha' X) Y dx$$

$$= L_X(\alpha \otimes \mu) \cdot Y \ , \text{ i.e.}$$

$$(\mathrm{ad}\ X)^*(\alpha \otimes \mu) = L_X(\alpha \otimes \mu).$$

Fix now $\alpha \otimes \mu \in \mathfrak{g}^*$ and take the co-adjoint orbit of G through this element. A typical element on this orbit is $\eta_*(\alpha \otimes \mu)$ and hence a typical element of the tangent space at $\eta_*(\alpha \otimes \mu)$ to the orbit is $(\mathrm{ad}\ X)^* \eta_*(\alpha \otimes \mu) = L_X \eta_*(\alpha \otimes \mu)$ for an arbitrary $X \in \mathfrak{g}$.

We now compute the weak symplectic form ω_α of the orbit $G \cdot (\alpha \otimes \mu)$ given by the Kostant-Kirillov-Souriau theorem.

$$\omega_\alpha(\eta_*(\alpha \otimes \mu))(L_X \eta_*(\alpha \otimes \mu), L_Y \eta_*(\alpha \otimes \mu))$$

$$= \eta_*(\alpha \otimes \mu) \cdot [Y,X] = (\alpha \otimes \mu)[\eta^* Y, \eta^* X]$$

$$= \int_{S^1} \alpha((Y \circ \eta)'(X \circ \eta) - (X \circ \eta)'(Y \circ \eta)) dx$$

$$= \int_{S^1} \alpha((Y' \circ \eta)(X \circ \eta) - (X' \circ \eta)(Y \circ \eta)) \eta' dx$$

$$= \int_{S^1} (\alpha \circ \eta^{-1})(Y' X - X' Y) dx \ .$$

Let $\alpha = \frac{1}{2} dx$. Then

$$\omega_{1/2}(\eta_*(\tfrac{1}{2}dx \otimes dx))(L_X \eta_*(\tfrac{1}{2}dx \otimes dx),\ L_Y \eta_*(\tfrac{1}{2}dx \otimes dx))$$

$$= \frac{1}{2}\int_{S^1} (Y' X - X' Y) dx, \quad X \equiv X \frac{\partial}{\partial x}\ , \ Y \equiv Y \frac{\partial}{\partial x}$$

which is the periodic version of the Fadeev-Zakharov symplectic structure.

The fact that this symplectic structure is identical to the one Adler obtains for pseudo-differential smoothing operators strengthens the belief that $\mathfrak{g} = H^s(TS^1)$ and Adler's Lie algebra fit together yielding eventually the KdV equation as a Hamiltonian system on an infinite dimensional Lie group contining these two, namely on the group of invertible Fourier integral operators.

Bibliography

Abraham, R., Marsden, J.: *Foundations of Mechanics*, Second Edition, Benjamin (1978).

Adler, M.: On the relationship of some results of Gelfand-Dikii and P. van Moerbeke, and a natural trace functional for formal asymptotic pseudo-differential operators, preprint, to appear in Inventiones Math. (1978).

Adler, M: On a trace functional for formal pseudo-differential operators and the Hamiltonian structure of Korteweg-deVries type equations (this volume).

Adler, M., van Moerbeke, P., Completely integrable Moody Hamiltonian Systems (In preparation) (1979).

Carlson, R.: Genericity of Simple Eigenvalues, preprint, University of Utah, Salt Lake City (1978).

Chorin, A., Hughes, T., McCracken, M., Marsden, J.: Product formulas and numerical algorithms, Comm. Pure and Applied Math. Vol XXXI, p. 205-256, (1978).

Deift, P., Trubowitz, E.: Inverse scattering on the line, to appear (1978).

Ebin, D., Marsden, J.: Groups of diffeomorphisms and the motion of an incompressible fluid, Annals of Mathematics, Vol. 92, No. 1, p. 102-163 (1970).

Fadeev, L., Zakharov, V.: The Korteweg-de Vries Equation, a completely integrable Hamiltonian system, Functional Analysis and its Applications, Vol. 5, No. 4, p. 18-27, (1971) (in Russian).

Kato, T.: Quasi-linear equations of evolution, with applications to partial differential equations, in Spectral Theory and Differential Equations, Lecture Notes in Mathematics, #448, Springer Verlag (1974).

________: The Cauchy problem for quasi-linear symmetric hyperbolic systems, Archive for Rational Mechanics and Analysis, Vol. 58, No. 3, p. 181-205, (1975).

_______ : Linear and quasi-linear equations of evolution of hyperbolic type, CIME, Bressanone (1977).

Kazhdan, J., Kostant, B., Sternberg, S.: Hamiltonian group actions and dynamical systems of Calogero type, Comm. Pure and Applied Math., Vol. XXXI, p. 481-507, (1978).

Kostant, B.: Dynkin diagrams and the generalized Toda lattice, preprint, (1978).

Magnus, W., Winkler, W.: *Hill's equation*, New York, Interscience, Wiley (1966).

McKean, H.: (this volume)

McKean, H., van Moerbeke, P.: The spectrum of Hill's equation, Inventiones Math. 30, p. 217-274 (1975).

McKean, H., Trubowitz, E.: Hill's operator and hyperelliptic function theory in the presence of infinitely many branch points, Comm. Pure and Applied Math. Vol. XXIX, p. 143-226, (1976).

Marsden, J.: A remark on semigroups depending smoothly on a parameter, unpublished preprint, Berkeley (1972).

__________ : On product formulas for nonlinear semigroups, Journal of Functional Analysis, Vol. 13, No. 1, p. 51-72, (1973).

__________ : Applications of Global Analysis for Mathematical Physics, Publish or Perish, (1974).

__________ : Qualitative Methods in Bifurcation Theory, Bull. Am. Math. Soc. 84 (1978), 1125-1148.

Marsden, J. and McCracken, M.: The Hopf bifurcation and its applications, Springer Applied Math. Sciences #19 (1976).

Meimann, N.: The theory of one-dimensional Schrödinger operators with a periodic potential, Journal of Math. Physics, Vol. 18, No. 4, p. 834-848 (1977).

Mumford, D., van Moerbeke, P.: The spectrum of difference operators and algebraic curves, preprint, Berkeley (1978).

Moser, J.: Finitely many mass points on the line under the influence of an exponential potential--an integrable system, Lecture Notes in Physics, Dynamical Systems, Theory and Applications, ed. J. Moser, p. 467-498 (1975).

Trubowitz, E.: Inverse problem for periodic potentials, Comm. Pure and Applied Math. Vol. XXX, p. 321-337, (1977).

Schechter, E.: Stability conditions for nonlinear products and semigroups, preprint, Duke University, (1978).

Simon, B.: On the genericity of non-vanishing instability intervals in Hill's equation, Ann. Inst. H. Poincaré, Vol. XXIV, No. 1, p. 91-93, (1976).

van Moerbeke, P. : The spectrum of Jacobi matrices, Inventiones Math. 37, p. 45-81, (1976).

SELF-DUAL YANG-MILLS FIELDS

by

J. H. Rawnsley

School of Theoretical Physics
Dublin Institute for Advanced Studies
Dublin 4, Ireland.

1. Introduction

In this lecture I want to talk about a system of non-linear partial differential equations - the Yang-Mills field equations [20] - which have been the object of much study recently. They are the Euler-Lagrange equations of the following variational problem: Let $\mathfrak{g}$ be a compact Lie algebra with an invariant inner product, and consider a 1-form α (the vector potential) on $\mathbb{R}^4$ with values in $\mathfrak{g}$. Define a 2-form Ω (the field strengths) by

$$\Omega = d\alpha + \frac{1}{2}[\alpha,\alpha]$$

and set

$$S(\alpha) = \int_{\mathbb{R}^4} |\Omega|^2 \, d^4x. \tag{1}$$

The Yang-Mills equations are obtained by varying α in this action functional.

If $\mathfrak{g} = \mathbb{R}$ then these equations reduce to Maxwell's equations, and the Yang-Mills theory may be considered as a generalization of electromagnetic theory to a theory with internal symmetry. Maxwell's equations are formulated in Minkowski space, whilst we shall look at the Yang-Mills equations in Euclidean space. This is because very little is known in general about the Minkowski space theory, whilst quite a lot is known about the Euclidean theory. Obviously, for physical applications, it will be necessary to develope the theory in Minkowski space. However, from now on, we shall assume our metrics are positive definite.

On the non-compact space $\mathbb{R}^4$ it is necessary to impose growth conditions on the fields being considered in order that the action (1) converges. We shall use the conformal invariance of the action functional and the Yang-Mills equations to obtain solutions on $\mathbb{R}^4$ by restriction from S^4. On S^4 the integrals will converge automatically. All known solutions arise this way.

On S^4 we have to contend with the global restrictions imposed by the topology. The functional (1) is invariant under the substitution (a gauge transformation)

$$\alpha \to g^{-1}\alpha g + g^{-1}dg$$

where $g : \mathbb{R}^4 \to G$ is a map into the group G with Lie algebra $\mathfrak{g}$. This suggests α should be considered as the local connection form of some connection in a principal G-bundle over S^4, and Ω as its curvature 2-form.

The principal G-bundles on S^4 are parametrized by $\pi_3(G)$ which is discrete ($\mathbb{Z}^r$ for r the number of simple ideals in $\mathfrak{g}$), so in formulating the variational problem the principal bundle should be held fixed. Thus we have a principal G-bundle P over S^4 and we consider the connection 1-forms α on P. The curvature Ω of α is a 2-form with values in the vector bundle $P(\mathfrak{g})$ associated to P by the adjoint representation of G in $\mathfrak{g}$. Then the action $S(\alpha)$ is defined by

$$S(\alpha) = \int_{S^4} |\Omega|^2 \mu,$$

where μ is the Riemannian volume. We want to find critical points of $S(\alpha)$ as α varies over the connections on P, and in particular we want to find those connections on P which minimize S.

Let us define

$$k = \frac{1}{8\pi^2} \int_{S^4} \Omega \dot{\wedge} \Omega \tag{2}$$

where $\dot{\wedge}$ denotes exterior product of forms combined with the inner product in $\mathfrak{g}$, so that $\Omega \dot{\wedge} \Omega$ is a scalar-valued 4-form on S^4. Note that $S(\alpha)$ may be written

$$S(\alpha) = \int_{S^4} \Omega \dot{\wedge} * \Omega$$

where $*$ denotes the Hodge duality operator on forms. $*$ maps 2-forms to 2-forms, and acting on 2-forms $(*)^2 = \mathrm{id}$. Thus any 2-form may be decomposed into a sum of two parts belonging to the $+1$ and -1 eigenspaces of $*$. In particular

$$\Omega = \Omega_+ + \Omega_-$$

and

$$S(\alpha) = \|\Omega_+\|^2 + \|\Omega_-\|^2,$$

$$8\pi^2 k = \|\Omega_+\|^2 - \|\Omega_-\|^2,$$

where

$$\|\beta\|^2 = \int \beta \dot{\wedge} * \beta$$

for any form β. It follows easily

$$S(\alpha) \geq 8\pi^2 |k|$$

with equality if and only if

$$\Omega_- = 0 \quad \text{if} \quad k > 0;$$

$$\Omega = 0 \quad \text{if} \quad k = 0$$

$$\Omega_+ = 0 \quad \text{if} \quad k < 0.$$

The first and last cases may be interchanged by reversing the orientation on S^4, whilst the second is trivial and will not concern us further. We shall therefore look at the case $k > 0$, and if there is a connection α in the principal bundle P with $\Omega_- = 0$ this will give the absolute minimum of S on connections on P. We call these connections self-dual. They are also called instantons in the physics literature.

The equation

$$*\Omega = \Omega \tag{3}$$

as a differential equation for α is of first order and is non-linear if G is not abelian.

It is a standard result of Chern-Weil theory that the number k depends only on P and not the connection used to calculate it. Then a solution of (3) gives a minimum of S on the connections on P and so will satisfy the Yang-Mills equations. The latter are of second order whilst (3) is of first order, and we shall limit ourselves to the study of (3).

Most of the general facts we shall use concerning self-dual connections are taken from [2]. For instance, the holonomy group of a self-dual connection is semi-simple. Also, since $H^2(S^4) = 0$, we may pass to the simply-connected covering group, where the holonomy group is a product of simple factors. Therefore without loss of generality we shall assume that G is simple. In this case the inner product on $\mathfrak{g}$ may be normalized so that a long root has unit length; then k is an integer. In fact for a simple group $\pi_3(G) \cong \mathbb{Z}$, and the map $P \to k$ sets up the isomorphism (recalling that $\pi_3(G)$ classifies the isomorphism classes of principal G-bundles on S^4).

Finally we summarize the problem we shall be considering. We take a positive integer k and fix a principal G-bundle P over S^4 where G is compact and simple, and P corresponds with k under the isomorphism $\pi_3(G) \cong \mathbb{Z}$. Then we look for self-dual connections in P.

2. The moduli space.

The first self-dual connection on S^4 was found by Belavin et al. [6], and had SU(2) as its structure group and $k = 1$. Then 't Hooft (unpublished) generalized their construction to produce a 5k-parameter family for each $k > 0$ for SU(2). This was extended by Jackiw et al, [12], by using conformal invariance to obtain a 13-parameter family for $k = 2$ and a $5k+4$ parameter family for $k \geq 3$. Since every compact simple, simply-connected group G has a subgroup isomorphic to SU(2) inducing an isomorphism $\pi_3(G) \cong \pi_3(SU(2))$, these solutions may be considered as self-dual connections for any compact, non-abelian Lie group. Hence self-dual connections exist for any group G and integer $k > 0$. However we may require that the connection be irreducible, which is necessarily true for SU(2) self-dual connections, but need not be true for general groups. It turns out that for low values of k there need not exist any irreducible self-dual connection for a given group G.

It is natural to ask whether the solutions found by Jackiw et al. exhaust all possible self-dual SU(2) connections, and what is the number of parameters in the general solution for any simple group G. These questions need careful formulation because if $\tau : P \to P$ is a bundle automorphism, and α is a self-dual connection in P then so is $\tau^*\alpha$, and the group of bundle automorphisms is infinite dimensional. We should obviously consider the space of equivalence classes, where two connections α, α' are equivalent if there is a bundle automorphism τ with $\alpha' = \tau^*\alpha$. The space of equivalence classes is called the moduli space of self-dual connections, which we shall denote by $M_k(G)$, and write $M_k^0(G)$ for the open subset (possibly empty) of equivalence classes of irreducible self-dual connections (this makes sense because $\tau^*\alpha$ is irreducible if and only if α is). The solutions of [12] are all inequivalent, so $M_1(SU(2))$ has dimension at least 5, $M_2(SU(2))$ at least 13, and $M_k(SU(2))$ at least $5k+4$ for $k \geq 3$.

Before we can actually assign a dimension to $M_k(G)$ it is necessary to see that it is a nice space in some sense. In fact little can be said about $M_k(G)$, but the following result holds for $M_k^0(G)$:

<u>Theorem</u>, [2]. There are integers n_G, m_G such that $M_k^0(G)$ is non-empty if and only is $k \geq n_G$, and for $k \geq n_G$ $M_k^0(G)$ is a smooth manifold of dimension $m_G k - \dim G$.

Here n_G and m_G are listed in [2], and also in [7]. For $G = SU(2)$, $n_G = 1$, $m_G = 8$, so that $M_k(SU(2))$ is a smooth manifold of dimension $8k - 3$. Parts of this theorem were also obtained in [13] and [18].

I do not want to spend much time on the proof of this theorem, despite its beauty, as I want to come to the construction of self-dual connections. The theorem is proven by means of deformation theoretical techniques analogous to those used by Kuranishi [14] to prove the existence of complete families of deformations of complex structures. The linearization of (3) around a known solution can be embedded in an elliptic complex. One cohomology group of this complex may be interpreted as the formal tangent space to $M_k^0(G)$, and the other cohomology groups all vanish. The Index Theorem then gives the formal dimension of $M_k^0(G)$. The hard part is to show every infinitesimal deformation is tangent to a curve of actual deformations in such a way that an open neighborhood of the origin in the formal tangent space is mapped into an open set in $M_k^0(G)$. The value of k for which self-dual connections first appear is found by induction over the maximal semi-simple subgroups. See [2] for full details.

It can be seen that $8k-3$ agrees with the number of parameters in the solutions of [12] only for $k = 1,2$. In the next section we describe how all self-dual connections may be constructed.

Some topological properties of the space of equivalence classes of connections and the relationship to the topology of the space of solutions of [12] are discussed in [3].

3. Construction of self-dual connections.

It is convenient to reformulate the problem in terms of Hermitian vector bundles. If we have a connection α in a principal G-bundle P, then a unitary representation of G gives rise to a Hermitian vector-bundle F, together with a metric connection

$$D : \Gamma(F) \to \Gamma(\wedge^1 \otimes F)$$

with

$$D(fs) = fDs + df \otimes s, \quad f \in C^\infty(S^4), \quad s \in \Gamma(F).$$

Here $\wedge^p$ denotes the bundle of p-covectors on S^4. Conversely, by considering the unitary frame bundle and reducing to the holonomy group, we can recover a principal bundle with connection with a compact structure group. If we began with an irreducible connection, and chose a faithful representation, we see the two pictures are equivalent. If $G = Sp(n)$ or $SO(n)$, then F will have a parallel symplectic or orthogonal structure compatible with its Hermitian structure.

We thus assume we have a Hermitian vector bundle F and a metric connection D in F. Then D extends to a covariant exterior derivative

$$D : \Gamma(\wedge^p \otimes F) \to \Gamma(\wedge^{p+1} \otimes F)$$

and the composite map D^2 is a section Ω of $\wedge^2 \otimes \mathrm{End}(F)$. As before we say D is self-dual if $*\Omega = \Omega$.

The integer k may be identified with the second Chern class $c_2(F)$. Then F, as a Hermitian vector bundle, is determined up to isomorphism by k and rank (F).

The first step toward constructing self-dual connections was taken by Atiyah

Ward in [4]. Using ideas of Penrose [16], they reformulated the problem so that it became amenable to the techniques of algebraic geometry. There is a fibering

$$\pi : P^3(\mathbb{C}) \to S^4$$

whose fibres are lines in $P^3(\mathbb{C})$ such that a 2-form ω on S^4 is self-dual if and only if $\pi^*\omega$ is of type (1,1) relative to the complex structure of $P^3(\mathbb{C})$.

If F is a Hermitian vector bundle on S^4, let E denote its pull-back to $P^3(\mathbb{C})$, then any connection D in F induces a connection $\tilde{D}$ in E whose curvature $\tilde{\Omega}$ is the pull-back $\pi^*\Omega$ of the curvature Ω of D. Thus D is self-dual if and only if $\tilde{\Omega}$ is of type (1,1). Since $\tilde{D}$ is again metric, $\tilde{\Omega}$ will be then skew-Hermitian, so it is of type (1,1) if and only if its (0,2) component vanishes.

Let $\Lambda^{o,p}$ be the bundle of (o,p)-covectors on $P^3(\mathbb{C})$. Then $\tilde{D}$ splits into a sum $D' + D''$ of parts of type (1,0) and (0,1) respectively such that

$$D'' : \Gamma(E) \to \Gamma(\Lambda^{0,1} \otimes E)$$

satisfies

$$D''(fs) = fD''s + \bar{\partial}f \otimes s, \quad f \in C^\infty(P^3(\mathbb{C})), \quad s \in \Gamma(E). \tag{4}$$

D'' extends to a map

$$D'' : \Gamma(\Lambda^{o,p} \otimes E) \to \Gamma(\Lambda^{o,p+1} \otimes E)$$

and the composite $(D'')^2$ is the (0,2) component of $\tilde{\Omega}$. Hence D is self-dual if and only if $(D'')^2 = 0$.

It easily follows, from for example [15], that if $(D'')^2 = 0$ E has a unique holomorphic structure such that the space $\mathcal{O}(E)(U)$ of holomorphic sections of E on U is the space of solutions of

$$D''s = 0 , \quad s \in \Gamma(E|U).$$

Hence a self-dual connection D in F gives rise to a holomorphic vector bundle E on $P^3(\mathbb{C})$. Such a bundle is called an instanton bundle.

The mapping π is most easily described by introducing the algebra of quaternions $\mathbb{H}$. $\mathbb{H}$ is a non-commutative 4-dimensional algebra with identity spanned by 1, i, j, k, where

$$i^2 = j^2 = k^2 = -1 , \quad ij = -ji = k.$$

If

$$q = a_1 1 + a_2 i + a_3 j + a_4 k$$

is in $\mathbb{H}$ we put

$$\bar{q} = a_1 1 - a_2 i - a_2 j - a_3 k, \quad |q|^2 = q\bar{q} = a_1^2 + a_2^2 + a_3^2 + a_4^2 ,$$

then

$$\overline{q_1 q_2} = \bar{q}_2 \bar{q}_1 , \quad |q_1 q_2| = |q_1|\ |q_2| .$$

The space $\mathbb{H}^k$ of k-tuples

$$\begin{pmatrix} q_1 \\ \cdot \\ \cdot \\ \cdot \\ q_k \end{pmatrix}$$

of elements of $\mathbb{H}$ is an $\mathbb{H}$-vector space with $\mathbb{H}$ acting on the right. Then $\mathbb{H}$-linear operators are given by $k \times k$ matrices acting on the left. The space of lines (one-dimensional subspaces) of $\mathbb{H}^2$ is denoted by $P^1(\mathbb{H})$, and is called the quaternionic projective line. It may be considered as the space of equivalence classes of elements $\binom{q_1}{q_2} \in \mathbb{H}^2 - 0$ modulo the relation

$$\begin{pmatrix} q_1 \\ q_2 \end{pmatrix} \sim \begin{pmatrix} q_1 q \\ q_2 q \end{pmatrix} , \quad q \in \mathbb{H} - o.$$

$P^1(\mathbb{H})$ may be identified as a manifold with S^4. We first identify $\mathbb{R}^5$ with $\mathbb{H} \times \mathbb{R}$ so that

$$S^4 = \{(q,r) \in \mathbb{H} \times \mathbb{R} \ \big| \ |q|^2 + r^2 = 1\}.$$

Then the map which sends the line in $P^1(\mathbb{H})$ which is spanned by $\binom{q_1}{q_2}$ to the point

$$\left(\frac{2q_1\bar{q}_2}{|q_1|^2 + |q_2|^2} , \frac{|q_1|^2 - |q_2|^2}{|q_1|^2 + |q_2|^2}\right)$$

in S^4 is a diffeomorphism.

Those points of S^4 with $r \neq 1$ may be mapped stereographically to $\mathbb{H}$ by

$$(q,r) \to q(1 - r)^{-1},$$

and this region of S^4 corresponds, in $P^1(\mathbb{H})$, with the subset where $q_2 \neq 0$. Then the stereographic projection, as a map on $P^1(\mathbb{H})$ is

$$\binom{q_1}{q_2} \cdot (\mathbb{H} - 0) \to \frac{2q_1\bar{q}_2}{|q_1|^2 + |q_2|^2} \cdot \frac{|q_1|^2 + |q_2|^2}{2|q_2|^2} = q_1 q_2^{-1} .$$

Thus stereographic projection in S^4 corresponds with taking quaternion affine coordinates in $P^1(\mathbb{H})$.

Any element q of $\mathbb{H}$ may be uniquely expressed in the form

$$q = z_1 + jz_2$$

with z_1 and z_2 in $\mathbb{C}$, where we identify $\mathbb{C}$ with the subspace of $\mathbb{H}$ spanned by 1 and i. Then we may further identify $\mathbb{H}^2$ with $\mathbb{C}^4$ by means of the map

$$\begin{pmatrix} q_1 \\ q_2 \end{pmatrix} \to \begin{pmatrix} z_1 \\ z_2 \\ z_3 \\ z_4 \end{pmatrix}, \quad q_1 = z_1 + jz_2,\ q_2 = z_3 + jz_4.$$

We have thus made $\mathbb{H}^2$ into a complex vector space with $\mathbb{C}$ acting on the right by restricting the action of $\mathbb{H}$. Let $P^3(\mathbb{C})$ denote the space of complex lines in $\mathbb{H}^2$. Clearly any complex line is contained in a quaternionic line, so we have the map

$$\pi : P^3(\mathbb{C}) \to S^4.$$

Let σ denote the map of $\mathbb{C}^4$ which corresponds with right multiplication by j on $\mathbb{H}^2$. Then

$$\sigma \begin{pmatrix} z_1 \\ z_2 \\ z_3 \\ z_4 \end{pmatrix} = \begin{pmatrix} -\bar{z}_2 \\ \bar{z}_1 \\ -\bar{z}_4 \\ \bar{z}_3 \end{pmatrix},$$

and σ induces a corresponding map of $P^3(\mathbb{C})$ to itself. Clearly σ preserves the fibres of π. The converse is true: A line in $P^3(\mathbb{C})$ is the image of a two-dimensional subspace of $\mathbb{C}^4$, a real line is one which is mapped into itself by σ. The fibres of π are precisely the set of real lines.

The map σ makes it possible to determine which holomorphic vector bundles on $P^3(\mathbb{C})$, amongst all holomorphic vector bundles on $P^3(\mathbb{C})$, are instanton bundles. If E is an instanton bundle, then the connection $\tilde{D}$ is flat on each real line, so that if z is in $P^3(\mathbb{C})$ then we can parallelly translate from z to $\sigma(z)$, and, combined with the Hermitian structure, obtain a map $\tau : \sigma^*\bar{E} \to E^*$. Here $\bar{E}$ denotes the C^∞ bundle which is conjugate to E, and E^* the dual of E. Since σ is antilinear, $\sigma^*\bar{E}$ is again a holomorphic bundle and $\tau : \sigma^*\bar{E} \to E^*$ is a holomorphic isomorphism. Also E is holomorphically trivial on each fibre of π because it is parallel on each fibre, and for the connection $\tilde{D}$ covariant constant sections are certainly holomorphic (the converse is also true on the fibres of π, see [2]).

The instanton bundles are then characterized as holomorphic vector bundles on $P^3(\mathbb{C})$ which are holomorphically trivial on each fibre of π, and which have a holomorphic isomorphism $\tau : \sigma^*\bar{E} \to E^*$. τ must satisfy certain conditions of positivity and triviality, see [2] for details. If F is a symplectic vector bundle, then $E^* \cong E$ and so τ can be regarded as a map $\sigma^*\bar{E} \to E$, and in this case $\tau^2 = -1$. If F is orthogonal, a map τ is obtained with $\tau^2 = 1$. This characterization of instanton bundles for $G = SU(2)$ was the main result of Atiyah and Ward [4].

The connection $\tilde{D}$ is recoverable from the holomorphic and Hermitian structures of E as follows: Because E is locally framed by holomorphic sections, there is a unique linear map

$$D'' : \Gamma(E) \to \Gamma(\wedge^{0,1} \otimes E)$$

satisfying (4). Define D' by

$$(D's,t) = d(s,t) - (s,D''t), \quad s,t \in \Gamma(E),$$

and put

$$\tilde{D} = D' + D''.$$

The reader may verify that $\tilde{D}$ is a metric connection in E, called the canonical connection.

In addition to being a complex manifold, $P^3(\mathbb{C})$ is a complex algebraic variety and a theorem of Serre [19] says that any holomorphic bundle on $P^3(\mathbb{C})$ is algebraic. Thus the problem of finding self-dual connections on S^4 becomes one of finding algebraic bundles E on $P^3(\mathbb{C})$ with an algebraic isomorphism $\tau : \sigma^*\bar{E} \to E^*$. Because the pull-back preserves characteristic classes E will satisfy

$$c_1(E) = c_3(E) = 0, \quad c_2(E) = k.$$

The second step in constructing self-dual connections was taken by Atiyah et al. [1] who applied results of Horrocks [11] and Barth and Hulek[5] concerning algebraic bundles on projective spaces to give a description of instanton bundles in terms of linear algebra. By working back through the Atiyah-Ward transformation it is then possible to give a construction of the principal bundle and self-dual connection on S^4 directly from this linear algebraic data.

Let L denote the holomorphic line bundle on $P^3(\mathbb{C})$ whose space of holomorphic sections is four-dimensional (L is determined up to isomorphisms). For p in $\mathbb{Z}$ let L^p denote the p-th power of L, and for any holomorphic vector bundle E on $P^3(\mathbb{C})$ let $E(p)$ denote $E \otimes L^p$. Horrocks proved that the module $\bigoplus_{p \in \mathbb{Z}} H^1(E(p))$ for $\bigoplus_{p \in \mathbb{Z}} H^0(L^p) = \mathbb{C}[z_1,\dots,z_4]$ determines E up to isomorphism. If $H^1(E(-2))$ vanishes this module is generated by $H^1(E(-1))$.

By analyzing this result of Horrocks' further, Barth and Hulek showed that if $H^1(E(-2)) = 0$, $E \cong E^*$ and E is holomorphically trivial on some line then there is a sequence

$$W^*(-1) \overset{a}{\to} V \overset{b}{\to} W(1)$$

of holomorphic vector bundles with a injective, $b \circ a = 0$, the kernel of b is a sub-bundle, $b \circ a = 0$ and

$$E \cong \operatorname{Ker} b/\operatorname{Im} a.$$

Here V is the trivial bundle with fibre $H^1(E \otimes \Omega^1)$ where Ω^1 is the bundle of holomorphic cotangents, and W is the trivial bundle with fibre $H^1(E(-1))$. The isomorphism $E \cong E^*$ induces a bilinear form on V with respect to which $b = a^t$. This bilinear form is symplectic if E is symplectic and orthogonal if E is.

There is an exact sequence

$$0 \to \Omega^1 \to C^4(-1) \to C \to 0 \tag{5}$$

where the first map is

$$\sum_{i=1}^{4} a_i dz_i \to \begin{pmatrix} a_1 \\ a_2 \\ a_3 \\ a_4 \end{pmatrix} ,$$

and the second map is

$$\begin{pmatrix} a_1 \\ a_2 \\ a_3 \\ a_4 \end{pmatrix} \to \sum_{i=1}^{4} a_i z_i .$$

Tensoring (5) with E gives an exact sequence

$$0 \to E \otimes \Omega^1 \to C^4 \otimes E(-1) \to E \to 0$$

whose long exact cohomology sequence contains the segment

$$0 \to H^0(E) \to H^1(E \otimes \Omega^1) \overset{\tilde{b}}{\to} C^4 \otimes H^1(E(-1)). \tag{6}$$

$\tilde{b}$ may be regarded as a map $V \to W(1)$ and gives b. The dual map $W^*(-1) \to V$ gives a. Note that $\tilde{b}$ is injective if and only if $H^0(E) = 0$.

In the case of instanton bundles we certainly have E holomorphically trivial on a line since it is trivial on all the real lines. If we are in the symplectic or orthogonal cases, then $E \cong E^*$, so that to apply the above construction it is necessary to know that $H^1(E(-2))$ vanishes. This follows by showing that $H^1(E(-2))$ may identified with the kernel of $\Delta + R/6$ acting on $\Gamma(F)$, where R is the scalar curvature of S^4. Since $R > 0$ this operator is injective and so $H^1(E(-2)) = 0$. Further details may be found in [17].

Let us assume, for simplicity, we are in the symplectic case. Then $\tau : \sigma^*\bar{E} \to E$ induces antilinear maps of $H^1(E \otimes \Omega^1)$ and $H^1(E(-1))$ which we also denote by σ. On $H^1(E(-1))$, $\sigma^2 = 1$, whilst on $H^1(E \otimes \Omega^1)$, $\sigma^2 = -1$. Further, if $\langle \cdot , \cdot \rangle$ denotes the symplectic structure on $H^1(E \otimes \Omega^1)$ then σ preserves $\langle \cdot , \cdot \rangle$ and $\langle \cdot , \sigma \cdot \rangle$ is a positive definite Hermitian inner product. If $a : W^*(-1) \to V$ is described by the linear family

$$A(z) = \sum_{i=1}^{4} z_i A_i : H^1(E(-1)) \to H^1(E \otimes \Omega^1)$$

then

$$\sigma(A(z)w) = A(\sigma z)\sigma w \ , \quad w \in H^1(E(-1)).$$

If E has rank $2n$, and $c_2(E) = k$ then the Hirzebruch-Riemann-Roch formula shows

$$\dim H^1(E(-1)) = k, \dim H^1(E \otimes \Omega^1) = 2k + 2n.$$

Let J be the $2(n+k) \times 2(n+k)$ matrix which in terms of $(n+k) \times (n+k)$ blocks is given by

$$J = \begin{pmatrix} 0 & -1 \\ 1 & 0 \end{pmatrix}$$

then we can choose bases for $H^1(E(-1))$ and $H^1(E \otimes \Omega^1)$ such that after identifying $H^1(E(-1))$ with C^k and $H^1(E \otimes \Omega^1)$ with C^{2k+2n}, σ is the ordinary conjugation on C^k and $\sigma v = J\bar{v}$ on C^{2k+2n}. The family $A(z)$ consists of $(2k + 2n) \times k$ complex matrices satisfying

(i) $A(z)^t J A(z) = 0 \ , \forall z \in C^4$;

(ii) $J\overline{A(z)} = A(\sigma z) \ , \forall z \in C^4$;

(iii) $\mathrm{rk} A(z) = k \ , \forall z \in C^4 - 0$.

If $z \in C^4 - 0$ let $\hat{z}$ denote the corresponding point in $P^3(C)$. Put

$$U_{\hat{z}} = A(z)C^k \ , \ U^0_{\hat{z}} = \mathrm{Ker}\, A(z)^t J.$$

then we obtain bundles U, U^0 with $U \subset U^0$ and

$$E = U^0/U$$

is an instanton bundle.

As a C^∞ bundle E is isomorphic to the orthogonal complement $\tilde{E}$ of U in U^0. $\tilde{E}$ is a sub-bundle of the trivial bundle $P^3(C) \times C^{2k+2n}$, so has a connection induced by orthogonal projection from the trivial connection in the product bundle. The first step in obtaining the Yang-Mills potential from A is to see that this connection in $\tilde{E}$ corresponds with the canonical connection in E. Recall that the canonical connection is determined by the requirements that it be metric, and that $D''s = 0$ for any local holomorphic section s.

Consider the connection in $\tilde{E}$. Let $P_{\hat{z}} : C^{2k+2n} \to \tilde{E}_{\hat{z}}$ be the orthogonal projection, then

$$\tilde{D}v = P\,dv$$

for any section v of $\tilde{E}$ (which we regard as a function on $P^3(C)$ with values in C^{2k+2n}). Then

$$\tilde{D}'' = P\bar{\partial}v,$$

so that $\tilde{D}''v = 0$ means

$$\bar{\partial}v \perp \tilde{E}.$$

It is easy to see that $C^{2k+2n} = U_{\hat{z}} + \tilde{E}_{\hat{z}} + U_{\sigma\hat{z}}$ is an orthogonal direct sum for each $\hat{z}$ in $P^3(C)$, so that $\tilde{D}''v = 0$ for v a local section of $\tilde{E}$ implies

$$\bar{\partial}v = A(z)w_1 + J\overline{A(z)}w_2$$

for some local sections w_1, w_2 of $C^k \otimes \Lambda^{0,1}$. But v satisfies

$$A(z)^*v = 0 = A(z)^tJv,$$

so

$$A(z)^tJ\bar{\partial}v = 0$$

and from (i) it then follows

$$A(z)^t\overline{A(z)}w_2 = 0.$$

Since $A(z)$ is injective this implies $w_2 = 0$ and so

$$\bar{\partial}v = A(z)w_1.$$

Applying $\bar{\partial}$ gives

$$A(z)\,\bar{\partial}w_1 = 0$$

and again, by injectivity, $\bar{\partial}w_1 = 0$ so that $w_1 = \bar{\partial}w$ for some local section w. Thus

$$\bar{\partial}(v - A(z)w) = 0.$$

Hence v projects into E to give a holomorphic section s.

Conversely, any local holomorphic section s of E has locally a lift to a holomorphic section of U^0 which can be split as $v + u$ with v in $\tilde{E}$ and u in U. Then $\bar{\partial}v = -\bar{\partial}u$ so

$$P\bar{\partial}v = -P\bar{\partial}u = 0,$$

and hence

$$\tilde{D}''v = 0.$$

Thus, as claimed, $\tilde{D}$ in $\tilde{E}$ corresponds with the canonical connection in E. From (i) and (ii) it is easy to see that the subspace $\tilde{E}_{\hat{z}}$ of C^{2k+2n} depends only on $\pi\hat{z}$, so we may define a vector bundle F on S^4 by

$$F_x = \tilde{E}_{\hat{z}} \quad \text{if} \quad x = \pi\hat{z}.$$

Then $\pi^*F = \tilde{E}$, and F is a subbundle of $S^4 \times C^{2k+2n}$. Further the connection D in F induced by orthogonally projecting the trivial connection in $S^4 \times C^{2k+2n}$ will pull back to $\tilde{D}$ in $\tilde{E}$. Thus D is a self-dual connection in F by the Atiyah-Ward result, and E is the instanton bundle associated to it.

In [5] it is also determined when two families $A(z)$, $A'(z)$ give rise to isomorphic bundles, and therefore to isomorphic self-dual connections. Namely: if there is an element S of $Sp(n+k)$ and an element T in $GL(k,\mathbb{R})$ with

$$A'(z) = SA(z)T^{-1}.$$

Moreover the action of $Sp(n+k) \times GL(k,\mathbb{R})$ modulo its centre, which acts trivially, is free on those families for which $a : C^k \otimes C^4 \to C^{2k+2n}$ is onto. From (6) this last condition holds if and only if $H^0(E) = 0$. Using [17] one may easily see that $H^0(E)$ is isomorphic to the space of covariant constant sections of F. For $n = 1$ there can be no such sections, so the group action is automatically free for $Sp(1) = SU(2)$ self-dual connections.

To analyze condtions (i), (ii), (iii) further it is convenient to reformulate them in terms of the quaternions. Condition (ii) implies

$$A_2 = J\bar{A}_1 \text{ , } A_4 = J\bar{A}_3 .$$

If we write

$$A_i = \begin{pmatrix} B_i \\ C_i \end{pmatrix} \quad i = 1, 2, 3, 4,$$

with B_i, C_i being $(k+n) \times k$ complex matrices, then

$$B_2 = -\bar{C}_1, \; C_2 = \bar{B}_1, \; B_4 = -\bar{C}_3, \; C_4 = \bar{B}_3 .$$

Consider

$$\sum_{i=1}^{4}(B_i z_i + jC_i z_i) = B_1 z_1 - \bar{C}_1 z_2 + B_3 z_3 - \bar{C}_3 z_4 + j(C_1 z_1 + \bar{B}_1 z_2 + C_3 z_3 + \bar{B}_3 z_4)$$

$$= (B_1 + jC_1)(z_1 + jz_2) + (B_3 + jC_3)(z_3 + jz_4).$$

If we put $R_1 = B_1 + jC_1$, $R_2 = B_3 + jC_3$, $q_1 = z_1 + jz_2$, $q_2 = z_3 + jz_4$, then $A(z)$ is completely determined by the family $R(q) = R_1 q_1 + R_2 q_2$. It is left as an exercise for the reader to show that (i), (ii), (iii) are then equivalent to

$$R(q)^* R(q) \in GL(k, \mathbb{R}), \ \forall q \in \mathbb{H}^2 - 0. \tag{7}$$

Here R^* denotes the quaternionic Hermitian adjoint of R.

$Sp(n + k)$ may be identified with the group of $(n+k) \times (n + k)$ quaternion matrices S satisfying $S^*S = 1$ and then two families R, R' arise from families A, A' giving isomorphic bundles if and only if there is S in $Sp(n + k)$ and T in $GL(k, \mathbb{R})$ with

$$R'(q) = SR(q)T^{-1}, \ \forall q \in \mathbb{H}^2. \tag{8}$$

It is not known how to characterize those families $R(q)$ which give irreducible self-dual connections, except for $n = 1$, where they all do. For $n = 1$ we can thus give a model for $M_k(SU(2))$ as all families $R(q)$ of $(k + 1) \times k$ quaternion matrices satisfying (7) modulo the action of $Sp(k + 1) \times GL(k, \mathbb{R})$ given by (8).

The principal bundle P over S^4 may be constructed from $R(q)$ as follows: Let

$$P_{(q_1, q_2)} = \{X \text{ a } (k + n) \times n \text{ quaternion matrix satisfying}$$

$$R(q)^* X = 0, \ X^* X = 1\}$$

for each $(q_1, q_2) \in \mathbb{H}^2 - 0$. Then $P_{(q_1, q_2)}$ depends only on the quaternion line spanned by $\binom{q_1}{q_2}$, so we obtain a bundle P over $P^1(\mathbb{H}) = S^4$. $Sp(n)$ acts on P on the right by

$$X \cdot g = Xg, \ X \in P, \ g \in Sp(n).$$

If $s : U \to P$ is a local section, then

$$\alpha_s = s^* ds$$

is the associated connection form of the self-dual connection.

Some further simplification may be made. It follows from (7) that $R_1^* R_1$ is real and positive definite, so there is T in $GL(k, \mathbb{R})$ with $T^2 = R_1^* R_1$. Then $R_1 T^{-1}$ forms the first k columns of some element S^{-1} of $Sp(n+k)$. In other words

$$R_1 = S^{-1} \overbrace{\begin{pmatrix} T \\ 0 \end{pmatrix}}^{k} \begin{matrix} \}k \\ \}n. \end{matrix}$$

Hence we may replace $R(q)$ by the equivalent family $SR(q)T^{-1}$, which we relabel $R(q)$ and

$$R_1 = \begin{pmatrix} 1 \\ 0 \end{pmatrix}, \quad R_2 = \begin{pmatrix} P \\ Q \end{pmatrix}. \tag{9}$$

Here P is $k \times k$ and Q is $n \times k$. Two families R, R' of this form $(R_1 = R_1' = (\begin{smallmatrix} 1 \\ 0 \end{smallmatrix}))$ are equivalent if there is S in $Sp(n+k)$ and T in $GL(k,\mathbb{R})$ with

$$\begin{pmatrix} 1 \\ 0 \end{pmatrix} = S \begin{pmatrix} 1 \\ 0 \end{pmatrix} T^{-1}, \quad \begin{pmatrix} P' \\ Q' \end{pmatrix} = S \begin{pmatrix} P \\ Q \end{pmatrix} T^{-1}.$$

It follows T is in $O(k)$, and if $S' = S\begin{pmatrix} T^{-1} & 0 \\ 0 & 1 \end{pmatrix}$ then S' is in $Sp(n+k)$ and

$$S'\begin{pmatrix} 1 \\ 0 \end{pmatrix} = \begin{pmatrix} 1 \\ 0 \end{pmatrix}.$$

Thus

$$S' = \begin{pmatrix} 1 & 0 \\ 0 & U \end{pmatrix}$$

with U in $Sp(n)$. Hence

$$S = \begin{pmatrix} T & 0 \\ 0 & U \end{pmatrix}$$

and

$$P' = TPT^{-1}, \quad Q' = UQT^{-1}. \tag{10}$$

If we substitute (9) into (7) we have

$$(q_1 + Pq_2)^*(q_1 + Pq_2) + \bar{q}_2 Q^* Q q_2 \in GL(k,\mathbb{R}). \tag{11}$$

Putting $q_2 = 0$ gives us no information, so we may assume $q_2 \neq 0$ and let $x = q_1 q_2^{-1}$. Then (11) becomes

$$(x + P)^*(x + P) + Q^*Q \in GL(k,\mathbb{R}). \tag{12}$$

For $x = 0$ we obtain

$$P^*P + Q^*Q \in GL(k,\mathbb{R}). \tag{13}$$

Thus (12) implies $\bar{x}P + P^*x$ is real for all x in $\mathbb{H}$. That is

$$\bar{x}P + P^*x = \bar{P}x + \bar{x}P^t$$

and hence

$$P = P^t. \tag{14}$$

Then

$$(x + P)^*(x + P) + Q^*Q = |x|^2 + \bar{x}P + \bar{P}x + P^*P + Q^*Q$$

is certainly real, so the only condition left in (12) is that it be invertible. This is the same as

$$v \neq 0,\ Pv = \lambda v \Rightarrow Qv \neq 0. \tag{15}$$

We thus obtain a model for the moduli space as all pairs (P,Q) satisfying (13), (14), (15) modulo the action of $Sp(n) \times O(k)$ given by (10). For $n = 1$ this action is free.

A section s of the principal bundle will be given by a pair of matrices X, Y, which are $k \times n$ and $n \times n$ respectively, with

$$(x + P)^* X + Q^*Y = 0, \quad X^*X + Y^*Y = 1$$

on the coordinate neighbourhood where $q_2 \neq 0$. At those points x where $x + P$ is invertible we may solve for X to give

$$X = -(x + P)^{-1*} Q^* Y.$$

Put

$$Z = -(x + P)^{-1*} Q^*,$$

then

$$X = ZY\ ,\ Y^* (Z^* Z + 1)Y = 1.$$

Thus

$$Y Y^* = (1 + Z^* Z)^{-1}.$$

For $n = 1$, this is a scalar equation and we can take $Y = (1 + |Z|^2)^{-1/2}$. In this case

$$\alpha_s = \frac{1}{2}(1 + |Z|^2)^{-1}(\bar{Z}dZ - d\bar{Z}Z).$$

Let me finish by examining the moduli spaces for $n = 1$, for $k = 1$ and 2.

k = 1.

In this case P and Q are scalars and (13, (14), (15) reduce to $Q \neq 0$. Two pairs (P,Q), (P',Q') are equivalent if

$$P' = P, \; Q' = \pm UQ$$

with U in Sp(1). Thus in each equivalence class there is a unique representative with Q real and strictly positive. P is arbitrary, so $M_1(SU(2)) \cong \mathbb{R}^4 \times (0,\infty)$ which is contractible, agreeing with [10].

k = 2.

Let

$$P = \begin{pmatrix} a & b \\ b & c \end{pmatrix}, \quad A = (d,e).$$

Then from (13) we have

$$\bar{a}b + \bar{b}c + \bar{d}e \in \mathbb{R},$$

or

$$\overline{(a - c)}b + \bar{b}(c - a) + \bar{d}e - \bar{e}d = 0.$$

This may be rewritten as

$$\bar{b}(c - a) + \bar{d}e = f$$

with f real.

The subset where $b = 0$ has codimension 4. If we assume $b \neq 0$ then we can solve for $c - a$ to give

$$c - a = \bar{b}^{-1}(f - \bar{d}e).$$

Thus if we neglect condition (15), the set of pairs (P,Q) satisfying (13), (14) differs from a space with the topology of $\mathbb{R}^4 \times S^3$ by a codimension 4 subset and so is connected and simply-connected. It can be seen that imposing (15) removes a codimension 4 subset, and so the space of matrices satisfying (13), (14), (15) is connected and simply-connected.

The moduli space is the quotient of this space of matrices by the action of $Sp(1) \times O(2)$, which is free after removing the centre. The centre lies in the connected component of the identity, so that the moduli space $M_2(SU(2))$ is the quotient of a connected and simply-connected space by a group with two connected components acting freely. It follows that $M_2(SU(2))$ is connected and has fundamental group $\mathbb{Z}_2$ agreeing again with [10].

By similar arguments, J. D. S. Jones (unpublished) has also established that for $k = 3$ $M_3(SU(2))$ is connected and has fundamental group $\mathbb{Z}_2$, but these methods have proven too cumbersome to handle for larger values of k, and it is an open question whether $M_k(SU(2))$ is connected for $k \geq 4$.

Other applications of this quaternion form of the self-dual connections appear in [8] and [9].

REFERENCES

[1] Atiyah, M. F., Hitchin, N. J., Drinfeld, V. G., Manin, Yu. I. Phys. Letters A 65(1978)185-187.

[2] Atiyah, M. F., Hitchin, N. J., Singer, I. M. Self-duality in four-dimensional Riemannian geometry. To appear in Proc. Roy. Soc.

[3] Atiyah, M. F., Jones, J. D. S. Commun. Math. Phys. 61 (1978)97-118.

[4] Atiyah, M. F., Ward, R. S. Commun. Math. Phys. 55(1977)117-124.

[5] Barth, W., Hulek, K. Monads and moduli of vector bundles. Preprint, Erlangen, 1978.

[6] Belavin, A. A., Polyakov, A. M. , Schwartz, A. S., Tyupkin, S. Phys. Letters B 59(1976)85-87.

[7] Bernard, C. W., Christ, N. H., Guth. A. H., Weinberg, E. J. Instanton parameters for arbitrary gauge groups. Preprint, Columbia, 1977.

[8] Christ, N. H., Weinberg, E. J., Stanton, N. K. General self-dual Yang-Mills solutions. Preprint, Columbia, 1978.

[9] Corrigan, E. F., Fairlie, D. B., Templeton, S., Goddard, P. A Green's function for the general self-dual gauge field. Preprint, École Normale Supérieure, Paris, 1978.

[10] Hartshorne, R. Commun. Math. Phys. 59(1978)1-15.

[11.] Horrocks, G. Proc. London Math. Soc. 14(1964)689-713.

[12] Jackiw, R., Nohl, C., Rebbi, C. Phys. Rev. D 15(1977)1642-1654.

[13] Jackiw, R., Rebbi, C. Phys Letters B 57(1977)189-192.

[14] Kuranishi, M. New proof for the existence of locally complete families of complex structures, 'Proceedings of the conference on complex analysis, Minneapolis, 1964', Springer-Verlag, New York(1965).

[15] Nijenhuis, A., Woolf, W. Annals of Math. 77(1963)424-489.

[16] Penrose, R. Reports on Mathematical Physics 12(1977)65-76.

[17] Rawnsley, J. H. On the Atiyah-Hitchin-Drinfeld-Manin vanishing theorem for cohomology groups of instanton bundles. Preprint, Dublin Institute for Advanced Studies, 1978.

[18] Schwarz, A. S. Phys. Letters B 67(1977)172-174.

[19] Serre, J.-P. Annals Inst. Fourier (Grenoble)6(1956)1-42.

[20] Yang, C. N., Mills, R. Phys. Rev. 96(1954)191-195.

ABOUT ISOSPECTRAL DEFORMATIONS OF DISCRETE LAPLACIANS

by

Pierre van Moerbeke

Berkeley and Louvain

In this lecture, I like to discuss the relevance of algebraic geometry and group theory to problems of isospectral deformations of the periodic (discrete) Laplacian in one and two dimensions. This development grew out of the explicit linearization of the Korteweg-de Vries equation by inverse spectral methods [3,11] and a parallel theory related to Jacobi matrices had its origin in the periodic Toda problem [3,13]; both examples show that the isospectral deformations of a periodic second order differential or difference operator are in effect linear flows on the Jacobi variety of some hyperelliptic curve. This means, there are, roughly speaking, as many isospectral periodic second order differential or difference operators as there are points on the Jacobi variety. This theory was generalized to differential operators of any order by Kricever [10], inspired by special examples of Zaharov-Shabat [19], and to difference operators of any order by P. van Moerbeke and D. Mumford [14]. In the latter case, a one-to-one correspondence was established between curves of a certain type and classes of isospectral difference operators. These ideas led Mumford [16] to show the absence of "isospectral" deformations for Laplace-like two-dimensional periodic difference operators by relating the Picard variety with the class of such "isospectral" operators and by showing that for a generic class of such operators, the Picard variety is trivial. "Isospectral" here means that the spectrum is given for all Floquet multipliers.

Therefore isospectral flows appear only in the case of one-dimensional operators. It was found by the authors above that the flows defined by Lax-type commutators (and only those) are linearizable on the Jacobi variety of some related curve. These flows derive from a symplectic structure using the Kostant-Kirillov method of orbits. This leads to a calculus with "pseudo-difference" operators as explained in Dikii-Gelfand [21], van Moerbeke-Mumford [14], and finally M. Adler [20], who found an attractive and unifying algebraic formulation. This work and work by Kostant-Symes suggest a relationship between the Kac-Moody extension of the classical groups, their Bruhat decomposition, specific families of hyperelliptic curves, and isospectral deformations; see Ratiu [22] and Adler and van Moerbeke [2].

This paper has two parts: One will be devoted to the isospectral deformation of periodic Jacobi matrices from an algebraic-geometrical point of view and their relation with the Kac-Moody extension of $sl(N)$; the theorems generalizing these ideas to general difference operators will also be stated. In another part, I discuss Mumford's result on the absence of isospectral deformations for two-dimensional difference operators.

PART I: Isospectral Deformations of One-dimensional Difference Operators.

§1. Periodic Jacobi Matrices and Hyperelliptic Curves.

Consider a hyperelliptic curve X (possibly singular) of genus g with its hyperelliptic involution τ and two division points P and Q such that $P^\tau = Q$; this means that for some meromorphic function h,

$(h) = -NP + NQ$. Let $\mathcal{D} = \sum_{i=1}^{g} \mu_i \in \text{Jac}(X)$; assume regularity about $\mathcal{D}$, i.e.,

$$\dim \mathcal{L}(\mathcal{D} + kP - (k+1)Q) = 0, \ \forall\, k \in \mathbb{Z}$$

Theorem 1: There is a one-to-one correspondence between regular points on the Jacobi variety of hyperelliptic curves X with division points and periodic Jacobi matrices (symmetric and tridiagonal)

$$C = \begin{pmatrix} a_1 & b_1 & \cdots 0 \cdots & b_N \\ b_1 & a_2 & b_2 & \vdots \\ 0 & b_2 & \ddots & \vdots \\ \vdots & & \ddots & b_{N-1} \\ b_N & \cdots & b_{N-1} & a_N \end{pmatrix} \quad \text{with } a_i, b_i \in \mathbb{C},\ b_i \neq 0$$

modulo conjugation by diagonal matrices $(\pm 1, \ldots, \pm 1)$.

Proof: Start with a regular point $\mathcal{D} = \sum_{i=1}^{g} \mu_i \in \text{Jac}(X)$. From the Riemann-Roch theorem, from the fact that allowing one extra pole increases $\dim \mathcal{L}$ by at most one, and from the regularity of $\mathcal{D}$ (in that order) one has

$$1 \leq \dim \mathcal{L}(\mathcal{D} + dP - kQ) \leq \dim \mathcal{L}(\mathcal{D} + (k-1)P - kQ) + 1 = 1.$$

Let $f_k (k \in Z)$ be the unique function (up to scalars) in $\mathcal{L}(\mathcal{D} + kp - kQ)$. Clearly h^α is the unique function in $\mathcal{L}(\mathcal{D} + N\alpha P - N\alpha Q)$ $(\alpha \in Z)$. Normalize f_k such that $hf_k = f_{k+N}$ for all $k \in Z$ and $f_0 = 1$. Then the argument proceeds as follows:

1) Any function

$$f \in \mathcal{L}(\mathcal{D} + rP - sQ) \quad r \geq s \quad r,\ s \in \mathbb{Z}$$

is a linear combination of $f_k (s \leq k \leq r)$. The function f_k has a pole of exact order equal to

$$r + \# \text{ times } P \text{ appears in } \mathscr{D}.$$

The function f has at P a pole or order, at worse, this number. Therefore for some constant c_r

$$f - c_r f_r \in \mathscr{L}(\mathscr{D} + (r - 1)P - sQ);$$

the same argument applies over and over again so as to find constants c_k such that

$$f - \sum_{k=s}^{r} c_k f_k \in \mathscr{L}(\mathscr{D} + (s - 1)P - sQ);$$

finally because of $\mathscr{D}$'s regularity:

$$f = \sum_{k=s}^{r} c_k f_k .$$

With a little more effort f_k can be shown to have a pole and a zero of exact order k at P and Q.

2) As an immediate consequence of 1), any function

$$u \in \mathscr{R} = \{f \text{ meromorphic on } X, \text{ holomorphic on } X \setminus P \setminus Q\}$$
$$= \mathbb{C}[z, h, h^{-1}] / F(h, h^{-1}, z)$$

admits an expansion in the following sense

$$u f_k = \sum_{i=-M'}^{M} c_{k,k+i} f_{k+i} \quad \text{with} \quad c_{k,k+i} \in \mathbb{C}$$

where[+] $(u)_\infty = -M'Q - MP$. In particular, since $(z)_\infty = -P - Q$

[+] $(u)_\infty$ denotes the divisor of u on X restricted to $P \cup Q$.

$$z\ f_k = b_{k-1}f_{k-1} + a_k f_k + c_k f_{k+1} \qquad c_k,\ b_k \in \mathbb{C}^*.$$

Clearly, the f_k's can be renormalized so as to have $b_k = c_k$. Because of the fact that $hf_k = f_{k+N}$, the function z maps into an infinite periodic Jacobi matrix (of period N) (or, what is equivalent information, into C) and h into the matrix S (shift operator) zero everyhwere except for the Nth subdiagonal (to the right) containing 1's.

Conversely, start with an infinite periodic Jacobi matrix

$$C_\infty = \begin{bmatrix} \ddots & \ddots & b_N & & & & & & & \\ \ddots & \ddots & & & & & & & & \\ b_N & a_1 & b_1 & & & & & & & \\ & b_1 & a_2 & b_2 & & & & & & \\ & & b_2 & \ddots & \ddots & & & & & \\ & & & \ddots & \ddots & b_{N-1} & & & & \\ & & & & b_{N-1} & a_N & b_N & & & \\ & & & & & b_N & a_1 & b_1 & & \\ & & & & & & b_1 & \ddots & \ddots & \\ & & & & & & & \ddots & & \end{bmatrix}$$

Looking for the common eigenvectors $f = (\ldots, f_{-1}, f_0, f_1, \ldots)^T$ of C and S,

$$\text{(i)} \qquad \begin{aligned} Cf &= zf \\ Sf &= hf \end{aligned}$$

amounts to considering the spectrum of C_h

$$C_h\overline{f} = z\overline{f}\ ,$$

where $\overline{f} = (f_1, f_2, \ldots, f_{N-1}, h)^T$ and

$$C_h = \begin{pmatrix} a_1 & b_1 & 0 & & b_N h^{-1} \\ b_1 & a_2 & b_2 & & \\ 0 & b_2 & \ddots & \ddots & \\ & & \ddots & \ddots & b_{N-1} \\ b_N h & & & b_{N-1} & a_N \end{pmatrix}$$

Therefore, the couples (z, h) which are spectral points of (i) must satisfy

$$\det(C_h - zI) = F(h, h^{-1}, z) \equiv (-1)^{N+1}(A(h + h^{-}) - R(z)) = 0,$$

where $A = \prod_i^N b_i$ and $R(z) = z^N + \ldots$ is a polynomial of degree N. This polynomial equation defines a hyperelliptic curve X (of genus $g \neq N - 1$) of couples

$$(z, h = \frac{1}{2A}(R(z) \pm \sqrt{R^2(z) - 4A^2}))$$

with branch points given by the $2N$ roots of $R(z) = \pm 2A$. The hyperelliptic involution τ maps (z, h) into (z, h^{-1}). This curve may be singular. The meromorphic function h nas no poles nor zeroes except at $z = \infty$, where

$$h(z) = \frac{1}{2A}R(z)[1 \pm (1 - \frac{2A^2}{R^2(z)} + \ldots)] \qquad \text{for } z \text{ near } \infty.$$

$$= \frac{R(z)}{A} + \ldots \qquad \text{for } z \text{ near } \infty \text{ on the } + \text{ sheet.}$$

$$= \frac{A}{R(z)} + \ldots \qquad \text{for } z \text{ near } \infty \text{ on the } - \text{ sheet.}$$

So, $(h) = -NP + NQ$, where P and Q are the two points covering ∞ on the $+$ and $-$ sheet respectively.

Since $C_h\bar{f} = z\bar{f}$ is a system of linear equations in $f_1, \ldots, f_{N-1}$

with coefficients, linear in z, h, and h^{-1}, the f_k's are rational expressions in z, h, and h^{-1} and therefore meromorphic functions on X with a divisor structure to be explained next.

Let Δ_{ij} stand for the minor of $C_h - zI$ corresponding to the ij^{th} entry. Then

$$f_k = \frac{\Delta_{1,k}}{\Delta_{1,i}} f_i = \frac{\Delta_{2,k}}{\Delta_{2,i}} f_i = \dots = \frac{\Delta_{N,k}}{\Delta_{N,i}} f_i \qquad 1 \le i, k \le N.$$

In particular

$$f_k = \frac{\Delta_{N,k}}{\Delta_{NN}} h = \frac{\Delta_{kk}}{\Delta_{k,N}} h.$$

By inspection of the matrix $C_h - zI$, the following minors have the form

$$\Delta_{N1} = \prod_1^{N-1} b_i + (-1)^N b_N h^{-1} Q_1(z),$$

$$\Delta_{1N} = \prod_1^{N-1} b_i + (-1)^N b_N h Q_1(z)$$

and

$$\Delta_{NN} = (-z)^{N-1} + \dots = \text{polynomial of degree } N - 1,$$

where $Q_1(z) = (-z)^{N-2} + \dots$ is a polynomial of degree $N - 2$. Then, readily,

$$\begin{aligned}(f_1)_\infty &= (\Delta_{N1})_\infty + (h) - (\Delta_{NN})_\infty \\ &= -(2N - 2)Q - NP + NQ + (N - 1)P + (N - 1)Q = Q - P.\end{aligned}$$

To find the divisor structure of the remaining f_k's, consider $C_h - zI$ shifted by one, i.e.,

$$\tilde{C}_h - zI = \begin{bmatrix} a_2 - z & b_2 & & & b_1 h^{-1} \\ b_2 & a_3 - z & \ddots & & \\ & \ddots & \ddots & \ddots & \\ & & & a_N - z & b_N \\ b_1 h & & & b_N & a_1 - z \end{bmatrix}$$

Then

$$(\tilde{C}_h - zI) \begin{bmatrix} \tilde{f}_1 \\ \tilde{f}_2 \\ \vdots \\ \tilde{f}_{N-1} \\ h \end{bmatrix} = 0.$$

where $\tilde{f}_k = f_{k+1}/f_1$. The same argument as above leads to

$$(\tilde{f}_1)_\infty = Q - P;$$

therefore

$$(f_2)_\infty = (\tilde{f}_1)_\infty + (f_1)_\infty = 2Q - 2P.$$

In general,

$$(f_k)_\infty = kQ - kP.$$

We now turn to the zeroes and poles of f_k on the affine part $X_0 = X \setminus P \setminus Q$. Define $\mathcal{D}$ to be the minimal positive divisor such that

$$(f_k) + \mathcal{D} \geq kQ - kP.$$

Such a divisor exists, because it suffices to check the minimality of for $1 \leq k \leq N$ (by periodicity). Next we show that 1) order $\mathcal{D} = g$ and 2) $\mathcal{D}$ is regular.

1) The proof that order $\mathcal{D} = g$ is informal, using the relationship

ii)
$$\Delta_{NN}\Delta_{11} = \Delta_{N1}\Delta_{1N}.;$$

a rigorous proof can be found in P. van Moerbeke and D. Mumford [9]. Since $\Delta_{kk}(1 \leq k \leq N)$ is a polynomial in z only of degree $N - 1$, we have that, whenever Δ_{kk} vanishes at μ_i, it also does at μ_i^τ. So, let*

$$(\Delta_{NN})_0 = \sum_1^{N-1} \mu_i + \sum_1^{N-1} \mu_i^\tau$$

and

$$(\Delta_{11})_0 = \sum_1^{N-1} \nu_i + \sum_1^{N-1} \nu_i^\tau.$$

In view of (ii), Δ_{N1} and Δ_{1N} share the roots of Δ_{NN} and Δ_{11}. Because of the fact that

$$\Delta_{N1}^\tau = \Delta_{1N},$$

whenever ξ is a zero of Δ_{N1}, ξ^τ is a zero of Δ_{1N}. Moreover, since

$$(\Delta_{N1})_\infty = (2N - 2)Q \text{ and } (\Delta_{1N})_\infty = (2N - 2)P,$$

both Δ_{N1} and Δ_{1N} have $2N - 2$ zeroes on X_0. Whenever μ_i and μ_i^τ are zeroes of Δ_{N1} or Δ_{1N}, $h(\mu_i) = h(\mu_i^\tau)$, i.e., μ_i is a branch point of X. Therefore after, possibly, relabeling the roots μ_i and ν_i,

---

* $(\Delta)_0$ denotes the divisor of Δ on $X_0 = X \setminus P \setminus Q$.

$$(\Delta_{1N})_0 = \sum_1^{N-1} \mu_i + \sum_1^{N-1} \nu_i$$

and

$$(\Delta_{N1})_0 = \sum_1^{N-1} \mu_i^\tau + \sum_1^{N-1} \nu_i^\tau .$$

Therefore

$$(f_1)_0 = (\frac{\Delta_{N1}}{\Delta_{NN}} h)_0 = (\Delta_{NI})_0 = \sum_1^{N-1} \nu_i^\tau - \sum_1^{N-1} \mu_i .$$

Since moreover

$$b_N + (a_1 - z)f_1 + b_1 f_2 = 0,$$

we have that

$$(f_2)_0 = \sum_1^{N-1} \xi_i - \sum_1^{N-1} \mu_i$$

for some points $\xi_1, \ldots, \xi_{N-1}$ on X_0. Let $\mathscr{D} = \sum_1^{N-1} \mu_i$; then using the general linear relation between the f_k's.

$$b_{k-1} f_{k-1} + (a_k - z)f_k + b_k f_{k+1} = 0,$$

we have that for all $k \in \mathbb{Z}$

$$(f_k)_0 \geq - \mathscr{D}$$

and $\mathscr{D}$ is minimal; clearly order $(\mathscr{D}) = N - 1 = g$.

2) $\mathscr{D}$ is a regular divisor.

Firstly, one shows that $\mathscr{D}$ is general, i.e., $\dim \mathscr{L}(\mathscr{D}) = 1$. Consider an integer $k > g - 2$; then by the Riemann-Roch theorem and the fact that $\dim \Omega(-\mathscr{D} - kP) = 0$*,

---

*Because a holomorphic differential can have at most $2g - 2$ zeroes.

$$\dim \mathscr{L}(\mathscr{D} + kP) = \dim \Omega(-\mathscr{D} - kP) + g + k - g + 1 = k + 1.$$

Moreover

$$\mathscr{L}(\mathscr{D} + jP) \supsetneq \mathscr{L}(\mathscr{D} + (j - 1)P),$$

because f_j belongs to the first space and not to the second. Therefore by lowering j down to 0,

$$\dim \mathscr{L}(\mathscr{D}) = 1$$

which shows the first part of this statement. The proof of the regularity proceeds by induction. Since $\mathscr{L}(\mathscr{D}) = 1$ and since $1 \notin \mathscr{L}(\mathscr{D} - Q)$,

$$\dim \mathscr{L}(\mathscr{D} - Q) = 0.$$

Assume that

$$\dim \mathscr{L}(\mathscr{D} + kP - (k + 1)Q) = 0;$$

then

$$\dim \mathscr{L}(\mathscr{D} + (k + 1)P - (k + 1)Q) \leq \dim \mathscr{L}(\mathscr{D} + kP - (k + 1)Q) + 1 = 1$$

implies equality since f_{k+1} belongs to the first space. Since f_{k+1} belongs to $\mathscr{L}(\mathscr{D} + (k + 1)P - (k + 1)Q)$ but not to $\mathscr{L}(\mathscr{D} + (k + 1)P - (k + 2)Q)$, we have that

$$\dim \mathscr{L}(\mathscr{D} + (k + 1)P - (k + 2)Q) = 0.$$

This finishes the proof of the correspondence between isospectral classes of Jacobi matrices and hyperelliptic curves with division points.

§2. Flows on the Jacobi Variety.

As a result of Theorem 1, the Jacobi variety (except for a lower dimensional manifold) can be parametrized by periodic Jacobi matrices with a given periodic and antiperiodic spectrum. Therefore the linear flows on $\mathrm{Jac}(X) = \mathbb{C}^g/L$ can be regarded as isospectral flows on the periodic Jacobi matrices: They can be expressed as Lax-type commutation relations. Let[*] $\omega_k = z^{k-1}y^{-1}dz$ $(1 \le k \le g)$ be a basis for the space of holomorphic differentials on X.

Theorem 2: Every linear flow on $\mathrm{Jac}(X)$

$$\sum_{i=1}^{g} \int_{v_i(0)}^{v_i(t)} \omega_k = a_k t \quad 1 \le k \le g$$

is associated with a polynomial $T(z)$ of degree $\le g$ such that

$$a_k = \mathrm{Res}_P(\omega_k T(z)).$$

This flows translates into a system of differential equations given by[**]

$$\dot{C} = [C, T(C)^+ - T(C)^-].$$

Proof: Let $T(z)$ be a polynomial of degree $1 \le K \le g$; from the previous considerations

$$T = Tf_0 = \sum_{i=-K}^{0} c_{0,i} f_i + \sum_{i=1}^{K} c_{0,i} f_i = g_- + g_+ .$$

Then, since

---

[*] $y = \pm\sqrt{R^2(z) - 4A^2}$.

[**] $T(C)^+$ $(T(C)^-)$ denotes the upper (resp. lower) triangular part of $T(C)$ including the diagonal of $T(C)$.

$$(g_+) = -KP + Q - \mathscr{D} + \text{a positive divisor on } X_0$$

and

$$(g_-) = -KQ - \mathscr{D} + \text{a positive divisor on } X_0$$

we have that

$$(g_+ + \frac{1}{t}) = -KP - \mathscr{D} + \sum_{i=1}^{K} P_i(t) + \mathscr{D}(t)$$

$$(g_- - \frac{1}{t}) = -KQ - \mathscr{D} + \sum_{i=1}^{K} Q_i(t) + \mathscr{D}'(t).$$

For t small enough, $\mathscr{D}(t)$ and $\mathscr{D}'(t)$ will be near $\mathscr{D}$, $\sum_{i=1}^{K} P_i(t)$ near KP and $\sum_{i=1}^{K} Q_i(t)$ near KQ. Using Abel's theorem on $g_+ + \frac{1}{t}$ and $g_- - \frac{1}{t}$:

$$\sum_{i=1}^{g} \int_{\nu_i}^{\nu_i(t)} \omega = - \sum_{i=1}^{K} \int_{P}^{P_i(t)} \omega \quad \text{and} \quad \sum \int_{\nu_i}^{\nu_i(t)} \omega = - \sum_{i=1}^{K'} \int_{Q}^{Q_i(t)} \omega$$

for every holomorphic differential ω. Moreover by a contour integration argument (van Moerbeke and Mumford [14], Lemma 4, §3):

$$\lim_{t \downarrow 0} \frac{1}{t} \sum_{i=1}^{K} \int_{P}^{P_i(t)} \omega = -\text{Res}_P(\ g^+)$$

and

$$\lim_{t \downarrow 0} \frac{1}{t} \sum_{i=1}^{K} \int_{Q}^{Q_i(t)} \omega = \text{Res}_Q(\omega g^-).$$

Therefore

$$\sum_{i=1}^{g} \int_{\nu_i}^{\nu_i(t)} \omega = -t\ \text{Res}_P(\omega u) + O(t^2)$$

and

$$\sum_{i=1}^{g} \int_{\nu_i}^{\nu_i'(t)} \omega = t \operatorname{Res}_Q(\omega u) + O(t^2).$$

For t small enough, $\mathscr{D}(t)$ and $\mathscr{D}'(t)$ will both be regular, as the regular divisors form an open subset in Jac(X). Let $f_k(t)$ and $f_k'(t)$ be the meromorphic functions associated to $\mathscr{D}(t)$ and $\mathscr{D}'(t)$ respectively. Then

$$((1 + tg^+)f_k(t)) \geq -(K + k)P + kQ - \mathscr{D}$$
$$((1 - tg^-)f_k(t)) \geq -kP + (k - K)Q - \mathscr{D}$$

implying that

$$(1 + tg_+)f_k(t) = \sum_{i=k}^{k+K} a_{ki}^+(t) f_i(0)$$
$$(1 - tg_-)f_k(t) = \sum_{i:k-K}^{k} a_{ki}^-(t) f_i(0).$$

Define $a_{ki}^+ = 0$ for $i < k$ and $a_{ki}^- = 0$ for $i > k$; the difference of these two equations reads

$$\begin{aligned}\Sigma(a_{ki}^+ - a_{ki}^-)f_i(0) &= f_k(t) - f_k'(t) + t(g_+f_k(t) + g_-f_k'(t)) \\ &= tTf_k(0) + tT(f_k(t) - f_k(0)) + (f_k(t) - f_k'(t))(1 - tg_-) \\ &= t\sum_i T(C)_{ki} f_i(0) + O(t^2)\end{aligned}$$

uniformly over any open set V such that $\overline{V} \subset X_0 \subset D$. Because of the independence of the $f_i(0)$,

$$a_{ki}^+(t) - a_{ki}^-(t) = tT(C)_{ki} + O(t^2).$$

Multiply $f_k(t)$ with some function of t (which can always be done) such that $a_{kk}^+(t) = 1 + tT(C)_{kk} + O(t^2)$. Now we write $(1 + tg_+)z\, f(t)$ in two different ways. On the one hand

$$\begin{aligned}(1 + tg_+)z\, f(t) &= (1 + tg_+)C(t)f(t) \\ &= C(t)(1 + tg_+)f(t) \\ &= C(t)(I + tT(C)^+ + O(t^2))f(0)\end{aligned}$$

and on the other hand

$$z(1 + tg_+)f(t) = z(I + tT(C)^+ + 0(t^2))f(0)$$
$$= (I + tT(C)^+ + 0(t^2))C(0)f(0).$$

Therefore

$$(I + tT(c)^+ + 0(t^2))^{-1}C(t)(I + tT(C)^+ + 0(t^2))f(0) = C(0)f(0).$$

This implies that

$$\dot{C} = [T(C)^+, C] = [T(C)^+ - T(C)^-, C].$$

Finally observe that any flow on Jac(X) can be obtained in this way, because for $1 \leq i \leq g$,

$$\mathrm{Res}_P(z^i\omega_k) \neq 0 \quad \text{for} \quad k = g - i + 1$$
$$= 0 \quad \text{for} \quad 1 \leq k \leq g - i.$$

§3. Relation with the Kac-Moody Lie Algebras.

Consider the Lie algebra sl(N) with the norm $\| A\| = \sum_{i,j} |a_{ij}|$, $A \in sl(N)$ and the Kac-Moody extension of it

$$\ell = \{ \sum_{|k|<\infty} A_k t^k | A_k \in sl(N), \sum_{k\in Z} \|A_k\| < \infty\}.$$

Consider the infinite matrix

$$= \begin{bmatrix} \cdots A_{-2} & A_{-1} & A_0 & A_1 & A_2 & \cdots & & \\ & \cdots A_{-2} & A_{-1} & A_0 & A_1 & A_2 \cdots & & \\ & & \cdots A_{-2} & A_{-1} & A_0 & A_1 & A_2 & \cdots \end{bmatrix}$$

Then ℓ can be described alternatively as follows:

$$\ell = \{A \mid \sum_{k\in Z} \| A_k \| < \infty\}.$$

Let $H^k = R^N$, for all $k \in Z$ and consider $H = \oplus H^k$, i.e., the ℓ^1-closure of infinite sequences, almost everywhere zero, of elements in R^N:

$$H \ni v = (\ldots, v_{-1}, v_0, v_1, \ldots) \text{ where } v_k = (v_k^1, \ldots, v_k^N).$$

Consider the shift $S: H \to H$

$$Sv = (\ldots, v'_{-1}, v'_0, v'_1 \ldots) \text{ where } v'_k = (v_{k-1,N}, v_{k,1}, \ldots, v_{k,N-1}).$$

Still another way to describe ℓ is to consider

$$gl_N(\infty, R) = \{\text{bounded operators } \xi \text{ on } H \text{ such that } \xi S^N = S^N \xi\}$$

equipped with the bracket operation; then

$$\begin{aligned} \ell &= [gl_N(\infty, R), gl_N(\infty, R)] \\ &= \{\sum_{k\in Z} A_k S^{Nk} \mid \Sigma \|A_k\| < \infty, A_k \in sl(N)\}. \end{aligned}$$

Now ℓ admits the following decomposition:

$$\ell = \mathcal{M} + \mathcal{K},$$

where

$$\mathcal{M} = \{\sum_{k=-\infty}^{0} A_k S^{Nk} \mid A_k \in sl(N),\ A_0 \text{ lower triangular of trace } 0; \sum_{k=-\infty}^{0} \|A_k\| < \infty\}$$

and

$$\mathcal{K} = \{\sum_{k\in Z} A_k S^{Nk} \mid A_k^\tau = -A_{-k};\ \sum \|A_k\| < \infty\ A_k \in sl(N)\}.$$

Consider the duals of these Lie algebras in

$$\ell^* = \{ \sum_{k\in Z} B_k S^{Nk},\ B_k \in sl(N);\ \sup_{k\in Z} \|B_k\| < \infty \}$$

for the bilinear pairing

$$< \sum_{k\in Z} A_k S^{Nk}, \sum_{m\in Z} B_m S^{Nm} > = \sum_{k+m=0} <A_k, B_m> \quad \text{with } <A, B> = \mathrm{Tr}AB.$$

Then $\ell^* = \mathcal{M}^* + \mathcal{K}^*$, where

$$\mathcal{M}^* = \{ \sum_{k=0} B_k S^{Nk} | B_k \in sl(N);\ B_0 \text{ is upper triangular;} \ \sup_k \|B_k\| < \infty \}$$

and

$$\mathcal{K}^* = \sum_{k:-\infty}^{0} B_k S^{Nl} | B_k \in sl(N),\ B_0 \text{ is strictly lower triangular;} \ \sup_k \|B_k\| < \infty \}.$$

The algebras ℓ, $\mathcal{M}$, and $\mathcal{K}$ give rise to closed subgroups L, N, and K in GL(H) the space of bounded invertible operators on H by exponentiation and closure. The coadjoint action of $n \in N$ on $\mathcal{M}^*$ is given by

$$Ad^*_{N^{n^{-1}}} \nu = (n\nu n^{-1})^+$$

where + denotes the upper triangular part. Consider now the orbit of the element

$$\nu = B_{-1}S^{-N} + B_0 + B_1 S^N \in \mathcal{M}^*$$

under the action of N, where

$$B_0 = \begin{bmatrix} 0 & 1 & & \\ & 0 & 1 & \\ & & \ddots & \ddots \\ \bigcirc & & & 1 \end{bmatrix} \quad B_1 = \begin{bmatrix} & & \\ & \bigcirc & \\ 1 & & \end{bmatrix} \quad \text{and} \quad B_{-1} = \begin{bmatrix} & & 1 \\ & \bigcirc & \\ & & \end{bmatrix}$$

If $n = \sum_{k=-\infty}^{0} A_k S^{Nk}$, then $n^{-1} = \sum_{k=-\infty}^{0} \overline{A_k} S^{Nk}$ with $A_0 \in SL(N)$, lower triangular, and

$$A_0\overline{A_0} = 1,\ A_{-1}\overline{A_0} + A_0\overline{A_1} = 0,\ A_2\overline{A_0} + A_{-1}\overline{A_{-1}} \div A_0\overline{A_{-2}} = 0,\ \text{etc.} \ldots$$

An elementary calculation shows that the orbit $0(\nu)$ consists of the set of infinite matrices

$$\begin{bmatrix} \ddots & \ddots & & & & & & \\ & a_N & b_N & & & & & \\ & & a_1 & b_1 & & & & \\ & & & a_2 & b_2 & & & \\ & & & & \ddots & \ddots & & \\ & & & & & \ddots & b_{N-1} & \\ & & & & & & a_N & b_N \\ & & & & & & & a_1 \quad b_1 \\ & & & & & & & \ddots \quad \ddots \end{bmatrix}$$

such that $\sum_1^N a_i = 0$ and $\prod_1^N b_i = 1$; this orbit $0(\nu)$ is $2(N-1)$-dimensional.

According to a method analogous to the Kostant-Kirillov construction of symplectic structures on orbits, define the Poisson bracket

$$\{H_1, H_2\} = \langle C, [\partial H_1, \partial H_2]\rangle$$

between functions H_1 and H_2 on ℓ^* such that $\partial H_i \in \mathcal{M}$, with $C \in 0(\nu)$.

Then each flow defined in §3 derives, according to this symplectic structure, from the Hamiltonial $H = TrP(C)$ for some polynomial P of degree at most N. Details about this and extensions to other classical groups can be found in Adler, van Moerbeke [2], and Ratiu [12].

§4. General Difference Operators and Algebraic Curves.

Let f be an infinite column vector $f = (\dots f_{-1}, f_0, f_1, \dots)^T$. Let D operate on f as the shift $Df_k = f_{k+1}$. Consider the difference operator C defined by

$$(Cf)_M = \sum_{k=-M'}^{M} c_{n,n+k} f_{n+k} = \Big(\sum_{k=-M'}^{M} c_{n,n+k} D^k \Big) f_n, \quad c_{i,j} \in \mathbb{C};$$

C acts on f as an infinite band matrix (c_{ij}) does on f, zero outside the band $-M' \leq i - j \leq M$; C is said to have support $[-M', M]$. Assume C to be periodic of period N, i.e., $c_{i+N,j+N} = c_{ij}$; this amounts to the commutation relation $CS = SC$, where $S = D^N$. Let $(M, N) = n$ and $(M', N) = n'$; let $M_2 n = M$, $M_1' n' = M'$, and $N_1 n = N$, $N_1' n' = N$.

A difference operator C will be called <u>regular</u> if the n quantities

$$\sigma_i = c_{i,i+M} c_{i+M,i+2M} \cdots c_{i+(N_1-1)M,i+N_1 M} \qquad 1 \leq i \leq n$$

are all different from zero and different from each other and the same for the n quantities

$$\sigma_i' = c_{i,i-M'} c_{i-M',i-2M'} \cdots c_{i-(N_1'-1)M',i-N_1'M'} \qquad 1 \leq i \leq n'.$$

They involve only boundary elements, i.e., elements on the outer diagonals. Note that $\sigma_{i+n} = \sigma_i$ and $\sigma'_{i+n'} = \sigma'_i$.

A square matrix C_h of order N will be constructed as in §1: If $N > M + M'$, consider the matrix of order N taken from C, having c_{11} for upper left and c_{NN} for lower right corner, move the upper and lower triangular corners (see Figure 1) respectively to the upper right and lower left corner of the square matrix after multiplication by h^{-1} and h. In general, we write:

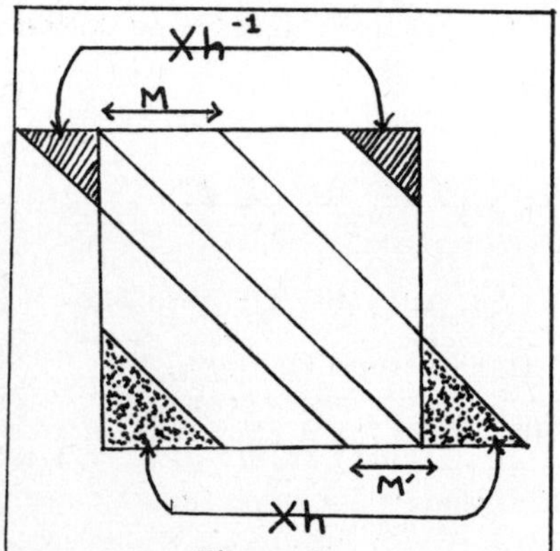

Figure 1.

$$(C_h)_{i,j} = \sum_{k=-\infty}^{+\infty} h^k \cdot c_{i,j+kN}$$

In fact C_h contains all the information contained in C. Also observe that $C_h D_h = (CD)_h$ for any two difference operators C and D. The determinant of $C_h - zI$ is readily seen to be a polynomial expression in z, h and h^{-1}, which has the form

$$F(h, h^{-1}, z)$$

$$\det(C_h - zI)$$

$$= A_0 h^M + A_1(z)h^{M-1} + \dots + A_M(z) + A_{M+1}(z)h^{-1} + \dots + A_{M+M'}h^{-M'} = 0;$$

where

$$A_0 = (-1)^{M(N-M)} \prod_{k=1}^{N} c_{k,k+M} = (-1)^{M(N-M)} \prod_{i=1}^{n} \sigma_i \neq 0$$

and

$$A_{M+M'} = (-1)^{M'(N-M')} \prod_{k=1}^{N'} c_{k,k-M'} = (-1)^{M'(N-M')} \prod_{i-1}^{n'} \sigma_i \neq 0.$$

Now let $\mathscr{D}$ be a positive divisor of degree g on a curve X. $\mathscr{D}$ is <u>general</u> if $\dim \mathscr{L}(\mathscr{D}) = 1$, i.e., $\dim \Omega(-\mathscr{D}) = 0$. $\mathscr{D}$ will be called <u>regular</u> with regard to two infinite sequences of smooth points $\{P_i\}_{i\in Z}$ and $\{Q_i\}_{i\in Z}$ if*

$$\dim \mathscr{L}(\mathscr{D} + \sum_{i=1}^{k} P_i - \sum_{i=0}^{k} Q_i) = 0 \quad \text{for} \quad k \in Z.$$

The points of $Jac(X)$ are all regular, except for a finite number of translates of the Θ-divisor ($g - 1$-dimensional subvariety of $Jac(X)$).

We now state the two main theorems (van Moerbeke-Mumford [14]).

<u>Theorem 3</u>: <u>There is a one-to-one correspondence between the two sets of data</u>:

a) <u>A regular difference operator</u> C <u>of support</u> $[-M', M]$ <u>and period</u> N, <u>modulo conjugation by diagonal periodic operators</u>.

b) <u>A curve</u> X, $(n + n')$ <u>points on</u> X, <u>a divisor</u> $\mathscr{D}$ <u>on</u> X <u>and</u> two functions h, z <u>on</u> X <u>subject to several conditions</u>.

---

*Define
$$\sum_{i=1}^{k} P_i = \sum_{1}^{k} P_i \qquad k \geq 1$$
$$= 0 \qquad k = 0$$
$$= -P_n - P_{n-1} - \dots - -P_{n+k+1} \qquad \text{for} \quad k < 0;$$

moreover
$$\sum_{i=0}^{k} Q_i = Q_{n'} + \sum_{i=1}^{k} Q_i \qquad k \geq 1$$
$$= Q_{n'} \qquad k = 0$$
$$= 0 \qquad k = -1$$
$$= -Q_{n'-1} - Q_{n'-2} - \dots - Q_{n'+k+1} \qquad \text{for} \quad k \leq -1$$

X may be singular, but always has genus:

$$g = \frac{(N-1)(M+M')-(n+n')+2}{2};$$

the $(n + n')$ points $P_1, \ldots, P_n, Q_1, \ldots, Q_{n'}$ are smooth and have a definite ordering. We define P_i (resp. Q_i) for all $i \in Z$ by $P_{i+n} = P_i$ (resp. $Q_{i+n'} = Q_i$). $\mathcal{D}$ has degree g and is regular for these sequences. The functions h and z have zeroes and poles as follows:

$$(h) = -N_1 \sum_{i=1}^{n} P_i + N_1' \sum_{i=1}^{n'} Q_i$$

and

$$(z) = -M_1 \sum_{i=1}^{n} P_i - M_1' \sum_{i=1}^{n'} Q_i + \text{a positive divisor not containing the}$$

P_i's and Q_i's.

Finally, $z^{N_1} h^{-M_1}$ (resp. $z^{N_1'} h^{M_1'}$) should take on distinct values at the P_i's (resp. the Q_i's).

As a result of §1, the Jacobi variety (except for a lower dimensional manifold) can be parametrized by difference operators of a given order with the same h-spectrum. Therefore the linear flows on Jac(X) (with regard to the group structure) can be regarded as isospectral flows on the space of difference operators. Let $\mathcal{A}$ denote the ring of meromorphic functions on X, holomorphic on $X_0 \equiv X \setminus \cup P_i \setminus \cup Q_i$. Since z, h, h^{-1} are affine coordinates on X_0, $\mathcal{A}$ is the ring of polynomials in z, h, h^{-1}. Let $\{\omega_k\}$ be a basis for the space of holomorphic differentials.

Define the following Poisson bracket between two differentiable functionals F and G on the space of N-periodic band matrices*:

$$\{F, G\} = \langle[(\tfrac{\partial F}{\partial C})^{[+]}, (\tfrac{\partial G}{\partial C})^{[+]}] - [(\tfrac{\partial F}{\partial C})^{[-]}, (\tfrac{\partial G}{\partial C})^{[-]}], C\rangle.$$

Theorem 4: Every linear flow on Jac(X)

$$\sum_{i=1}^{g} \int_{v_i(0)}^{v_i(t)} \omega_k = a_k t \qquad 1 \le k \le g$$

is associated with a polynomial $u = P(z, h, h^{-1})$ in $\mathcal{A}$ such that

$$a_k = \sum_{i=1}^{n} \mathrm{Res}_{P_i} (\omega_k u).$$

This flow is equivalent to the system of differential equations, given by

$$\dot{C} = [C[u]^{+}, C]$$

where $C[u] = P(C, S, S^{-1})$. Every linear flow on Jac(X) is a Hamiltonian flow with regard to the Poisson bracket { , } . In particular, in Poisson bracket notation, the typical flow

$$\dot{C} = [C, (S^{-k}C^{\ell})^{[+]}]$$

can be written as

---

*For any difference operators C, define

$$\begin{aligned} (C^{[+]})_{ij} &= C_{ij} && \text{if } i < j \\ &= \tfrac{1}{2}C_{ii} && \text{if } i = j \\ &= 0 && \text{if } i > j \end{aligned}$$

and $C^{[-]} = C - C^{[+]}$.

$$\dot{c}_{ij} = \{F, c_{ij}\}$$

where

$$F = \frac{1}{\ell+1} \operatorname{Tr}(S^{-k} C^{\ell+1}).$$

PART II. Mumford's Theorem on Isospectral Deformations of Two-dimensional Difference Operators.

Consider a two-dimensional (Laplace-like) difference operator T of order two, acting on a vector $(e_\alpha, \alpha \in Z^2)$ according to

$$(Te)_\alpha = T_{\alpha,\alpha} e_\alpha + \sum_{i=1}^{4} T_{\alpha,\alpha+\varepsilon_i} e_{\alpha+\varepsilon_i}$$

where $\varepsilon_1 = (1, 0)$, $\varepsilon_2 = (0, 1)$, $\varepsilon_3 = (-1, 0)$ and $\varepsilon_4 = (0, -1)$. Let $L_0 \subset Z^2$ be a two-dimensional sublattice; primitive periods $a = (a_1, 0)$ and $b = (b_1, b_2)$ can always be found such that $a_1 > 0$ and $b_1 > b_2 > 0$. For any $\gamma \in L_0$, let S^γ be the shift

$$(S^\gamma e)_\alpha = e_{\alpha+\gamma}.$$

The difference operator T is L_0-periodic in the sense that for all $\gamma \in L_0$,

$$T_{\alpha+\gamma,\beta+\gamma} = T_{\alpha,\beta}$$

which amounts to the commutation relations

$$TS^\gamma = S^\gamma T \quad \text{for all} \quad \gamma \in L_0.$$

As in the one-dimensional case, this set-up defines a ring of commuting operators, which therefore admits common eigenvectors. Let x and z^{γ} denote the respective eigenvalues of T and S^{γ}. In particular, let $z_1 = z^a$ and $z_2 = z^b$. Mumford shows that for a generic class of such L_0-periodic difference operators, T does not admit isospectral deformations given the common spectrum of T, S^a and S^b, i.e., the triples (x, z_1, z_2) for which T, S^a, and S^b admit common eigenvectors. In other words, T does not allow deformations given the spectrum of T for every Floquet multiplier.

My aim is to present this theorem and to show how it hinges on a deep fact about algebraic surfaces; namely, generic algebraic surfaces unlike curves, have a trivial Picard variety; the Picard or Jacobi variety of a curve is nontrivial. To provide a (superficial) explanation of this phenomenon, observe that a divisor on the curve is merely a collection of points, while a divisor on a surface (i.e., locally, the zero set of a meromorphic function) is an algebraic subvariety, or, in other words, a curve on the surface.

I have attempted to presuppose as little as possible about algebraic geometry, except for the notion of the Spec of some polynomial ring and some basic facts about cohomology of sheaves; they can easily be picked up, for instance in Griffith and Harris [4]. Except for these few facts, most statements will be shown by hand. I thank Nick Ercolani for many helpful conversations.

The problem of isospectral deformations of two dimensional Laplacians gave rise to a great deal of interest in the last few years; Singer [17], Guillemin-Kazhdan [5], and Weinstein [18] have shown results in the same direction as Mumford's one, but for continuous Laplacians (in various different contexts).

Before stating this theorem precisely, first a few remarks about L_0: various other sets of primitive periods will be used. A set of the form $\mu = (m, -m)$ and $\lambda = (\lambda_1, \lambda_2)$ with $\lambda_1 \geq -\lambda_2$ (respectively $\nu = (n, n)$ and $\sigma = (\sigma_1, \sigma_2)$ with $\sigma_1 \geq \sigma_2$) can always be found; let $\ell \equiv \lambda_1 + \lambda_2$ (respectively $s = \sigma_1 - \sigma_2$). Then

$$(\lambda_1, \lambda_2) = \alpha(a_1, 0) + \beta(b_1, b_2)$$

$$(-m, m) = \alpha'(a_1, 0) + \beta'(b_1, b_2) \qquad \alpha, \beta, \alpha', \beta' \in Z.$$

and

$$(a_1, 0) = \beta'(\lambda_1, \lambda_2) - \beta(-m, m)$$

$$(b_1, b_2) = -\alpha'(\lambda_1, \lambda_2) + \alpha(-m, m)$$

with $\alpha\beta' - \alpha'\beta = 1$, $\gcd(\alpha, \beta) = 1$, $\gcd(\alpha', \beta') = 1$, $\gcd(\alpha, \alpha') = 1$, and $\gcd(\beta, \beta') = 1$. This combined with the relations $\lambda_2 = \beta b_2$ and $m = \beta' b_2$ leads to $b_2 = \gcd(\lambda_2, m)$. Moreover, since

$$\det\begin{pmatrix} \lambda_1 & \lambda_2 \\ -m & m \end{pmatrix} = \det\begin{pmatrix} a_1 & 0 \\ b & b \end{pmatrix}$$

we have that $\ell m = a_1 b_2$. Similarly, $b_2 = \gcd(\sigma_2, n)$ and $sn = a_1 b_2$.

For $\alpha = (\alpha_1, \alpha_2) \in Z^2$, let $|\alpha| = \alpha_1 + \alpha_2$. We now proceed to the definition of the boundary operators: T^I and T^{III} are one-dimensional difference operators defined on e_α with $|\alpha| = |\alpha_0|$:

$$(T^I e)_{\alpha_0} = (T^\ell S^\lambda e)_{\alpha_0} \quad \text{with}^* \quad e_\alpha = 0 \quad \text{for} \quad |\alpha| > |\alpha_0|$$

---

* $(T^\ell S^\lambda e)_{\alpha_0}$ is a linear combination of e 's with $|\alpha| > |\alpha_0|$.

$$(T^{III}e)_{\alpha_0} = (T^{\ell}S^{-\lambda}e)_{\alpha_0} \quad \text{with}^{**} \quad e_\alpha = 0 \text{ for } |\alpha| < |\alpha_0|$$

T^I and T^{III} both commute with S^μ. Similarly, let T^{II} and T^{IV} be the one-dimensional difference operators T^sS^σ and $T^sS^{-\sigma}$ acting on e_α's with $\alpha_1 - \alpha_2 = 0$ and restricted to such e_α's; also T^{II} and T^{IV} both commute with S^ν.

The quantities

$$\eta_\alpha^I = \prod_{i=0}^{m-1} \frac{T_{\alpha+i(\varepsilon_1-\varepsilon_2),\alpha+i(\varepsilon_1-\varepsilon_2)-\varepsilon_2}}{T_{\alpha+i(\varepsilon_1-\varepsilon_2),\alpha+i(\varepsilon_1-\varepsilon_2)-\varepsilon_1}} \qquad \text{defined } \operatorname{mod}(\varepsilon_1 - \varepsilon_2, \lambda)$$

$$\eta_\alpha^{II} = \prod_{i=0}^{n-1} \frac{T_{\alpha+i(-\varepsilon_1-\varepsilon_2),\alpha+i(-\varepsilon_1-\varepsilon_2)-\varepsilon_1}}{T_{\alpha+i(-\varepsilon_1-\varepsilon_2),\alpha+i(-\varepsilon_1-\varepsilon_2)+\varepsilon_2}} \qquad \text{defined } \operatorname{mod}(-\varepsilon_1 - \varepsilon_2, \sigma)$$

$$\eta_\alpha^{III} = \prod_{i=0}^{m-1} \frac{T_{\alpha+i(-\varepsilon_1+\varepsilon_2),\alpha+i(-\varepsilon_1+\varepsilon_2)+\varepsilon_2}}{T_{\alpha+i(-\varepsilon_1+\varepsilon_2),\alpha+i(-\varepsilon_1+\varepsilon_2)+\varepsilon_1}} \qquad \text{defined } \operatorname{mod}(-\varepsilon_1 + \varepsilon_2, \lambda)$$

$$\eta_\alpha^{IV} = \prod_{i=0}^{n-1} \frac{T_{\alpha+i(\varepsilon_1+\varepsilon_2),\alpha+i(\varepsilon_1+\varepsilon_2)+\varepsilon_1}}{T_{\alpha+i(\varepsilon_1+\varepsilon_2),\alpha+i(\varepsilon_1+\varepsilon_2)-\varepsilon_2}} \qquad \text{defined } \operatorname{mod}(\varepsilon_1 + \varepsilon_2, \sigma)$$

will also be needed in the sequel.

THEOREM: (Mumford)

Let T be a Laplace difference operator which is L_0-periodic. Assume

1) *T, S^a, and S^b have no simultaneous 2-dimensional eigenspaces.*
2) *The pairs $\{T^I, S^\mu\}$, $\{T^{II}, S^\nu\}$, $\{T^{III}, S^\mu\}$ and $\{T^{IV}, S^\nu\}$ have no simultaneous 2-dimensional eigenspaces. Moreover they are regular difference operators in the sense of Theorem 3.*
3) *The quantities η_α^I, η_α^{II}, η_α^{III}, and η_α^{IV} are all different within their classes I, II, III, and IV.*

** $(T^{\ell}S^{-\lambda})_{\alpha_0}$ is a linear combination of e_α's with $|\alpha| < |\alpha_0|$.

<u>A generic L_0-periodic T has these properties. Then such a difference operator T has no isospectral deformations.</u>

<u>PROOF</u>: A first step is to show that the spectrum (x, z_1, z_2) of T, S^a, and S^b satisfies an algebraic relation, obtained in a similar way as in the one-dimensional case, by reducing the problem to a finite matrix one. Consider the points $\alpha \in \mathscr{R} \equiv [a_1-1, 0] \times [0, b_2-1] \subset Z^2$ ordered according to the rows from bottom to top and each row from left to right; as follows:

$0, \varepsilon_1, 2\varepsilon_1, \ldots, (a_1-1)\varepsilon_1, \varepsilon_2, \varepsilon_2 + \varepsilon_1, \ldots, \varepsilon_2 + (a_1-1)\varepsilon_1,$

$2\varepsilon_2, 2\varepsilon_2 + \varepsilon_1, \ldots, (b_2-1)\varepsilon_2, (b_2-1)\varepsilon_2 + \varepsilon_1, \ldots, (b\ -1)\varepsilon_2 + (a_1-1)\varepsilon_1.$

Consider now the column vector $\overline{e} = (e_\alpha, \alpha \in \mathscr{R})$. The problem now amounts to solving the spectral problem for the finite matrix $T^{z_1 z_2}\overline{e} = x\overline{e}$ for $z_1, z_2 \in C$; the square matrix $T^{z_1 z_2}$ of order $a_1 b_2 = \ell m$ is constructed according to the same recipe as in the one-dimensional case, as follows:

$$T^{z_1 z_2} = \begin{bmatrix} A_1 & B_1 & 0 & \cdots & & z_2^{-1} B'_{b_2} \\ B'_1 & A_2 & B_2 & & \bigcirc & \vdots \\ 0 & B'_2 & A_3 & \ddots & & \\ \vdots & & \ddots & \ddots & \ddots & B_{b_2-1} \\ z_2 B_{b_2} & \cdots & & & B'_{b_2-1} & A_{b_2} \end{bmatrix}$$

where

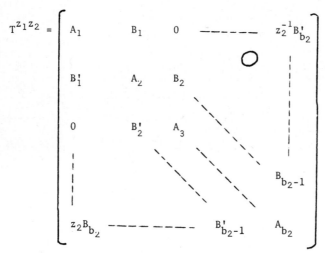

$$A_{k+1} = \begin{bmatrix} T_{k\varepsilon_2,k\varepsilon_2} & T_{k\varepsilon_2,\,k\varepsilon_2+\varepsilon_1} & 0 & & z_1^{-1}T_{k\varepsilon_2+a_1\varepsilon_1,k\varepsilon_2+(a_1-1)\varepsilon_1} \\ T_{k\varepsilon_2+\varepsilon_1,k\varepsilon_2} & T_{k\varepsilon_2+\varepsilon_1,k\varepsilon_2+\varepsilon_1} & T_{k\varepsilon_2+\varepsilon_1,k\varepsilon_2+2\varepsilon_1} & & \\ 0 & T_{k\varepsilon_2+2\varepsilon_1,k\varepsilon_2+\varepsilon_1} & T_{k\varepsilon_2+2\varepsilon_1,k\varepsilon_2+2\varepsilon_1} & \ddots & \\ & & \ddots & \ddots & T_{k\varepsilon_1+(a_1-2)\varepsilon_1,k\varepsilon_2+(a_1-1)\varepsilon_2} \\ z_1T_{k\varepsilon_2+(a_1-1)\varepsilon_1,k\varepsilon_2+a_1\varepsilon_1} & & & T_{k\varepsilon_2+(a_1-1)_1,\,k\varepsilon_2+(a_1-2)\varepsilon_1} & T_{k\varepsilon_2+(a_1-1)\varepsilon_1,k\varepsilon_2+(a_1-1)\varepsilon_1} \end{bmatrix}$$

for $0 \leq k \leq b_2 - 1$ and where

$$B_{k+1} = \mathrm{diag}(T_{k\varepsilon_2+j\varepsilon_1,(k+1)\varepsilon_2+j\varepsilon_1}; \quad 0 \leq j \leq a_1 - 1)$$

$$B'_{k+1} = \mathrm{diag}(T_{(k+1)\varepsilon_2+j\varepsilon_1,k\varepsilon_2+j\varepsilon_1}; \quad 0 \leq j \leq a_1 - 1)$$

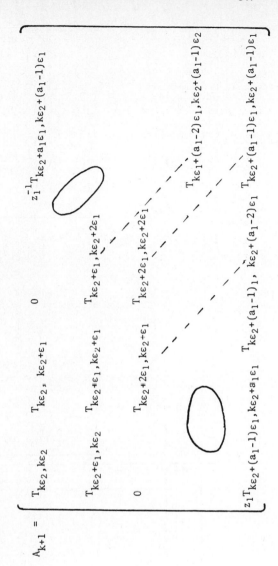

for $0 \leq k \leq b_2 - 2$; moreover

$$B'_{b_2} =$$

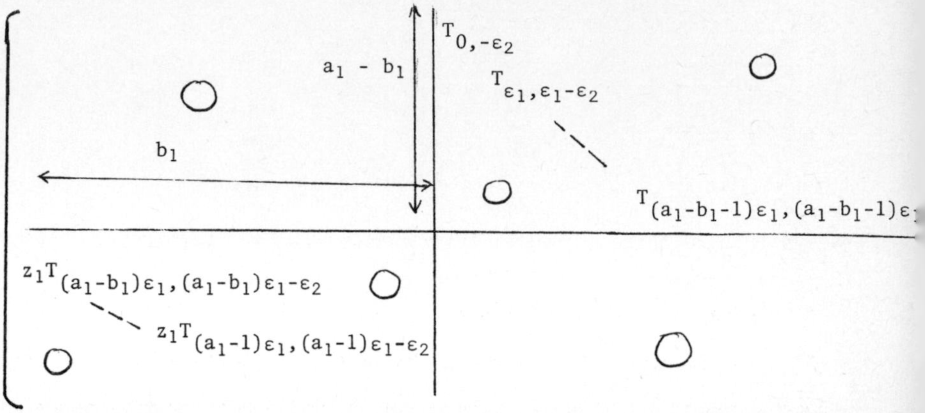

$$B_{b_2} =$$

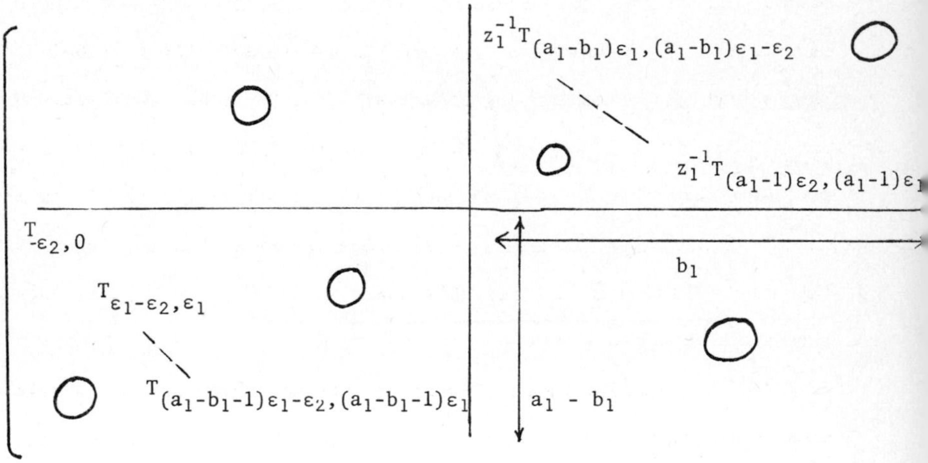

So x, z_1, and z_2 are eigenvalues of T, S^a, and S^b provided they satisfy the algebraic relation:

$$0 = \det(T^{z_1 z_2} - xI) = P(x, z_1, z_1^{-1}, z_2, z_2^{-1})$$

$$= Az_1^{b_2} + A'z_1^{-b_2} + Bz_2^{-a_1} z_1^{b_1} + B'z_2^{a_1} z_1^{-b_1} + (-1)^{a_1 b_2} x^{a_1 b_2}$$

$$+ \text{lower order terms in } z_1, z_1^{-1}, z_2, z_2^{-1}, \text{ and } x;$$

it defines an algebraic surface X_0 in $\mathbb{C} \times \mathbb{C}^* \times \mathbb{C}^*$.
The strategy now is as follows: The vector $\overline{e}$ defines ℓm meromorphic functions on X_0; they in turn define a line bundle or divisor on X_0 and vice versa: The sections of this line bundle L define in a unique way the ℓm meromorphic functions and therefore T. An infinitesimal deformation of T which preserves X_0 or what is the same, an infinitesimal deformation of the line bundle L on X_0 does not alter the topological structure of L. Therefore it would suffice to show there is only one line bundle within a given topological class; that is to say the Picard variety of X_0 is trivial. But the surface X_0 is highly singular at $x = \infty$; the divisor defined by T runs through $x = \infty$, so that X_0 is to be blown up, up to a point where this divisor sits completely within the smooth part of the blown up surface X. For this new surface X, $\mathrm{Pic}^o(X)$ is shown to be trivial.

Rather than blowing up X_0 we exhibit an appropriate algebraic extension $\mathcal{U}$ of the torus $\Omega \equiv \mathbb{C} \times \mathbb{C}^* \times \mathbb{C}^*$ such that X can be embedded in and such that X does not admit multiplicities, except maybe along subvarieties which the divisor defined by T does not intersect. Moreover, rather than computing $\mathrm{Pic}^o(X)$ directly on X, we compute it on an equivalent divisor, which is very simply defined; on that divisor, the question is reduced to a purely combinatorial one. These techniques of so-called "toroidal embedding" (referring to embedding of Ω in $\mathcal{U}$) have widely been used by algebraic geometers; the reader will find a general exposition of these ideas in Kempf, Knudsen, Mumford, and St.-Donat [8]. In order to show that X has no multiplicities, it suffices to show that T, S_1, and S_2, possibly rewritten in the new coordinates do not admit double eigenvalues (i.e., eigenvalues with at least two independent eigenvectors). The ideas of "toroidal embedding" will become clear in the course of this exposition, which will be broken up in various steps.

§1. Construction of the Toroidal Cradle.

Consider the vectors $x_i\,(0 \le i \le 4)$ in $\mathbb{R}^3$

$$x_0 = (1, 0, 0)$$
$$x_1 = (-1, a_1, b_1 + b_2)$$
$$x_2 = (-1, a_1, b_1 - b_2)$$
$$x_3 = (-1, -a_1, -b_2 - b_1)$$
$$x_4 = (-1, -a_1, +b_2 - b_1)$$

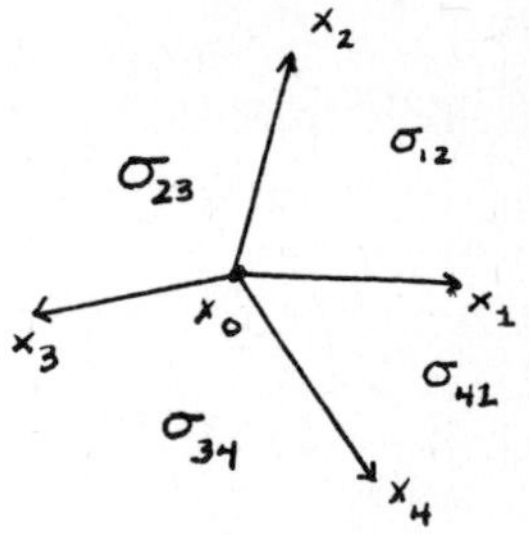

Figure 2.

and the 4 rational cones (semigroups) defined by them:

$$\sigma_{12} = \{\lambda_0 x_0 + \lambda_1 x_1 + \lambda_2 x_2 \mid \lambda_i \ge 0,\ \lambda_i \in Q\}$$
$$\sigma_{23} = \{\lambda_0 x_0 + \lambda_2 x_2 + \lambda_3 x_3 \mid \lambda_i \ge 0,\ \lambda_i \in Q\}$$
$$\sigma_{34} = \{\lambda_0 x_0 + \lambda_3 x_3 + \lambda_4 x_4 \mid \lambda_i \ge 0,\ \lambda_i \in Q\}$$
$$\sigma_{41} = \{\lambda_0 x_0 + \lambda_4 x_4 + \lambda_1 x_1 \mid \lambda_i \ge 0,\ \lambda_i \in Q\}$$

They have in common faces and edges. Each $\sigma_{i,i+1}$ has three faces and three edges. Let Σ be the union of these four cones. Consider now the dual figure in $\mathbb{Z}^3$. Let $I = (i, j, k) \in Z^3$ be called a <u>character</u>; the isomorphism

$$\mathbb{Z}^3 \to \mathbb{Z} \times L_0 : (i, j, k) \curvearrowright (i, \rho) \equiv (i, ja + kb)$$

will be used throughout; isomorphic sets will be denoted by $\cong$. Also, see*

Then each x_i defines a half-space as follows**:

$\check{\sigma}_0 = \{I \mid \ <I, x_0> \geq 0\} = \{I \mid i \geq 0\}$

$\check{\sigma}_1 = \check{\sigma}_0 \cap \{I \mid <I, x_1> \geq 0\} \cong \{(i, \rho) \mid i \geq 0, \rho \in L_0 \text{ s.t. } \rho - i\varepsilon_{1,2} \in$  $\}$

with generators $\{(\ell, \lambda), (\ell-1, \lambda), \ldots, (0, \lambda), (0, \pm\mu)\}$

$\check{\sigma}_2 = \check{\sigma}_0 \cap \{I \mid <I, x_2> \geq 0\} \cong \{(i, \rho) \mid i \geq 0, \rho \in L_0 \text{ s.t. } \rho - i\varepsilon_{1,4} \in$  $\}$

with generators $\{(s, \sigma), (s-1, \sigma), \ldots, (0, \sigma), (0, \pm\nu)\}$

$\check{\sigma}_3 = \check{\sigma}_0 \cap \{I \mid <I, x_3> \geq 0\} \cong \{(i, \rho) \mid i \geq 0, \rho \in L_0 \text{ s.t. } \rho - i\varepsilon_{3,4} \in$  $\}$

with generators $\{(\ell, -\lambda), (\ell-1, -\lambda), \ldots, (0, -\lambda), (0, \pm\mu)\}$

$\check{\sigma}_4 = \sigma_0 \cap \{I \mid <I, x_4> \geq 0\} \cong \{(i, \rho) \mid i \geq 0, \rho \in L_0 \text{ s.t. } \rho - i\varepsilon_{2,3} \in$  $\}$

with generators $\{(s, -\sigma), (s-1, -\sigma), \ldots, (0, -\sigma)(0, \pm\nu)\}$

Define the semigroups

$\check{\sigma}_{12} = \check{\sigma}_1 \cap \check{\sigma}_2$, $\check{\sigma}_{23} = \check{\sigma}_2 \cap \check{\sigma}_3$, $\check{\sigma}_{34} = \check{\sigma}_3 \cap \check{\sigma}_4$ and $\check{\sigma}_{41} = \check{\sigma}_4 \cap \check{\sigma}_1$. Then

$\check{\sigma}_{12} = \check{\sigma}_0 \cap \{I \mid <I, \lambda_1 x_1 + \lambda_2 x_2> \geq 0 \quad \lambda_1, \lambda_2 \geq 0\} \cong \{(i, \rho) \mid i \geq 0, \rho \in L_0, \rho - i\varepsilon_1 \in$  $\}$

$\check{\sigma}_{23} = \check{\sigma}_0 \cap \{I \mid <I, \lambda_2 x_2 + \lambda_3 x_3> \geq 0 \quad \lambda_2, \lambda_3 \geq 0\} \cong \{(i, \rho) \mid i \geq 0, \rho \in L_0, \rho - i\varepsilon_4 \in$  $\}$

$\check{\sigma}_{34} = \check{\sigma}_0 \cap \{I \mid <I, \lambda_3 x_3 + \lambda_4 x_4> \geq 0 \quad \lambda_3, \lambda_4 \geq 0\} \cong \{(i, \rho) \mid i \geq 0, \rho \in L_0, \rho - i\varepsilon_3 \in$  $\}$

$\check{\sigma}_{41} = \check{\sigma}_0 \cap \{I \mid <I, \lambda_4 x_4 + \lambda_1 x_1> \geq 0 \quad \lambda_4, \lambda_1 \geq 0\} \cong \{(i, \rho) \mid i \geq 0, \rho \in L_o, \rho - i\varepsilon_2 \in$  $\}$

Let $\partial_i = \partial\sigma_i$ and $\partial_{ij} = \partial_i \cap \partial_j$. Then

---

* $\rho - i\varepsilon_{1,2}$ stands for $\rho - i\varepsilon_1$ or $\rho - i\varepsilon_2$; the little figure  stands for $\{\alpha = (\alpha_1, \alpha_2) \mid \alpha \in Z^2, |\alpha_1| \geq 0\}$, 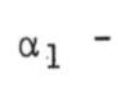 for $\{\alpha \mid \alpha \in Z^2, \alpha_1 - \alpha_2 \geq 0\}$, etc...; the other ones used in the text are self-explanatory.

** $<I, x>$ denotes the usual scalar product in $\mathbb{R}^3$.

$\partial_1 \cong \{(i, \rho) | i \geq 0,\ \rho \in L_0;\ \rho - i\varepsilon_{1,2} \in \ast\ \}$ generated by $\{(\ell, \lambda), (0, \pm\mu)\}$

$\partial_2 = \{(i, \rho) | i \geq 0,\ \rho \in L_0;\ \rho - i\varepsilon_{1,4} \in \ast\ \}$ generated by $\{(s, \sigma), (0, \pm\nu)\}$

$\partial_3 = \{(i, \rho) | i \geq 0,\ \rho \in L_0;\ \rho - i\varepsilon_{3,4} \in \ast\ \}$ generated by $\{(\ell, -\lambda), (0, \pm\mu)\}$

$\partial_4 = \{(i, \rho) | i \geq 0,\ \rho \in L_0;\ \rho - i\varepsilon_{2,3} \in \ast\ \}$ generated by $\{(s, -\sigma), (0, \pm\nu)\}$

and

$\partial_{12} = \{(i, \rho) | i \geq 0,\ \rho \in L_0, \rho - i\varepsilon_1 = 0\}$ generated by (a_1, a).

$\partial_{23} = \{(i, \rho) | i \geq 0,\ \rho \in L_0, \rho - i\varepsilon_4 = 0\}$

$\partial_{34} = \{(i, \rho) | i \geq 0,\ \rho \in L_0, \rho - i\varepsilon_3 = 0\}$ generated by $(-a_1, -a)$

$\partial_{41} = \{(i, \rho) | i \geq 0,\ \rho \in L_0, \rho - i\varepsilon_2 = 0\}$

Every $I = (i, j, k) \cong (i, \rho) \in Z \times L_0$ maps into a monomial $x^i z_1^j z_2^k = x^i z^\rho$.
Every rational subcone σ of Σ containing x_0

$$\sigma = \{\lambda_0 x_0 + \sum_{j=1}^{p} \lambda_j y_j;\ \lambda_j \in Q,\ \lambda_j \geq 0\}, \qquad y_j \in \Sigma$$

defines, by duality, the semigroup

$$\check{\sigma} = \{I | I \in Z^3, \langle I, x\rangle \geq 0, \quad \forall\ x \in \sigma\};$$

this semigroup of characters I, maps into the polynomial ring

$$\mathbb{C}[\check{\sigma}] \equiv \mathbb{C}[\ldots, x^i z^\rho, \ldots]_{(i,\rho)\in\check{\sigma}} \subset \mathbb{C}[x, z_1, z_1^{-1}, z_2, z_2^{-1}],$$

and therefore, since $\mathbb{C}[\check{\sigma}] \subset \mathbb{C}[\check{\sigma}']$ implies $\operatorname{spec}\mathbb{C}[\check{\sigma}] \supset \operatorname{spec}\mathbb{C}[\check{\sigma}']$,

$$\Omega \equiv \operatorname{spec}\mathbb{C}[x, z_1, z_1^{-1}, z_2, z_2^{-1}] = \mathbb{C} \times \mathbb{C}^* \times \mathbb{C}^* \supset \operatorname{spec}\mathbb{C}[\check{\sigma}].$$

If $(i_j, \rho_j)(1 \leq j \leq k)$ is a set of generators* for the semigroup $\check{\sigma}$, then

---

*Gordan's Lemma asserts that such a semigroup is always finitely generated [8].

spec $\mathbb{C}[\check{\sigma}]$ defines the variety

$$V = \{(t_1, \ldots, t_k) \mid \prod_{j=1}^{k} t_j^{\alpha_j} = \prod_{j=1}^{k} t_j^{\alpha'_j} \quad \text{whenever} \quad \Sigma\alpha_j(i_j, \rho_j) = \Sigma\alpha'_j(i_j, \rho_j)\}$$

The ring of rational holomorphic function on V is given by $\mathbb{C}[\check{\sigma}]$. Observe also that spec $\mathbb{C}[\check{\sigma}]$ is invariant under the obvious action of $(a, b, c) \in \mathbb{C}^* \times \mathbb{C}^* \times \mathbb{C}^*$

$$(a, b, c)x^i z_1^j z_2^k = a^i b^j c^k \; x^i z_1^i z_2^k$$

In particular, consider the following four embeddings of the torus Ω

$\Omega \subset U_1 \equiv \text{spec } \mathbb{C}[\check{\sigma}_{12}]$

$\Omega \subset U_\ell \equiv \text{spec } \mathbb{C}[\check{\sigma}_{23}]$

$\Omega \subset U_3 \equiv \text{spec } \mathbb{C}[\check{\sigma}_{34}]$

$\Omega \subset U_4 \equiv \text{spec } \mathbb{C}[\check{\sigma}_{41}]$

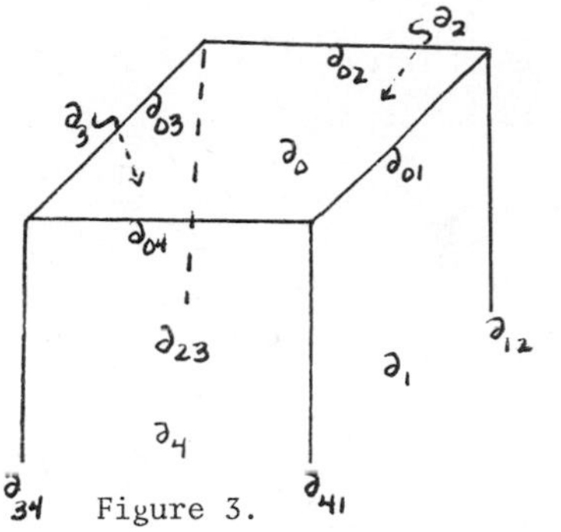

Figure 3.

and the toroidal cradle $\mathcal{U} = \bigcup_{i=1}^{4} \mathcal{U}_i$. The four semigroups $\sigma_{j,j+1} (i \leq j \leq 4)$ with their faces and edges are pictured in Figure 3. Their faces and edges correspond to subvarieties; let D_j in $\mathcal{U}_j \cap \mathcal{U}_{j-1}$ be defined by spec $\mathbb{C}[\partial_j]$, $D_{j,j+1}$ (j mod 4) by $\mathbb{C}[\partial_{j,j+1}]$ and $D_{0,j}$ $(j = 1, \ldots, 4)$ by $\mathbb{C}[\partial_{0j}]$; for instance

$$\text{spec } \mathbb{C}[\partial_1] = \text{spec } \mathbb{C}[x^\ell z^\lambda, z^{\pm\mu}],$$

$$\text{spec } \mathbb{C}[\partial_{12}] = \text{spec } \mathbb{C}[x^{a_1} z^a],$$

$$\text{spec } \mathbb{C}[\partial_{01}] = \text{spec } \mathbb{C}[z^\mu].$$

§2. The Spectral Surface Defined on the Toroidal Cradle.

Associate now to every point in $\mathcal{U} = \bigcup_{i=1}^{4} \mathcal{U}_i$ the space F_{α_0} of common eigenvectors of $T^i S^\rho (i \in Z,\ \rho \in L_0,\ i \geq 0)$, choosing the origin at $\alpha_0 \in Z^2$; compute $\dim F_{\alpha_0}$.

i) To every point in Ω, using the assumption, there corresponds at most one eigenvector $(e_\alpha,\ \alpha \in Z^2)$. Moreover $\dim(F_\alpha) = 1$ at a point of Ω if and only if this point belongs to the surface X defined by the algebraic equation above. In Ω, the spaces F_α are independent of α.

ii) Consider now the spectral problem along the subvariety D_1 in $\mathcal{U}_1 \cap \mathcal{U}_2$; it is defined by

$$x^i z^\rho = 0 \quad \text{for} \quad (i,\ \rho) \in \gamma_1,\ (i,\ \rho) \notin \partial_1.$$

In view of the generators, the whole spectral problem is now summarized by the set of equations

$$\begin{aligned} (T^i S^\lambda e)_\alpha &= x^i z^\lambda e_\alpha = 0 \quad 0 \leq i \leq \ell-1, \quad |\alpha| \geq |\alpha_0|. \\ (T^\ell S^\lambda e)_\alpha &= x^\ell z^\lambda e_\alpha \quad \text{with} \quad x^\ell z^\lambda \neq 0 \\ (S^{\pm\mu} e)_\alpha &= z^{\pm\mu} e_\alpha . \end{aligned}$$

We now prove that $e_\alpha = 0$ for all $\alpha \in Z^2$ such that $|\alpha| > |\alpha_0|$. For the sake of simplicity assume $\alpha_0 = 0$. From the fact that $(S^\lambda e)_\alpha = 0$ for all $|\alpha| \geq 0$, it follows that $e_\alpha = 0$ for all $|\alpha| \geq \ell$. Moreover, let $0 \leq i_0 < \ell$ be the smallest integer such that $e_\alpha = 0$ for all $|\alpha| > i_0$; this is to say that some $e_\alpha (|\alpha| = i_0)$ does not vanish. Then for $|\alpha| = 0$.

$$(T^{\ell - i_0} S^\lambda e)_\alpha = 0.$$

Let $\delta = \varepsilon_1 - \varepsilon_2 = -\varepsilon_3 + \varepsilon_4$; define

$$\mathscr{T}_\alpha \equiv \begin{bmatrix} T_{\alpha,\alpha+\varepsilon_3} & T_{\alpha,\alpha+\varepsilon_4} & 0 & 0 & 0 \\ 0 & T_{\alpha+\delta,\alpha+\varepsilon_3+\delta} & T_{\alpha+\delta,\alpha+\varepsilon_4+\delta} & 0 & \\ 0 & 0 & T_{\alpha+2\delta,\alpha+\varepsilon_3+2\delta} & T_{\alpha+2\delta,\alpha+\varepsilon_4+2\delta} & \\ & & & \ddots & T_{\alpha+(m-2)\delta,\alpha+\varepsilon_4+(m-2)\delta} \\ z^{-\mu}T_{\alpha+(m-1)\delta,\alpha+\varepsilon_4+(m-1)\delta} & & & 0 & T_{\alpha+(m-1)\delta,\alpha+\varepsilon_3+(m-1)\delta} \end{bmatrix}$$

and

$$\bar{e}^{\,\alpha} \equiv (e_\alpha,\ e_{\alpha+\delta},\ e_{\alpha+2\delta},\ \ldots,\ e_{\alpha+(m-1)\delta})^T.$$

Then, since $e_\alpha = 0$ for $|\alpha| > i_0$, $(S^\lambda T^{\ell-i_0} e)_\alpha$, with $|\alpha| = 0$, is a linear combination of e_α's with $|\alpha| = i_0$ only; therefore $S^\mu e_\alpha = z^\mu e_\alpha (|\alpha| = i)$ and $(S^\lambda T^{\ell-i} e)_\alpha = 0$ $(|\alpha| = 0)$ amount to

$$(1)\quad \mathscr{T}_\alpha \mathscr{T}_{\alpha+\varepsilon_3} \mathscr{T}_{\alpha+2\varepsilon_3} \cdots \mathscr{T}_{\alpha+(\ell-i_0-1)\varepsilon_3} \bar{e}^{\,\alpha+(\ell-i_0)\varepsilon_3+\lambda} = 0 \quad \text{for} \quad |\alpha| = 0$$

with vector $\bar{e} \neq 0$. But, since

$$(S^\lambda T^\ell e)_\alpha = x^\ell z^\lambda e_\alpha \neq 0 \quad \text{for some} \quad |\alpha| = i_0,$$

we have that

$$\mathscr{T}_\alpha \mathscr{T}_{\alpha+\varepsilon_3} \cdots \mathscr{T}_{\alpha+(\ell-1)\varepsilon_3} \bar{e}^{\,\alpha+\ell\varepsilon_3+\lambda} \neq 0 \quad \text{for} \quad |\alpha| = i$$

or, putting $\alpha = \alpha' - i_0\varepsilon_3$ with $|\alpha'| = 0$,

$$\mathscr{T}_{\alpha'-i_0\varepsilon_3} \mathscr{T}_{\alpha'-(i_0-1)\varepsilon_3} \cdots \mathscr{T}_{\alpha'-\varepsilon_3} (\mathscr{T}_{\alpha'} \mathscr{T}_{\alpha'+\varepsilon_3} \cdots \mathscr{T}_{\alpha'+(\ell-i_0-1)\varepsilon_3} \bar{e}^{\,\alpha'+(\ell-i_0)\varepsilon_3+\lambda}) \neq 0 \qquad |\alpha'| = 0$$

which clearly contradicts (1). Therefore $e_\alpha = 0$ for all $|\alpha| > 0$.

It follows that $(S^\lambda T^\ell e)_\alpha$ with $|\alpha| = 0$ is the m-periodic one-dimensional difference operator T^I acting on $\bar{e}^{\alpha}$ with $|\alpha| = 0$. By assumption this difference operator has no two-dimensional eigenspaces. Therefore along the subvariety D_1,

$$\dim F_{\alpha_0} \leq 1.$$

The locus points on D_1 where $\dim F_{\alpha_0} = 1$, is the curve associated with the commuting operators T^I and S^{μ}.

iii) Consider now the spectral problem along the orbit in $\mathscr{U}_1$ defined by

$$D_{12}: x^i z^{\rho} = 0 \quad \text{for} \quad (i, \rho) \in \check{\delta}_{12}, \ (i, \rho) \notin \partial_{12}$$

The locus of points on D_{12} where $\dim F_{\alpha_0} \geq 1$, will be given by its intersection with the curve defined by the one-dimensional operators T^I and S^{μ}. The finite matrix version reads as follows (cf. Theorem 3, Part I).

$$\mathscr{T}_0 \mathscr{T}_{\varepsilon_3} \cdots \mathscr{T}_{(\ell-1)\varepsilon_3} \bar{e}^{\lambda-\ell\varepsilon_1} \equiv \bar{e}^{\,0} = x^{\ell} z^{\lambda} \bar{e}^{\,0}$$

where $\mathscr{S}$ is the difference operator T^I, of period m, somewhat reshuffled, so as to have support $(-\lambda_2, \lambda_1)$. Since this difference operator is <u>regular</u> by assumption, the curve intersects D_{12} in g.c.d.$(\lambda_2, m) = b_2$ points $p_0, \ldots, p_{b_2-1}$ and D_{41} in g.c.d.(λ_1, m) points. To show that at these points $\dim F_{\alpha_0} = 1$, we need to unwind the definition of regularity. To begin with, the outer left subdiagonal elements of $\mathscr{S}$ are given by

$$\Pi_{\alpha} \equiv T_{\alpha-(\ell-1)\varepsilon_1, \alpha-\ell\varepsilon_1} T_{\alpha-(\ell-2)\varepsilon_1, \alpha-(\ell-1)\varepsilon_1} \cdots T_{\alpha, \alpha-\varepsilon_1},$$

with $\alpha = 0, \delta, 2\delta, \ldots, (m-1)\delta$. Regularity means that the elements

$$\begin{aligned} P_i &\equiv \Pi_{i\delta}\Pi_{(i+\lambda_2)\delta}\Pi_{(i+2\lambda_2)\delta} \cdots \Pi_{(i+(\frac{m}{b_2}-1)\lambda_2)\delta} \\ &= T_{\varepsilon_1+\alpha,\alpha} T_{2\varepsilon_1+\alpha,\varepsilon_1+\alpha} \cdots T_{a_1\varepsilon_1+\alpha,(a_1-1)\varepsilon_1+\alpha} \end{aligned}$$

with $0 \leq i \leq b_2 - 1$, are all different from each other; they are also the outer left subdiagonal elements of $\mathscr{S}^{m/b_2}$. From the one-dimensional theory, it also follows that the local parameter

$$(x^{\ell}y^{\lambda})^{m/b_2}(z^{\mu})^{-\beta} = x^{a_1}z^{a}$$

assumes the value P_i at the points $p_i\,(0 \leq i < b\ )$.

Moreover, since

$$\begin{aligned} (S^{-\mu}e)_{\alpha} &= 0 \quad \text{with} \quad \alpha_1 \geq \alpha_2 \geq 0 \\ (S^{\nu}e)_{\alpha} &= 0 \quad \text{with} \quad \alpha_1 \geq -\alpha_2 \geq 0 \\ (S^{a}e)_{\alpha} &= 0 \quad \text{with} \quad \alpha_1 \geq |\alpha_2|, \end{aligned}$$

we have that $e_{\alpha} = 0$ in the shaded region of Figure 4.

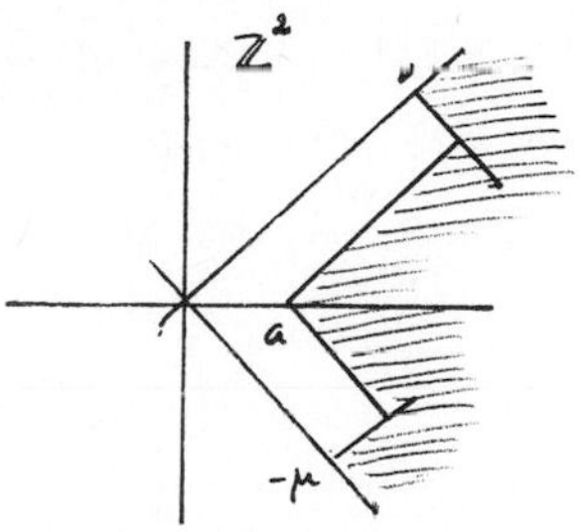

Figure 4.

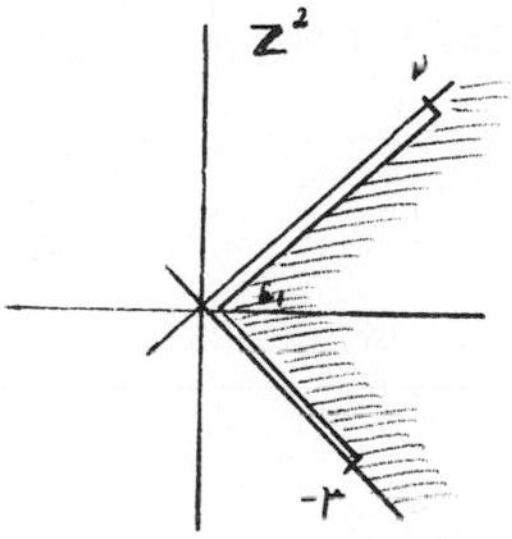

Figure 5.

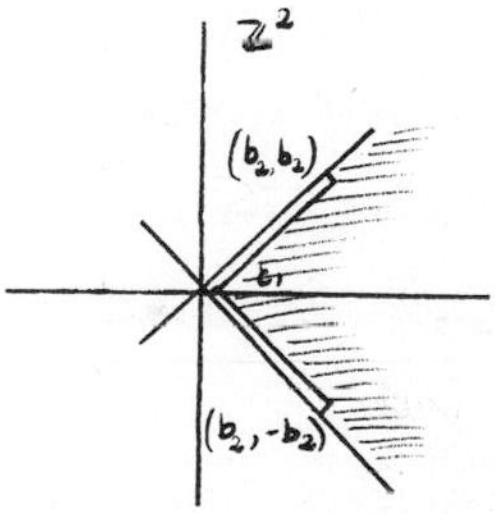

Figure 6.

The fact that $(S^{a}T^{i}e)_{\alpha} = 0$ for $\alpha_1 = |\alpha_2|$ $(1 \leq i \leq a_1)$ shows that $e_{\alpha} = 0$ for all α in the shaded region of Figure 5, except possibly for a finite number of α's such that $\alpha_1 = |\alpha_2|$. This is done by choosing $\alpha_1 = |\alpha_2|$ as large as possible and $i(1 \leq i < a\ )$ such that $(S^{a}T^{i}e)_{\alpha}$

involves one e_β only with β in the unshaded strip of Figure 4; then $(S^aT^ie)_\alpha = Ce_\beta = 0$, where $C \neq 0$; so, e_β must vanish. In this fashion all the e_β's with β in the strip can be made to vanish the one after the other, except possibly for e_α with $\alpha_1 = |\alpha_2| \leq \max(m, n)$.

Since, also $(S^{b}T^{b_1-b_2}e)_\alpha = 0$ for $\alpha_1 = \alpha_2 \geq 0$, we conclude that $e_\alpha = 0$ for $\alpha_1 = \alpha_2 \geq b_2$. Let $c \equiv -b + qa$, where q is the smallest integer such that $|c| \geq 0$. Then since $(S^cT^{|c|}e)_\alpha = 0$ for $\alpha_1 = -\alpha_2 \geq 0$, $e_\alpha = 0$ for $\alpha_1 = -\alpha_2 \geq b_2$.

Besides notice that for $\alpha_1 = |\alpha_2| = i$, $0 \leq i \leq b_2 - 1$,

$$(T^{a_1}S^ae)_\alpha = T_{\alpha,\alpha-\varepsilon_1}T_{\alpha-\varepsilon_1,\alpha-2\varepsilon_1}\cdots T_{\alpha-(a_1-1)\varepsilon_1,\alpha-a_1\varepsilon_1}e_\alpha + \text{linear combination of } e_{j(\varepsilon_1 \pm \varepsilon_2)} \text{ with } i < j < b_2$$

$$= x^{a_1}z^ae_\alpha \qquad \text{(where } \pm = \text{sign } \alpha_2).$$

Now at the point $p_j(0 \leq j \leq b_2 - 1)$, the local parameter $x^{a_1}z^a$ takes on the value P_j. The coefficient of e_α in $(T^{a_1}S^ae)_\alpha$ is P_j for $i \neq j$. Therefore $e_\alpha = 0$ for $\alpha = i(\varepsilon_1 + \varepsilon_2)$ with $i > j$ and $\alpha = i(\varepsilon_1 - \varepsilon_2)$ with $i < j - b_2$. At $\alpha = j(\varepsilon_1 + \varepsilon_2)$, e_α can be chosen $\neq 0$. Then e_β for $\beta = i(\varepsilon_1 + \varepsilon_2)$, $0 \leq i \leq j - 1$ is the unique solution of a triangular system of equations with determinant

$$\prod_{0 \leq i < j} (P_i - P_j) \neq 0;$$

the same holds for e_β with $\beta = i(\varepsilon_1 - \varepsilon_2)$, $j - b_2 < i \leq 0$. Therefore $\dim F_{\alpha_0} = 1$ at each of the points $p_i(0 \leq i \leq b_2 - 1)$.

iv) D_{01} is a subvariety of $\mathcal{U}_1 \cap \mathcal{U}_2$ defined by the equations

$$x^iz^\lambda = 0 \qquad 0 \leq i \leq \ell.$$

The spectral problem reads now

$$(T^i S^\lambda e)_\alpha = 0 \qquad 0 \leq i \leq \ell, \qquad |\alpha| \geq 0$$

and

$$(S^\mu e)_\alpha = z^\mu e_\alpha \quad \text{with} \quad z^\mu \neq 0.$$

From $(S^\lambda e)_\alpha = 0$ with $|\alpha| = 0$, it follows that $e_\alpha = 0$ for $|\alpha| \geq \ell$.
Then $(T^i S^\lambda e)_\alpha = 0$ with $|\alpha| = 0$ implies

$$\mathcal{T}_\alpha \mathcal{T}_{\alpha+\varepsilon_3} \cdots, \mathcal{T}_{\alpha+(i-1)\varepsilon_3} \bar{e}^{\alpha+i\varepsilon_3} = 0 \qquad \alpha = \lambda, \quad 1 \leq i \leq \ell.$$

If $\det(\mathcal{T}_{\alpha+(i-1)\varepsilon_3}) \neq 0$ for $1 \leq i \leq \ell$, then $e_\alpha \equiv 0$ for all $|\alpha| \geq 0$.
The only way to obtain a nonzero solution is to assume that
$\det(\mathcal{T}_{\alpha+(i-1)\varepsilon_3}) = 0$ for at least one value of i with $1 \leq i \leq \ell$; let
i_0 be this value. This imposes the value of y^μ:

$$y^\mu = \frac{\prod_{i=0}^{m-1} T_{\alpha+(i_0-1)\varepsilon_3+i\delta,\alpha+(i_0-1)\varepsilon_3+i\delta+\varepsilon_4}}{\prod_{i=0}^{m-1} T_{\alpha+(i_0-1)\varepsilon_3+i\delta,\alpha+(i_0-1)\varepsilon_3+i\delta+\varepsilon_3}} \equiv (-1)^m \eta^I_{\alpha+(i_0-1)\varepsilon_3}, \qquad \alpha = \lambda.$$

By assumption the ℓ quantities $\eta^I_{\alpha+j\varepsilon_3}$ $(\alpha = \lambda,\ 0 \leq j \leq \ell - 1)$ are
different from each other. Therefore $\bar{e}^{\alpha+i_0\varepsilon_3}$ is the only solution
modulo a multiplicative constant. Then $(T^{i_0+1} S^\lambda)_\alpha = 0$ with $|\alpha| = 0$
implies

$$\mathcal{T}_{\alpha+(i_0-1)\varepsilon_3} f + \mathcal{T}_{\alpha+(i_0-1)\varepsilon_3} \mathcal{T}_{\alpha+i_0\varepsilon_3} \bar{e}^{\alpha+(i_0+1)\varepsilon_3} = 0, \quad \alpha = \lambda$$

for some vector f. Since $\operatorname{rank}(\mathcal{T}_{\alpha+i_0\varepsilon_3}) = m$ and $\operatorname{rank}(\mathcal{T}_{\alpha+(i_0-1)\varepsilon_3}) = m - 1$,
the space of solutions $\bar{e}^{\alpha+(i_0+1)\varepsilon_3}$ of this equation is two dimensional.
Next, the equation $(T^{i_0+2} S^\lambda)_\alpha = 0$ with $|\alpha| = 0$ breaks up as follows:

$$\mathscr{T}_{\alpha+(i_0-1)\varepsilon_3} f' + \mathscr{T}_{\alpha+(i_0-1)\varepsilon_3}\mathscr{T}_{\alpha+i_0\varepsilon_3}\mathscr{T}_{\alpha+(i_0+1)\varepsilon_3} e^{\alpha+(i_0+2)\varepsilon_3} = 0$$

for some vector f' depending on the solution of previous step; the solution of this system of equations exists but is not unique; one proceeds in the same fashion for the other e's. Therefore the surface X intersects D_{01} in the ℓ distinct points $y^\mu = (-1)^m \eta_{\alpha+i\varepsilon_3}$ $(1 \leq i \leq \ell)$ where $\dim F_{\alpha_0} \geq 1$ for all α_0, with equality for $\alpha_0 = (i-1)\varepsilon_1$ (mod δ and λ).

For every point of $X \setminus \bigcup_{i=1}^{4} (X \cap D_{0i})$ we have that $\dim F_\alpha = 1$ and for every point of $X \cap D_{0i}$ there is at least an α such that $\dim F_\alpha = 1$.

§3. The Divisor on X Defined by T.

First we examine the behavior of e_α around D_i; for the sake of this argument, choose D_1. Let $w = x^{\ell-1}z^\lambda$; then $w = 0$ at D_1. Now

$$(2) \qquad (T^{\ell-1}S^\lambda e)_\alpha = we_\alpha \quad \text{for} \quad |\alpha| = 0.$$

In a small enough neighborhood of D_1, the functions e_α with $|\alpha| = 0$ are nonzero while at D_1, $e_\alpha = 0$ for $|\alpha| > 0$. From this fact and (2) it follows that the e_α's with $|\alpha| \geq 1$ must have an expansion in powers of w starting with $w^j (j \geq 1)$ with $j = 1$ for some of them. Choosing now the origin at a point of $|\beta| = 1$ and solving

$$(T^{\ell-1}S^\lambda e')_\beta = we'_\beta \quad \text{for} \quad |\beta| = 1,$$

would lead to the same conclusion as before, but now with regard to $|\beta| = 1$. However, off D_1, the solutions e_α are independent of the origin α_0 of the lattice, modulo some scalar. Therefore e_α is proportional to e'_α and again because of (2) e_α/e'_α has an expansion in w starting with w. The conclusion is that

$$(e_\alpha)_\infty = \mathscr{D}_\alpha \equiv (\alpha_1 + \alpha_2)D_1 + (\alpha_1 - \alpha_2)D_2 - (\alpha_1 + \alpha_2)D_3 - (\alpha_1 - \alpha_2)D_4.$$

Let $\mathscr{D}$ be the minimal effective divisor such that

$$(e_\alpha) \geq -\mathscr{D} + \mathscr{D}_\alpha .$$

Let $\mathscr{L}_\alpha$ be the line bundle defined by $\mathscr{D} - \mathscr{D}_\alpha$. Then in order to compute $\dim H^i(\mathscr{O}(\mathscr{L}_\alpha))$, cover X with the four affines $\bar{\mathscr{U}}_i$ $X \cap \mathscr{U}_i$, each one equipped with its space of holomorphic functions. The set of holomorphic functions $H^0(\mathscr{O}(\bar{\mathscr{U}}_i))$ on $\bar{\mathscr{U}}_i$ is a subspace of the linear span of some of the e_α's as follows:

$H^0(\mathscr{O}(\bar{\mathscr{U}}_1)) = \mathbb{C}[\check{\sigma}_{12}] \subset$ linear span of e_α with $\alpha_1 + \alpha_2 \geq 0$ and $\alpha_1 - \alpha_2 \geq 0$.

$H^0(\mathscr{O}(\bar{\mathscr{U}}_2)) = \mathbb{C}[\check{\sigma}_{23}] \subset$ linear span of e_α with $\alpha_1 - \alpha_2 \geq 0$ and $-\alpha_1 - \alpha_2 \geq 0$.

$H^0(\mathscr{O}(\mathscr{U}_3)) = \mathbb{C}[\check{\sigma}_{34}] \subset$ linear span of e_α with $-\alpha_1 - \alpha_2 \geq 0$ and $\alpha_2 - \alpha_1 \geq 0$.

$H^0(\mathscr{O}(\mathscr{U}_4)) = \mathbb{C}[\check{\sigma}_{41}] \subset$ linear span of e_α with $\alpha_2 - \alpha_1 \geq 0$ and $\alpha_1 + \alpha_2 \geq 0$.

To see this, assume $e_0 = 1$; then every holomorphic function on $\bar{\mathscr{U}}_1$ is given by a linear combination of $x^i z^\rho$ with $\rho - i\varepsilon_1 \in$  ; but $x^i z^\rho = x^i z^\rho e_0 = (T^i S^\rho e)_0$ is a linear combination of e_α's with $\alpha \in$  . Hence this function will never be holomorphic on the other patches $\mathscr{U}_i (i \neq 1)$, unless it is constant. Therefore $\dim H^0(\mathscr{O}(\mathscr{L}_\alpha)) = 1$. To compute $\dim H^1(\mathscr{O}(\mathscr{L}_\alpha))$, find holomorphic functions ϕ_{ij} on $\mathscr{U}_i \cap \mathscr{U}_j$ $(1 \leq i < j \leq 4)$ such that on $\bar{\mathscr{U}}_i \cap \bar{\mathscr{U}}_j \cap \bar{\mathscr{U}}_k$ $(1 \leq i \leq j \leq k \leq 4)$, $\phi_{ij} + \phi_{jk} - \phi_{ik} = 0$, modulo the ϕ_{ij}'s such that $\phi_{ij} = \phi_i - \phi_j$ on $\bar{\mathscr{U}}_{ij}$. Then the spaces of holomorphic functions $H^0(\mathscr{O}(\mathscr{U}_i \cap \mathscr{U}_j))$ on $\mathscr{U}_i \cap \mathscr{U}_j$ are given by

$H^0(\mathscr{O}(\mathscr{U}_1 \cap \mathscr{U}_2)) = \mathbb{C}[\check{\sigma}_2] \subset$ linear span of e_α, with $\alpha_1 - \alpha_2 \geq 0$

$H^0(\mathscr{O}(\mathscr{U}_1 \cap \mathscr{U}_3)) = H(\ (\Omega)) = \mathbb{C}[x, z_1, z_1^{-1}, z_2, z_2^{-1}] \subset$ linear span of all e_α's.

$$H^o(\mathcal{O}(\mathcal{U}_1 \cap \mathcal{U}_4)) = \mathbb{C}[\sigma_1] \quad \text{linear span of } e_\alpha, \text{ with } \alpha_1 + \alpha_2 \geq 0$$
$$H^o(\mathcal{O}(\mathcal{U}_2 \cap \mathcal{U}_3)) = \mathbb{C}[\sigma_3] \quad \text{linear span of } e_\alpha, \text{ with } \alpha_1 + \alpha_2 \leq 0$$
$$H^o(\mathcal{O}(\mathcal{U}_2 \cap \mathcal{U}_4)) = H^o(\mathcal{O}(\Omega)) = \mathbb{C}[x, z_1, z_1^{-1}, z_2, z_2^{-1}] \subset \text{linear span of all } e_\alpha\text{'s.}$$
$$H^o(\mathcal{O}(\mathcal{U}_3 \cap \mathcal{U}_4)) = \mathbb{C}[\sigma_4] \subset \text{linear span of } e_\alpha, \text{ with } \alpha_1 - \alpha_2 \leq 0.$$

Because of the independence of the e_α's, the only ϕ_{ij}, for which the relations above are satisfied, are given by $\phi_{ij} = c_{ij}$ = constant over each patch $\bar{\mathcal{U}}_i \cap \bar{\mathcal{U}}_j$. Therefore

$$\dim H^1(\mathcal{O}(\mathcal{L}_\alpha)) = \dim \check{H}^1(\bar{\mathcal{U}}_i), \mathbb{C}) = 0,$$

where $H^1(\bar{\mathcal{U}}_i), \mathbb{C})$ is complex Čech-cohomology over the $\bar{\mathcal{U}}_i$'s. Analogously.

$$\dim H^2(\mathcal{O}(\mathcal{L}_\alpha)) = 0.$$

§4. The Divisor on $\mathcal{U}$ Defined by X.

For the sake of later computation, the divisor X on $\mathcal{U}$ defined by the equation $P(x, z_1, z_1^{-1}, z_2, z_2^{-1}) = 0$ in Ω, will be replaced by an equivalent divisor on $\mathcal{U}$. The function $P(x, z_1, z_1^{-1}, z_2, z_2^{-2})$ is a meromorphic function on $\mathcal{U}$, whose zero is given by X and whose poles turn out to be given by $V \Sigma D_i$ where $V \equiv m\ell = a_1 b_2$ = area of the period paralleogram. This will now be shown. Since the rational function P has no poles in Ω, they must lie on the D_i's. Near D_1, consider the new variables $u \equiv z^\mu$, $v \equiv x^\ell z^\lambda$, and $w \equiv x^{\ell-1} z^\lambda$, so that $x = v/w$, $z^\lambda = w^\ell / v^{\ell-1}$ and $z^\mu = u$. Recall that $z_1 \equiv z^a$ and $z_2 \equiv z^b$. Then, according to the relations between the primitive periods (λ, μ) and (a, b),

$$z_1 = z^a = (z^\lambda)^{\beta'} (z^\mu)^{-\beta} = \frac{w^{\ell\beta'}}{v^{(\ell-1)\beta'}} u^{-\beta},$$
$$z_2 = z^b = (z^\lambda)^{-\alpha'} (z^\mu)^\alpha = \frac{w^{-\ell\alpha'}}{v^{-(\ell-1)\alpha'}} u^\alpha.$$

When $P(x, z_1, z_1^{-1}, z_2, z_2^{-1})$ is expressed in the u, v, w coordinates, the leading terms are

$$z_1^{-b_2} = \frac{w^{-\ell\beta' b_2}}{v^{-(\ell-1)\beta' b_2}} u^{\beta b_2} = w^{-\ell m} \cdot u^{\lambda_2} v^{(\ell-1)m}$$

$$z_1^{b_1} z_2^{-a_1} = \frac{w^{\ell\beta' b_1}}{v^{(\ell-1)\beta' b_1}} u^{-\beta b_1} \frac{w^{\ell\alpha' a_1}}{v^{(\ell-1)\alpha' a}} u^{-\alpha a_1} = w^{-\ell m} \cdot u^{-\lambda_1} v^{(\ell-1)m}$$

and

$$z^{a_1 b_2} = w^{-\ell m} v^{\ell m};$$

moreover the coefficient of $w^{-\ell m}$ in $P(x, z_1, z_1^{-1}, z_2, z_2^{-1})$ does not vanish identically. Therefore P has at each D_i a pole of order V, i.e.,

$$X \text{ is equivalent to } V \sum^{4} D_i .$$

§5. The Picard Variety of X.

The exact sequence of sheaves on X

$$0 \to \mathbb{Z} \to \mathcal{O}_X \to \mathcal{O}_X^* \to 0$$

implies the following exact sequence of cohomology:

$$\ldots \to H^1(X, \mathbb{Z}) \to H^1(X, \mathcal{O}_X) \to H^1(X, \mathcal{O}_X^*) \overset{c}{\to} H^2(X, \mathbb{Z}) \to \ldots ;$$

$H^1(X, \mathcal{O}_X^*)$ is the set of line bundles L (Picard group) and $c(L)$ is the Chern class of the bundle, i.e., the Poincare dual of a cycle in X; so c is a topological property of L. Consider now the line bundle $\mathcal{L}$ on X defined by $\mathcal{D}$ which in turn is defined by the difference operator T as described in §3. An infinitesimal isospectral deformation of T implies an infinitesimal deformation of the line bundle $\mathcal{L}$; in fact, it does not modify the topological type of the line bundle $\mathcal{L}$, so that $c(\mathcal{L})$ will be preserved under this isospectral deformation. So in order to show that T cannot be deformed isospectrally, it suffices to show that if $\mathcal{L}$ and $\mathcal{L}'$ have the same Chern class $c(\mathcal{L}) = c(\mathcal{L}')$, they are equal, i.e., the map c is injective; since c is a group homomorphism (in the sense $c(\mathcal{L}\otimes\mathcal{L}') = c(\mathcal{L}) + c(\mathcal{L}'))$ it suffices to show that $\ker c \equiv \mathrm{Pic}^{o}(X) = \{0\}$*. The exactness of the sequence above implies

$$\mathrm{Pic}^{o}(X) = \frac{H^1(X, \mathcal{O}_X)}{H^1(X, \mathbb{Z})}$$

and since $H^1(X, \mathbb{Z})$ is a discrete group, it suffices to show that $H^1(X, \mathcal{O}_X) = \{0\}$; let $H^1(\mathcal{O}_X)$ stand for $H^1(X, \mathcal{O}_X)$.

The variety X in $\mathcal{U}$ defines a line bundle χ; let χ^{-1} correspond to the divisor $-X$; the same symbol also denotes its sheaf of section: On open sets 0 of $\mathcal{U}$, define holomorphic functions f such that $(f|_0) \geq X \cap 0$. This sheaf is trivially embedded in $\mathcal{O}_U$ and every element $(0, f_0)$ of $\mathcal{O}_{\mathcal{U}}$ which comes from χ^{-1}, vanishes at X. Finally every holomorphic functoin on an open set of X can be extended to a holomorphic function on an open set of $\mathcal{U}$. Therefore

$$0 \to \chi^{-1} \to \mathcal{O}_{\mathcal{U}} \to \mathcal{O}_X \to 0$$

is an exact sequence; hence also the sequence

$$\text{(3)} \qquad \to H^1(\mathcal{O}_U) \to H^1(\mathcal{O}_X) \to H^2(\chi^{-1}) \to H^2(\mathcal{O}_U) \to$$

is exact. Finally remark that since the line bundle χ^{-1} is defined by the divisor $-X$ up to equivalence classes, we may as well consider the following

*Notice that $\mathrm{Pic}^{o}(X)$ is the Jacobi variety of X, when X is an algebraic curve.

equivalent sheaf: on open sets 0 of $\mathcal{U}$ define holomorphic functions f such that $(f|_0) \geq V \sum D_i \cap 0$ (see §4). Now we are in a position to show that $H^1(\mathcal{O}_X) = 0$, using this exact sequence. It would suffice to show that ϕ is injective and that $H^1(\mathcal{O}_{\mathcal{U}}) = 0$. We need to compute $H^i(\mathcal{O}_U)$ and $H^i(\chi^{-1})$. The same methods apply to both cases; so let us consider $H^i(\chi^{-1})$.

First introduce the following notation: For

$I = (i, j, k) \simeq (i, \rho) \in Z \times L_0$

$$\chi^I \equiv x^i z_1^j z_2^k = x^i z^\rho .$$

Since every holomorphic function is a linear combination of monomials χ^I, we have that

$$H^i(\chi^{-1}) = \bigoplus_I H^i(\chi^{-1})^I$$

where $H^i(\chi^{-1})^I$ is cohomology restricted to the monomial χ^I. A practical way to compute cohomology $H^i(\chi^{-1})^I$ is to use Čech-cohomology $H^i(\{0_j\}, \chi^{-1})^I$ for a locally finite covering of the space $\mathcal{U}$ by open sets 0_j, such that finite intersections of these open sets carry no (strictly positive) cohomology over the same sheaf (Leray's theorem, see Hartshorne [6], P. 215, or Griffith and Harris [4], Chapter 0, iii). They are chosen to be open affines $\mathcal{U}_\sigma = \operatorname{spec} \mathbb{C}[\check{\sigma}]$ coming from polyhedral cones of the type

$$\sigma = \{\lambda_0 x_0 + \lambda_1 y_1 + \lambda_2 y_2 | \lambda_i \geq 0, \quad \lambda_i \quad Q\}$$

where both y_1 and y_2 belong to one of the two-dimensional faces

$$\Sigma_i = \{\mu_0 x_0 + \mu_1 x_i + \mu_2 x_{i+1} | \mu_i \in Q, \Sigma\mu_i = 1, \mu_i \geq 0\}, \quad 1 \leq i \leq 4.$$

For open affines, it is known that

$$H^i(\mathcal{U}_\sigma, \chi^{-1})^I = \{0\} \quad \text{for} \quad i > 0.$$

(See Hartshorne [6] P. 215); it is a simple observation that

$$\begin{aligned} H^0(\mathcal{U}_\sigma, \chi^{-1})I &= \mathbb{C}\chi^I \quad \text{if} \quad (\chi^I)_{\mathcal{U}_\sigma} \geq V \,\Sigma\, D_j \cap \mathcal{U}_\sigma \\ &= 0 \quad \text{otherwise.} \end{aligned}$$

The latter condition amounts to the inequalities

$$\langle I, x_0 \rangle \geq 0$$

and

$$\langle I, y_k \rangle \geq \phi(y_k), \quad 1 \leq k \leq 2,$$

where ϕ is the function on $\bigcup_1^4 \Sigma_i$ which is linear on each Σ_i and such that

$$\phi(x_k) = V \quad 1 \leq k \leq 4$$

$$\phi(x_0) = 0.$$

To verify this statement, for example, when $y_1 = y_2 = x_i$ $(1 \leq i \leq 4)$, observe* that χ^{I_i}, has a simple zero along the subvariety D_i; therefore the condition

$$(\chi^I)_{\mathcal{U}_\sigma} \geq V \sum_1^4 D_j \cap \mathcal{U}_\sigma = V\, D_i \cap \mathcal{U}_\sigma \quad \text{for some} \quad 1 \leq i \leq 4$$

amounts to say that χ^I is holomorphic on $\mathcal{U}_\sigma$ and that

$$\chi^I (\chi^{I_i})^{-V}$$

is holomorphic on a sufficiently small neighborhood of $D_i \cap \mathcal{U}_\sigma$; this is

---

*Define $I_1 = (\ell - 1, \lambda)$, $I_2 = (s - 1, \sigma)$, $I_3 = (\ell - 1, -\lambda)$, and $I_4 = (s - 1, -\sigma)$.

expressed by the inequalities $<I, x_0> \geq 0$ and

$$<I - VI_i, x_i> \geq 0.$$

But since+ $<I_i, x_i> = 1$, they amount to $<I, x_0> \geq 0$ and

$$<I, x_i> \geq V\phi(x_i)$$

To carry out the Čech cohomology argument, cover each Σ_i with triangles (x_0, y_1, y_2) and consider the corresponding polyhedral cones σ and affines $\mathcal{U}_\sigma$. Any finite intersection of such affines (which is the affine obtained by intersection the corresponding triangles of Σ_i) is also an affine, with trivial positive cohomology by (4). The only affines $\mathcal{U}_\sigma$, for which $c \cdot \chi^I$ with $c \in \mathbb{C}^*$ is a global section, are those triangles $(x_0, y_1, y_2) \subset \cup \Sigma_i$ such that

$$(x_0, y_1, y_2) \subset A_I \equiv \{y | <I, y> \geq \phi(y), y \in \bigcup_1^4 \Sigma_i\}.$$

So, every Čech complex of complex multiples of χ^I with regard to the sheaf χ^{-1} corresponds to a Čech complex for the sheaf $\mathbb{C}$ on triangles of A_I; thus we have the isomorphism:

$$H^i(\mathcal{U}, \chi^{-1})^I \simeq H^i_{A_I}(\Sigma_i, \mathbb{C}).$$

Since $\cup\Sigma_i$ is homeomorphic to a disc, it is simply connected and connected; therefore $H^0(\cup\Sigma_i, \mathbb{C}) = \mathbb{C}$ and $H^1(\Sigma_i, \mathbb{C}) = 0$. Moreover

---

+Indeed $<I_1, x_1> = <(\ell - 1, \alpha, \beta), (-1, a_1, b_1 + b_2)>$

$$= -\ell + 1 + \alpha a_1 + \beta(b_1 + b_2)$$

$$= -\ell + 1 + \lambda_1 + \lambda_2$$

$$= 1.$$

$$H^0_{A_I}(\Sigma_i, \mathbb{C}) = \mathbb{C} \quad \text{if} \quad A_I = \cup\Sigma_i$$
$$= 0 \quad \text{otherwise.}$$

In fact, there is not I such that $A_I = \cup\Sigma_i$; if so it would imply that

$$<x_i, I> \geq V \qquad 1 \leq i \leq 4$$

which is a contradiction, whatever be the integer $V \geq 0$. Therefore according to Appendix 1, the following exact sequence holds

$$0 \to H^0(\cup\Sigma_i, \mathbb{C}) \to H^0(\cup\Sigma_i \setminus A_I, \mathbb{C}) \to H^1_{A_I}(\cup\Sigma_i, \mathbb{C}) \to 0,$$

so that

$$H^1(\mathcal{U}, \chi^{-1})^I \simeq H^1_{A_I}(\cup\Sigma_i, \mathbb{C}) = \frac{H^0(\cup\Sigma_i \setminus A_I, \mathbb{C})}{H^0(\cup\Sigma_i, \mathbb{C})}$$

is nontrivial, provided $\cup\Sigma_i \setminus A_I$ is not connected; the latter can only happen when I satisfies

$$<x_1, I> \geq V \quad \text{or} \quad <x_1, I> < V$$
$$<x_2, I> < V \quad \text{or} \quad <x_2, I> \geq V$$
$$<x_3, I> \geq V \quad \text{or} \quad <x_3, I> < V$$
$$<x_4, I> < V \quad \text{or} \quad <x_4, I> \geq V$$

Again this never occurs because, as a result of the first column,

(5)

$$V + i \leq <x_1, I> + i \leq ja_1 + k(b_1 + b_2) = -i - <x_3, I> \leq -V - i$$

is a contradiction, since $i \geq 0$. Therefore

$$H^1(\mathcal{U}, \chi^{-1}) = \{0\}.$$

All the arguments above can be adapted to compute $H^1(\mathcal{U}, \mathcal{O}_{\mathcal{U}})$: It suffices to replace V by 0. With this modification, no I can satisfy the inequalities above: (5), in which V is replaced by 0, can only be satisfied when $i = 0$; however that would contradict one of the two inequalities $<x_2, I> < 0$ or $<x_4, I> < 0$. Therefore

$$H^1(\mathcal{U}, \mathcal{O}_{\mathcal{U}}) = \{0\}.$$

Again, from the exact sequence in Appendix 1, since $H^1(\Sigma_i, \mathbb{C}) = H^2(\Sigma_i, \mathbb{C}) = 0$, the sequence

$$0 \to H^1(\cup\Sigma_i \setminus A_I, \mathbb{C}) \to H^2_{A_I}(\Sigma_i, \mathbb{C}) \to 0$$

is exact, so that

$$H^2(\mathcal{U}, \chi^{-1})^I \simeq H^2_{A_I}(\cup\Sigma_i, \mathbb{C}) = H^1(\cup\Sigma_i \setminus A_I, \mathbb{C})$$

would be different from zero provided $\cup\Sigma_i \setminus A_I$ would not be simply connected; this can only happen for those I such that

$$<I, x_j> < V.$$

Analogously,

$$\dim H^2(\mathcal{U}, \mathcal{O}_{\mathcal{U}})^I = 1 \quad \text{iff} \quad <I, x_j> < 0 \quad 1 \leq j \leq 4 \quad \text{and} \quad i \geq 0.$$

Therefore the spaces $H^2(\mathcal{U}, \chi^{-1})$ ($H^2(\mathcal{U}, \mathcal{O}_{\mathcal{U}})$ respectively) can be parametrized by the set of I's such that $i \geq 0$ and $<I, x_j> < V$ ($<I, x_j> < 0$ respectively) ($1 \leq j \leq 4$). The natural map

$$H^2(\mathcal{U}, \chi^{-1}) \overset{\phi}{\to} H^2(\mathcal{U}, \mathcal{O}_{\mathcal{U}})$$

induced by the exact sequence (3) amounts to multiplying the monomials $x^i z_1^j z_2^k$ in $H^2(\mathcal{U}, \chi^{-1})$ by x^N; this means: Add $(N, 0, 0)$ to I in $H^2(\mathcal{U}, \chi^{-1})$. Clearly

$$<I, x_j> < V \text{ if and only if } <I + (V, 0, 0), x_j> < 0 \quad 1 \leq j \leq 4.$$

So this map is injective. This finishes the proof of Mumford's theorem.

Appendix 1.

Consider a disc D^1 and a subset $A \subset D^1$. The following exact sequence holds:

$$0 \to H_A^0(D, \mathbb{C}) \to H^0(D, \mathbb{C}) \to H^0(D \setminus A, \mathbb{C}) \to H_A^1(D, \mathbb{C}) \to H^1(D, \mathbb{C}) \to$$
$$\to H^1(D \setminus A, \mathbb{C}) \to H_A^2(D, \mathbb{C}) \to H^2(D, \mathbb{C}) \to \ldots$$

where $H_A^i(D, \mathbb{C})$ is the ith cohomology space with regard to the sheaf $\mathbb{C}$ vanishing on $D \setminus A$. Then

$$H_A^0(D, \mathbb{C}) = \mathbb{C} \text{ if } A = D$$
$$= 0 \text{ otherwise.}$$

Appendix 2.

It is instructive to check that $\dim \mathrm{Jac}\, X = N - 1$ for nonsingular hyperelliptic curves with division points $NP - NQ \quad 0$ (as described in Part I), using this method. This will help to understand why Pic^0 is nontrivial in one dimension and trivial in two dimensions. Consider the vectors in $\mathbb{R}^2$

$$x_0 = (1, 0)$$
$$x_1 = (-1, N)$$
$$x_2 = (-1, -N)$$

and cover $\Omega = \mathbb{C} \times \mathbb{C}^* = \mathrm{Spec}\ \mathbb{C}[x, h, h^{-1}] = \mathrm{Spec}\ \mathbb{C}[\sigma_{x_0}]$ with the two affines

$$\mathcal{U}_1 = \mathrm{Spec}\ \mathbb{C}[\check{\sigma}_{x_1}], \quad \text{where } \check{\sigma}_{x_1} = \{I = (i, j) \mid \langle I, x_1 \rangle \geq 0\} \cap \check{\sigma}_{x_0}.$$
$$\mathcal{U}_2 = \mathrm{Spec}\ \mathbb{C}[\check{\sigma}_{x_2}], \quad \text{where } \check{\sigma}_{x_2} = \{I = (i, j) \mid \langle I, x_2 \rangle \geq 0\} \cap \check{\sigma}_{x_0}.$$

Consider in $\mathcal{U}_1$ the subvariety D_1 defined by

$$x^i h = 0 \quad \text{for} \quad 0 \leq i < N \quad \text{with coordinate} \quad x^N h$$

and in $\mathcal{U}_2$ the subvariety D_2 defined by

$$x^i h^{-1} = 0 \quad \text{for} \quad 0 \leq i < N \quad \text{with coordinate} \quad x^N h^{-1}.$$

Define

$$\Sigma_i = \{\lambda x_0 + (1 - \lambda) x_i \mid \lambda \in Q, \quad 0 \leq \lambda \leq 1\} \quad \text{for} \quad 1 \leq i \leq 2.$$

The curve

$$A(h + h^{-1}) - R(z) = 0$$

can now be embedded in $\mathcal{U} \equiv \mathcal{U}_1 \cup \mathcal{U}_2$; it intersects D_1 and D_2 in the points $hz^N = A$ and $h^{-1}z^N = A$ respectively. This divisor in $\mathcal{U}$ is equivalent to $ND_1 + ND_2$. Consider the sheaf χ^{-1} associated to the line bundle $-ND_1 - ND_2$ and the exact sequence of cohomology

$$\to H^0(\mathcal{O}_X) \to H^1(\chi^{-1}) \to H^1(\mathcal{O}_{\mathcal{U}}) \to H^1(\mathcal{O}_X) \to H^2(\chi^{-1}) \to$$

Then, using the same arguments as before,

$$(6) \quad g = \dim \mathrm{Jac}(X) = \dim \mathrm{Pic}^o(X) = \dim H^1(\mathcal{O}_X) = \dim \frac{H^1(\mathcal{O}_{\mathcal{U}})}{H^1(\chi^{-1})};$$

the last equality holds, because

$$H^0(\mathcal{O}_X) = H^2(\chi^{-1}) = \{0\}.$$

Indeed, on the one hand, there are no global holomorphic functions on the curve X, and on the other hand,

$$H^2(\chi^{-1})^I \simeq H^2_{A_I}(\Sigma_1 \cup \Sigma_2, \mathbb{C}) \simeq H^1(\Sigma_1 \cup \Sigma_2\ A_I, \mathbb{C})$$

where

$$A_I = \{y \in \Sigma_1 \cup \Sigma_2 | \langle y, I\rangle \geq (1 - \lambda)N\} \quad \text{with} \quad y = \lambda x_0 + (1 - \lambda)x_i.$$

The latter space $H^1 = \{0\}$, because $\Sigma_1 \cup \Sigma_2 \setminus A_I$ is, at worse, a union of intervals.

In order to compute the dimension of $\mathrm{Jac}(X)$, it suffices to compute the dimension of the quotient in (6). At first, for A_I defined as above

$$H^1(\chi^{-1})^I = H^1_{A_I}(\Sigma_1 \cup \Sigma_2, \mathbb{C}) \simeq \frac{H^0(\Sigma_1 \cup \Sigma_2 \setminus A_I, \mathbb{C})}{H^0(\Sigma_1 \cup \Sigma_2, \mathbb{C})} = \mathbb{C}^m, \quad m \geq 0,$$

whenever $\Sigma_1 \cup \Sigma_2 \setminus A_I$ has $m + 1$ components. Either $A_I = \Sigma_1 \cup \Sigma_2$ for $\langle I, x_i\rangle \geq N$ $(i = 1, 2)$ or $\Sigma_1 \cup \Sigma_2 \setminus A_I$ has two components, for $\langle I, x_i\rangle < N$ $(i = 1, 2)$; in all other cases it has one component. Therefore

$$H^1(\chi^{-1})^I = \mathbb{C} \quad \text{if and only if} \quad \langle I, x_i\rangle < N \quad i = 1, 2$$

and, similarly

$$H^1(\mathcal{O}_X)^I = \mathbb{C} \quad \text{if and only if} \quad \langle I, x_i\rangle < 0 \quad i = 1, 2.$$

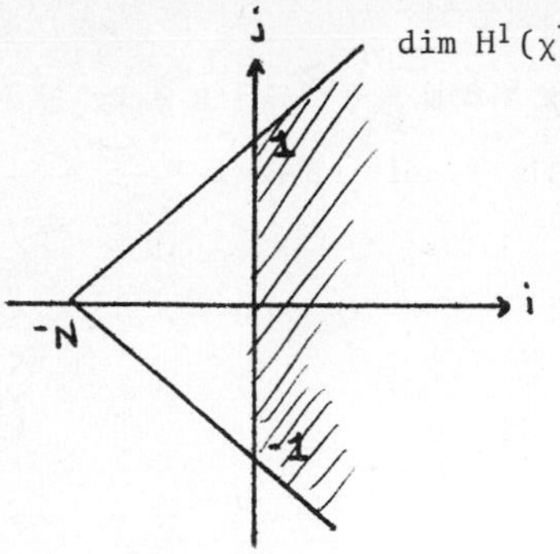

$\dim H^1(\chi^{-1})^I = 1$ in the shaded region

$= 0$ outside

Figure 7.

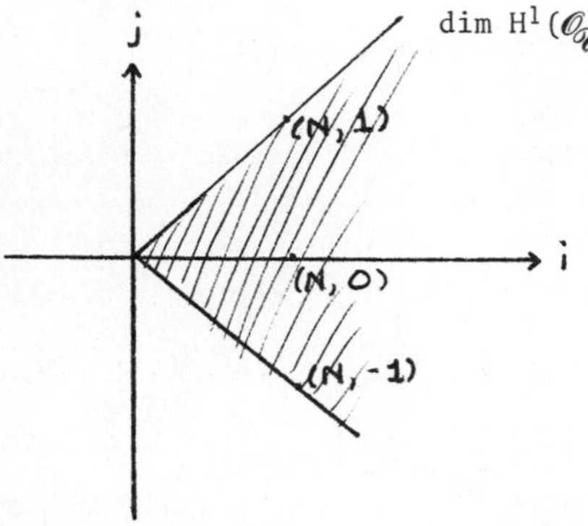

$\dim H^1(\mathcal{O}_{\mathcal{U}})^1 = 1$ in the shaded region

$= 0$ outside

Figure 8.

The injective map

$$H^1(\chi^{-1}) \to H^1(\mathcal{O}_X)$$

induced by the exact sequence above, amounts to multiplying a monomial in H^1 with x^N or, what is the same, adding $(N, 0)$ to I. Then the quotient $H^1(\mathcal{O}_{\mathcal{U}})/H^1(\chi^{-1})$ contains those I in the shaded region of Figure 8 which are not contained in the translate by $(N, 0)$ of Figure 7; it contains exactly $N - 1$ integer points; see Figure 9.

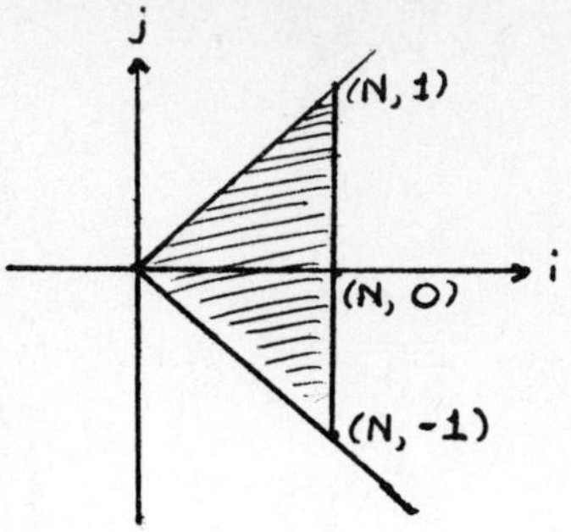

Figure 9.

Therefore $g = \dim \dfrac{H^1(\mathcal{O}_{\mathcal{U}})}{H^1(\chi^{-1})} = N - 1$, confirming the result of Part I.

REFERENCES

[1] Abraham, R., Marsden, J., Foundations of Mechanics, Benjamin, San Francisco, 1978.

[2] Adler, M., van Moerbeke, P., Algebraic curves and the Classical Kac-Moody Algebras (to appear).

[3] Dubrovin, B. A., Matveev, V. B., and Novikov, S. P., Uspehi Mat. Nauk 31 (1976); Russian Math. Surveys 31 (1976).

[4] Griffith, P., Harris, J., Algebraic Geometry (to appear).

[5] Guillemin, V., Kazhdan, D., Some Inverse Spectral Results for Negatively Curved Two-manifolds, preprint 1978.

[6] Hartshorne, R., Algebraic Geometry, Springer-Verlag, NY, 1977.

[7] Kac, M., van Moerbeke, P., On Periodic Toda Lattices, PNAS 72 (1975), 1627-29 and, A Complete Solution of the Periodic Toda Problem, PNAS 72 (1975) 2875-80.

[8] Kempf, G., Knudsen, F., Mumford, D., Saint-Donat, B., Toroidal Embeddings I, Berlin-Heidelberg-New York: Springer Vol 39 (1973).

[9] Kostant, B., Quantization and Unitary Representation. Lectures on Modern Analysis and Applications III, Berlin-Heidelberg-New York: Springer Vol. 170 (1970).

[10] Krichever, I. M., Algebra-Geometrical Construction of the Zaharov-Shabat Equations and their Periodic Solutions, Sov. Math. Dokl. 17 (1976) 394-397.

[11] McKean, M. P., van Moerbeke, P., The Spectrum of Hill's Equation, Inv. Math. 30 (1973) 174-217.

[12] McKean, M. P., van Moerbeke, P., Sur le Spectre de Quelques Opérateurs et les Variétés de Jacobi, Sem. Bourbaki, 1976-76 No. 474, 1-15.

[13] van Moerbeke, P., The Spectrum of Jacobi Matrices, Inv. Math. 37 (1976), 45-81.

[14] van Moerbeke, P., Mumford, D., The Spectrum of Difference Operators and Algebraic Curves, Acta Mathematica 1978 (to appear).

[15] Mumford, D., An Algebra-Gemoetrical Construction of Commuting Operators and of Solutions to the Toda Lattice Equation, Korteweg-de Vries Equation and Related Nonlinear Equations, Kyoto Conference on Algebraic Geometry.

[16] Mumford, D., On Isospectral Deformations of Laplace-like Difference Operators (to appear).

[17] Singer, I., On Deformations of Two Dimensional Laplacians (to appear).

[18] Weinstein, A., Eigenvalues of the Laplacian Plus a Potential, Internat. Congress of Math., Helsinki 1978.

[19] Zaharov, V. E., Shabat, A. B., A Scheme for Integrating the Nonlinear Equations of Math. Physics by the Method of the Inverse Scattering Problem I, Funct. Analysis and its Appl. 8 (1974) (translation 1975, P. 226).

[20] Adler, M., On a Trace Functional for Formal Pseudo-differential Operators and the Symplectic Structure of the Korteweg-de Vries Equations, Inv. Math. 1979.

[21] Dikii, L. A., Gel'fand, I. M., The Resolvent and Hamiltonian Systems, Funct. Anal.Phil. 11 (1977) 11-27.

[22] Ratiu, T., Thesis, Berkeley, (1979).

Bifurcations of periodic orbits in autonomous systems

Yieh-Hei Wan

Department of Mathematics
State University of New York
Buffalo, New York

(A)

Let X_μ be a smooth 1-parameter family of smooth vector fields on a smooth manifold M. Suppose γ is a periodic orbit of X_{μ_0} for some $\mu_0 \in R$ with period T. Denote by $\phi_\mu : V \to U$ the Poincaré map defined on an open subset V of the cross section U through p in γ for μ close to μ_0. The bifurcation analysis of the vector fields X_μ near γ for μ near μ_0 may be reduced to that of the diffeomorphisms ϕ_μ near the fixed point p. For instance, a fixed point of ϕ_μ corresponds to a periodic orbit of X_μ with period close to T. A periodic point of ϕ_μ with order n gives a subharmonic solution of X_μ with period close to nT. An invariant circle of ϕ_μ corresponds to invariant torus of X_μ. In this article, we carry out the bifurcation analysis of X_μ around the periodic orbit γ through the analysis of the associated Poincaré maps ϕ_μ.

When the spectrum of $D\phi_{\mu_0}$ lies away from the unit circle in C, the phase portraits remain the same for μ close to μ_0. Thus, bifurcations of phase portraits happen only if $D\phi_{\mu_0}$ has eigenvalues $\lambda(\mu_0)$ with absolute value 1 (i.e. $|\lambda(\mu_0)| = 1$). In this article, we consider two possibilities. (1) $D\phi_{\mu_0}(p)$ has exactly one *simple real* eigenvalue (thus, $\lambda(\mu_0) = \pm 1$), (2) $D\phi_{\mu_0}(p)$ has exactly a pair of *simple complex* eigenvalues $\lambda(\mu_0)$, $\overline{\lambda(\mu_0)}$ (thus, $\lambda(\mu_0) \neq \pm 1$). Since all bifurcations occur in center manifolds, we may, therefore, take $U = R$ in case (1) and $U = R^2 = C$ in case (2).

For simplicity in notation, let us set $\mu_0 = 0$, and $p = 0$.

(B)

In this part, case 1) is examined.

<u>Case 1(a)</u>: $\lambda(0) = 1$.

$\phi_\mu(x) = a(\mu) + b(\mu)x + c(\mu)x^2 = o(|x|^3)$ with $a(0) = 0$, $b(0) = 1$, $\frac{da}{d\mu}(0) > 0$ and $c(0) \neq 0$. Clearly, the equation $\phi_\mu(x) - x = 0$ can be solved as $\mu = -\frac{c(0)}{a'(0)}x^2 + o(|x|^3)$, and the dynamics of ϕ_μ as shown in Diagram 1.

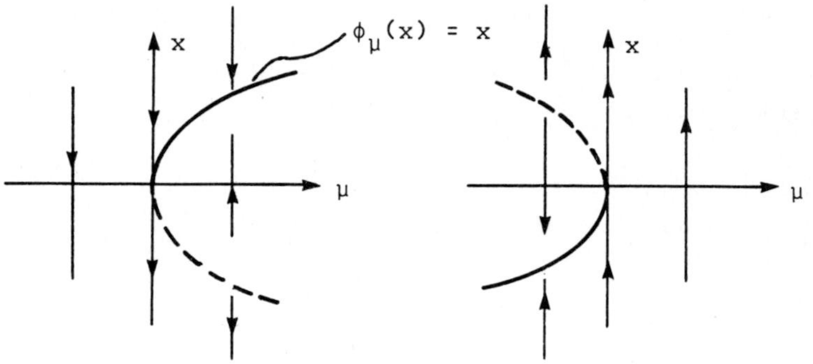

$c(0) < 0$ $c(0) > 0$

Diagram 1

Thus, one concludes that precisely two periodic orbits bifurcate from γ for $\mu > 0$ ($\mu < 0$), one stable, one unstable if $c(0) < 0$, ($c(0) > 0$).

<u>Case 1(b)</u>: $\lambda(0) = -1$.

By implicit function theorem, ϕ_μ can be put in the form: $a(\mu)x + b(\mu)x^2 + o(|x|^3)$ with $a(0) = -1$. Let

$x = y + \frac{b(\mu)}{a^2(\mu) - a(\mu)} y^2$, ϕ_μ in y coordinates, becomes
$a(\mu)y + c(\mu)y^3 + o(|y|^4)$. Assume now, $\frac{da}{d\mu}(0) > 0$ and $c(0) \neq 0$.
Clearly, $y = 0$ is the only fixed point of ϕ_μ near 0.
$\phi_\mu^2(y) = a^2(\mu)y + 2a(\mu)c(\mu)y^3 + o(|y|^4)$. Thus the equation
$\phi_\mu^2(y) - y = 0$ can be solved as $\mu = -\frac{c(0)}{a'(0)}y^2 + o(|y|^3)$, and the
dynamics of ϕ_μ^2 as shown in Diagram 2.

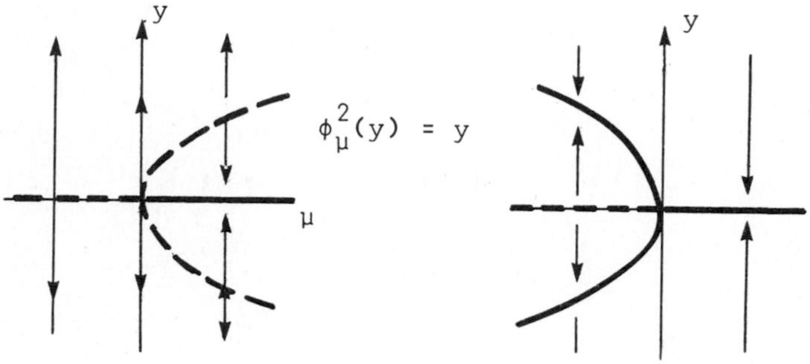

$c(0) < 0$ $c(0) > 0$

Diagram 2

Hence, one obtains that exact one stable (unstable) subharmonic solution of order 2 branches from γ for $\mu < 0$ $(\mu > 0)$ if $-c(0) < 0$ $(-c(0) > 0)$.

C)

In this part Case 2) with $\lambda^4(0) \neq 1$ is studied.

Case 2(a): $\lambda^3(0) \neq 1$, $\lambda^4(0) \neq 1$.

By suitable change of coordinates, ϕ_μ takes the normal form:
$\phi_\mu(z) = \lambda(\mu)z + \alpha_1(\mu)z^2\bar{z} + \alpha_2(\mu)\bar{z}^3 + o(|z|^4)$, $z \in C$, $(\alpha_2(\mu) = 0$ if $\lambda^5(0) \neq 1)$. It has been proved by Sacker, Ruelle and Takens [3] if $\frac{d|\lambda(\mu)|}{d\mu}(0) > 0$, and $\mathrm{Re}\overline{\lambda(0)}\alpha_1(0) < 0$ $(\mathrm{Re}\overline{\lambda(0)}\alpha_1(0) > 0)$ then attracting (repelling) invariant tori bifurcate from γ for

$\mu > 0$ $(\mu < 0)$. See Diagram 3

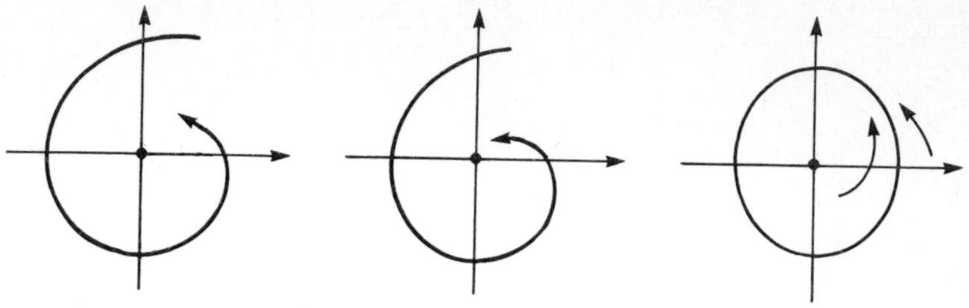

Diagram 3. $\mathrm{Re}(\overline{\lambda(0)}\alpha_1(0)) < 0$.

The periodic solution γ of X_{μ_0} or the fixed point p of ϕ_{μ_0} is said to be at *resonance* if $\lambda(\mu_0)$ is a root of unity (i.e. $\lambda^k(\mu_0) = 1$ for some positive integer k). Bifurcations of the vector fields X_μ become complicated at those points of resonance.

<u>Case 2(b)</u>: $\lambda(0) = (-1 + \sqrt{3}i)/2$ (i.e. $\lambda^3(0) = 1$).

ϕ_μ can be brought into the normal form $\phi_\mu(z) = \lambda(\mu)z + \alpha_1(\mu)z^2\bar{z} + \alpha_0(\mu)\bar{z}^2 + o(|z|^4)$. Assume that $\frac{d\,|\mu(\mu)|}{d\mu}(0) > 0$, $\mathrm{Re}(\overline{\lambda(0)}\alpha_1(0)) \neq 0$ and $\alpha_0(0) \neq 0$. From the works of Takens [4], Arnold [1], we know that $\phi_\mu = \lambda(0)\psi_\mu + o(|z|^4)$ where ψ_μ is the time 1 map of the vector field $X_\mu(z) = \gamma(\mu)z + \sigma_2(\mu)z^2\bar{z} + \sigma_3(\mu)\bar{z}^2$, with $e^{\gamma(\mu)} = \overline{\lambda(0)}\lambda(\mu)$, $\sigma_2(0) = \overline{\lambda(0)}\alpha_0(0)$, $\sigma_3(0) = \overline{\lambda(0)}\alpha_0(0)$. The phase portraits of X_μ are shown in diagram 4.

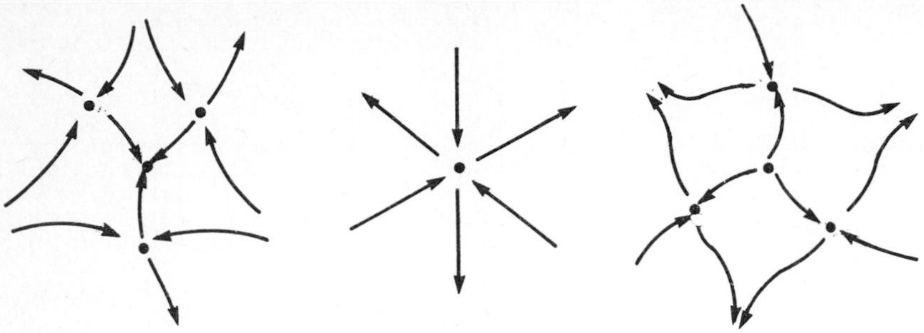

Diagram 4. $\lambda(0) = (-1 + \sqrt{3}i)/2$.

Hence, one gets that one (saddle) subharmonic solution of order 3 branches from γ on both sides of $\mu = 0$ (see also [2]). No invariant tori are expected to bifurcate from γ.

(D)

Some analysis in case 2) with $\lambda(0) = i$ (i.e. $\lambda^4(0) = 1$) are given here. Again, in this situation, ϕ_μ has the normal form:
$\phi_\mu(z) = \lambda(\mu)z + \alpha_1(\mu)z^2\bar{z} + \alpha_2(\mu)\bar{z}^3 + o(|z|^5)$. Write
$\lambda(\mu) = i(1 + \sigma(\mu)) = i(1 + \sigma_1\mu^k + o(\mu^{k+1}))$ for some integer $k \geq 1$. It can be shown as in [2] that (1) if $|\mathrm{Im}(\bar{\sigma}_1\overline{\lambda(0)}\alpha_2(0)| > |\bar{\sigma}_1\alpha_3(0)|$ then no subharmonic solutions of order 4 bifurcate (2) if $|\mathrm{Im}\ \bar{\sigma}_1\overline{\lambda(0)}\alpha_2(0)| < |\bar{\sigma}_1\alpha_3(0)|$ then subharmonic solutions of order 4 do branch from γ. The following result, analogue to that in case 2(a) is conjectured by Iooss and Joseph in [2] and justified in Wan [5]. If $\mathrm{Re}\ \sigma_1 \neq 0$, $\mathrm{Re}\ \overline{\lambda(0)}\alpha_2(0) < 0$, $(\mathrm{Re}\ \overline{\lambda(0)}\alpha_2(0) > 0)$ and $|\mathrm{Im}(\bar{\sigma}_1\overline{\lambda(0)}\alpha_2(0)| > |\bar{\sigma}_1\alpha_3(0)|$, then attracting (repelling) invariant tori bifurcating from γ are, one for each $\mathrm{Re}\ \sigma(\mu) > 0$, and none for each $\mathrm{Re}\ \sigma(\mu) \leq 0$ with small $|\mu|$ (see Diagram 5).

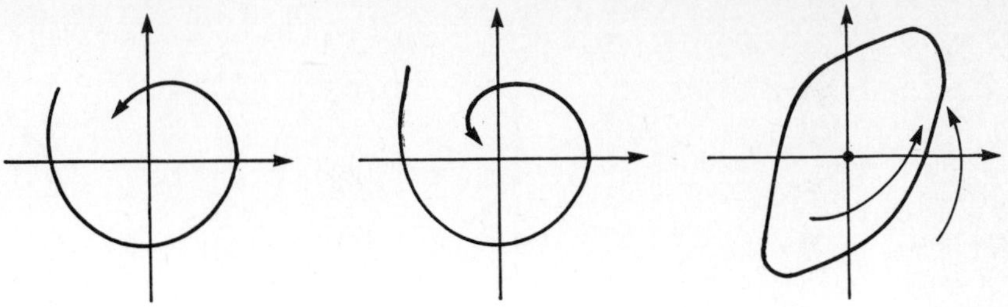

Re $\sigma(\mu) < 0$ Re $\sigma(\mu) = 0$ Re $\sigma(\mu) > 0$

Diagram 5.

Remark: Since $|\lambda(\mu)| = 1 + (\text{Re}\ \sigma_1)\mu^k + o(\mu^{k+1})$, the condition Re $\sigma_1 \neq 0$ is a natural extension of the condition $\frac{d|\lambda(\mu)|}{d\mu}(0) \neq 0$. One needs the weaker form in applications.

Now, I sketch the proof of the above result. Choosing a vector field $Y_\mu(z) = \gamma(\mu)z + \sigma_2(\mu)z^2\bar{z} + \sigma_3(\mu)\bar{z}^3$ with $\sigma_2(0) = \overline{\lambda(0)}\alpha_2(0)$, $\sigma_3(0) = \overline{\lambda(0)}\alpha_3(0)$ and $e^{\gamma(u)} = 1 + \sigma(\mu)$ so that $\phi_\mu = i\psi_\mu + o(|z|^5)$. Here, ψ_μ denotes the time 1 map of Y_μ. First, study the phase portrait of the vector field Y_μ. In particular, by Poincaré-Bendixon theorem and some stability computations, one obtains a (unique) hyperbolic periodic orbit in suitable range of μ. Second, show that ϕ_μ is a "small" perturbation of $i\psi_\mu$ so that phase portrait is preserved.

As pointed out by Takens (4) and Arnold [1], the bifurcations of X_μ with $\lambda(\mu_0) = i$, in general, are very complicated.

References

1. Arnold, V.I., "Loss of stability of self-oscillations close to resonance and versal deformations of equivariant vector fields," Functional Anal. and its Appl. Vol. 11, No. 2 (1977).

2. Iooss, G., and D. D. Joseph, "Bifurcation and stability of nT-periodic solutions branching from T-periodic solutions at points of resonance," Arch. Rational Mech. Anal. 66 (1977), 135-172.

3. Ruelle, D. and F. Takens, "On the nature of turbulence," Comm. Math. Phys. 20 (1971), 167-192, 23(1971)343-344.

4. Takens, F., "Forced oscillations and bifurcations. Application of global analysis I," Symposium, Utrecht State University (1973).

5. Wan, Y.H., "Bifurcations into invariant tori at points of resonance," Arch. Rational Mech. Anal. 68(1978)343-357.

Vol. 580: C. Castaing and M. Valadier, Convex Analysis and Measurable Multifunctions. VIII, 278 pages. 1977.

Vol. 581: Séminaire de Probabilités XI, Université de Strasbourg. Proceedings 1975/1976. Edité par C. Dellacherie, P. A. Meyer et M. Weil. VI, 574 pages. 1977.

Vol. 582: J. M. G. Fell, Induced Representations and Banach *-Algebraic Bundles. IV, 349 pages. 1977.

Vol. 583: W. Hirsch, C. C. Pugh and M. Shub, Invariant Manifolds. IV, 149 pages. 1977.

Vol. 584: C. Brezinski, Accélération de la Convergence en Analyse Numérique. IV, 313 pages. 1977.

Vol. 585: T. A. Springer, Invariant Theory. VI, 112 pages. 1977.

Vol. 586: Séminaire d'Algèbre Paul Dubreil, Paris 1975–1976 (29ème Année). Edited by M. P. Malliavin. VI, 188 pages. 1977.

Vol. 587: Non-Commutative Harmonic Analysis. Proceedings 1976. Edited by J. Carmona and M. Vergne. IV, 240 pages. 1977.

Vol. 588: P. Molino, Théorie des G-Structures: Le Problème d'Equivalence. VI, 163 pages. 1977.

Vol. 589: Cohomologie l-adique et Fonctions L. Séminaire de Géométrie Algébrique du Bois-Marie 1965–66, SGA 5. Edité par L. Illusie. XII, 484 pages. 1977.

Vol. 590: H. Matsumoto, Analyse Harmonique dans les Systèmes de Tits Bornologiques de Type Affine. IV, 219 pages. 1977.

Vol. 591: G. A. Anderson, Surgery with Coefficients. VIII, 157 pages. 1977.

Vol. 592: D. Voigt, Induzierte Darstellungen in der Theorie der endlichen, algebraischen Gruppen. V, 413 Seiten. 1977.

Vol. 593: K. Barbey and H. König, Abstract Analytic Function Theory and Hardy Algebras. VIII, 260 pages. 1977.

Vol. 594: Singular Perturbations and Boundary Layer Theory, Lyon 1976. Edited by C. M. Brauner, B. Gay, and J. Mathieu. VIII, 539 pages. 1977.

Vol. 595: W. Hazod, Stetige Faltungshalbgruppen von Wahrscheinlichkeitsmaßen und erzeugende Distributionen. XIII, 157 Seiten. 1977.

Vol. 596: K. Deimling, Ordinary Differential Equations in Banach Spaces. VI, 137 pages. 1977.

Vol. 597: Geometry and Topology, Rio de Janeiro, July 1976. Proceedings. Edited by J. Palis and M. do Carmo. VI, 866 pages. 1977.

Vol. 598: J. Hoffmann-Jørgensen, T. M. Liggett et J. Neveu, Ecole d'Eté de Probabilités de Saint-Flour VI – 1976. Edité par P.-L. Hennequin. XII, 447 pages. 1977.

Vol. 599: Complex Analysis, Kentucky 1976. Proceedings. Edited by J. D. Buckholtz and T. J. Suffridge. X, 159 pages. 1977.

Vol. 600: W. Stoll, Value Distribution on Parabolic Spaces. VIII, 216 pages. 1977.

Vol. 601: Modular Functions of oneVariable V, Bonn 1976. Proceedings. Edited by J.-P. Serre and D. B. Zagier. VI, 294 pages. 1977.

Vol. 602: J. P. Brezin, Harmonic Analysis on Compact Solvmanifolds. VIII, 179 pages. 1977.

Vol. 603: B. Moishezon, Complex Surfaces and Connected Sums of Complex Projective Planes. IV, 234 pages. 1977.

Vol. 604: Banach Spaces of Analytic Functions, Kent, Ohio 1976. Proceedings. Edited by J. Baker, C. Cleaver and Joseph Diestel. VI, 141 pages. 1977.

Vol. 605: Sario et al., Classification Theory of Riemannian Manifolds. XX, 498 pages. 1977.

Vol. 606: Mathematical Aspects of Finite Element Methods. Proceedings 1975. Edited by I. Galligani and E. Magenes. VI, 362 pages. 1977.

Vol. 607: M. Métivier, Reelle und Vektorwertige Quasimartingale und die Theorie der Stochastischen Integration. X, 310 Seiten. 1977.

Vol. 608: Bigard et al., Groupes et Anneaux Réticulés. XIV, 334 pages. 1977.

Vol. 609: General Topology and Its Relations to Modern Analysis and Algebra IV. Proceedings 1976. Edited by J. Novák. XVIII, 225 pages. 1977.

Vol. 610: G. Jensen, Higher Order Contact of Submanifolds of Homogeneous Spaces. XII, 154 pages. 1977.

Vol. 611: M. Makkai and G. E. Reyes, First Order Categorical Logic. VIII, 301 pages. 1977.

Vol. 612: E. M. Kleinberg, Infinitary Combinatorics and the Axiom of Determinateness. VIII, 150 pages. 1977.

Vol. 613: E. Behrends et al., L^p-Structure in Real Banach Spaces. X, 108 pages. 1977.

Vol. 614: H. Yanagihara, Theory of Hopf Algebras Attached to Group Schemes. VIII, 308 pages. 1977.

Vol. 615: Turbulence Seminar, Proceedings 1976/77. Edited by P. Bernard and T. Ratiu. VI, 155 pages. 1977.

Vol. 616: Abelian Group Theory, 2nd New Mexico State University Conference, 1976. Proceedings. Edited by D. Arnold, R. Hunter and E. Walker. X, 423 pages. 1977.

Vol. 617: K. J. Devlin, The Axiom of Constructibility: A Guide for the Mathematician. VIII, 96 pages. 1977.

Vol. 618: I. I. Hirschman, Jr. and D. E. Hughes, Extreme Eigen Values of Toeplitz Operators. VI, 145 pages. 1977.

Vol. 619: Set Theory and Hierarchy Theory V, Bierutowice 1976. Edited by A. Lachlan, M. Srebrny, and A. Zarach. VIII, 358 pages. 1977.

Vol. 620: H. Popp, Moduli Theory and Classification Theory of Algebraic Varieties. VIII, 189 pages. 1977.

Vol. 621: Kauffman et al., The Deficiency Index Problem. VI, 112 pages. 1977.

Vol. 622: Combinatorial Mathematics V, Melbourne 1976. Proceedings. Edited by C. Little. VIII, 213 pages. 1977.

Vol. 623: I. Erdelyi and R. Lange, Spectral Decompositions on Banach Spaces. VIII, 122 pages. 1977.

Vol. 624: Y. Guivarc'h et al., Marches Aléatoires sur les Groupes de Lie. VIII, 292 pages. 1977.

Vol. 625: J. P. Alexander et al., Odd Order Group Actions and Witt Classification of Innerproducts. IV, 202 pages. 1977.

Vol. 626: Number Theory Day, New York 1976. Proceedings. Edited by M. B. Nathanson. VI, 241 pages. 1977.

Vol. 627: Modular Functions of One Variable VI, Bonn 1976. Proceedings. Edited by J. P. Serre and D. B. Zagier. VI, 339 pages. 1977.

Vol. 628: H. J. Baues, Obstruction Theory on the Homotopy Classification of Maps. XII, 387 pages. 1977.

Vol. 629: W. A. Coppel, Dichotomies in Stability Theory. VI, 98 pages. 1978.

Vol. 630: Numerical Analysis, Proceedings, Biennial Conference, Dundee 1977. Edited by G. A. Watson. XII, 199 pages. 1978.

Vol. 631: Numerical Treatment of Differential Equations. Proceedings 1976. Edited by R. Bulirsch, R. D. Grigorieff, and J. Schröder. X, 219 pages. 1978.

Vol. 632: J.-F. Boutot, Schéma de Picard Local. X, 165 pages. 1978.

Vol. 633: N. R. Coleff and M. E. Herrera, Les Courants Résiduels Associés à une Forme Méromorphe. X, 211 pages. 1978.

Vol. 634: H. Kurke et al., Die Approximationseigenschaft lokaler Ringe. IV, 204 Seiten. 1978.

Vol. 635: T. Y. Lam, Serre's Conjecture. XVI, 227 pages. 1978.

Vol. 636: Journées de Statistique des Processus Stochastiques, Grenoble 1977, Proceedings. Edité par Didier Dacunha-Castelle et Bernard Van Cutsem. VII, 202 pages. 1978.

Vol. 637: W. B. Jurkat, Meromorphe Differentialgleichungen. VII, 194 Seiten. 1978.

Vol. 638: P. Shanahan, The Atiyah-Singer Index Theorem, An Introduction. V, 224 pages. 1978.

Vol. 639: N. Adasch et al., Topological Vector Spaces. V, 125 pages. 1978.

Vol. 640: J. L. Dupont, Curvature and Characteristic Classes. X, 175 pages. 1978.

Vol. 641: Séminaire d'Algèbre Paul Dubreil, Proceedings Paris 1976–1977. Edité par M. P. Malliavin. IV, 367 pages. 1978.

Vol. 642: Theory and Applications of Graphs, Proceedings, Michigan 1976. Edited by Y. Alavi and D. R. Lick. XIV, 635 pages. 1978.

Vol. 643: M. Davis, Multiaxial Actions on Manifolds. VI, 141 pages. 1978.

Vol. 644: Vector Space Measures and Applications I, Proceedings 1977. Edited by R. M. Aron and S. Dineen. VIII, 451 pages. 1978.

Vol. 645: Vector Space Measures and Applications II, Proceedings 1977. Edited by R. M. Aron and S. Dineen. VIII, 218 pages. 1978.

Vol. 646: O. Tammi, Extremum Problems for Bounded Univalent Functions. VIII, 313 pages. 1978.

Vol. 647: L. J. Ratliff, Jr., Chain Conjectures in Ring Theory. VIII, 133 pages. 1978.

Vol. 648: Nonlinear Partial Differential Equations and Applications, Proceedings, Indiana 1976–1977. Edited by J. M. Chadam. VI, 206 pages. 1978.

Vol. 649: Séminaire de Probabilités XII, Proceedings, Strasbourg, 1976–1977. Edité par C. Dellacherie, P. A. Meyer et M. Weil. VIII, 805 pages. 1978.

Vol. 650: C*-Algebras and Applications to Physics. Proceedings 1977. Edited by H. Araki and R. V. Kadison. V, 192 pages. 1978.

Vol. 651: P. W. Michor, Functors and Categories of Banach Spaces. VI, 99 pages. 1978.

Vol. 652: Differential Topology, Foliations and Gelfand-Fuks-Cohomology, Proceedings 1976. Edited by P. A. Schweitzer. XIV, 252 pages. 1978.

Vol. 653: Locally Interacting Systems and Their Application in Biology. Proceedings, 1976. Edited by R. L. Dobrushin, V. I. Kryukov and A. L. Toom. XI, 202 pages. 1978.

Vol. 654: J. P. Buhler, Icosahedral Golois Representations. III, 143 pages. 1978.

Vol. 655: R. Baeza, Quadratic Forms Over Semilocal Rings. VI, 199 pages. 1978.

Vol. 656: Probability Theory on Vector Spaces. Proceedings, 1977. Edited by A. Weron. VIII, 274 pages. 1978.

Vol. 657: Geometric Applications of Homotopy Theory I, Proceedings 1977. Edited by M. G. Barratt and M. E. Mahowald. VIII, 459 pages. 1978.

Vol. 658: Geometric Applications of Homotopy Theory II, Proceedings 1977. Edited by M. G. Barratt and M. E. Mahowald. VIII, 487 pages. 1978.

Vol. 659: Bruckner, Differentiation of Real Functions. X, 247 pages. 1978.

Vol. 660: Equations aux Dérivée Partielles. Proceedings, 1977. Edité par Pham The Lai. VI, 216 pages. 1978.

Vol. 661: P. T. Johnstone, R. Paré, R. D. Rosebrugh, D. Schumacher, R. J. Wood, and G. C. Wraith, Indexed Categories and Their Applications. VII, 260 pages. 1978.

Vol. 662: Akin, The Metric Theory of Banach Manifolds. XIX, 306 pages. 1978.

Vol. 663: J. F. Berglund, H. D. Junghenn, P. Milnes, Compact Right Topological Semigroups and Generalizations of Almost Periodicity. X, 243 pages. 1978.

Vol. 664: Algebraic and Geometric Topology, Proceedings, 1977. Edited by K. C. Millett. XI, 240 pages. 1978.

Vol. 665: Journées d'Analyse Non Linéaire. Proceedings, 1977. Edité par P. Bénilan et J. Robert. VIII, 256 pages. 1978.

Vol. 666: B. Beauzamy, Espaces d'Interpolation Réels: Topologie et Géometrie. X, 104 pages. 1978.

Vol. 667: J. Gilewicz, Approximants de Padé. XIV, 511 pages. 1978.

Vol. 668: The Structure of Attractors in Dynamical Systems. Proceedings, 1977. Edited by J. C. Martin, N. G. Markley and W. Perrizo. VI, 264 pages. 1978.

Vol. 669: Higher Set Theory. Proceedings, 1977. Edited by G. H. Müller and D. S. Scott. XII, 476 pages. 1978.

Vol. 670: Fonctions de Plusieurs Variables Complexes III, Proceedings, 1977. Edité par F. Norguet. XII, 394 pages. 1978.

Vol. 671: R. T. Smythe and J. C. Wierman, First-Passage Perculation on the Square Lattice. VIII, 196 pages. 1978.

Vol. 672: R. L. Taylor, Stochastic Convergence of Weighted Sums of Random Elements in Linear Spaces. VII, 216 pages. 1978.

Vol. 673: Algebraic Topology, Proceedings 1977. Edited by P. Hoffman, R. Piccinini and D. Sjerve. VI, 278 pages. 1978.

Vol. 674: Z. Fiedorowicz and S. Priddy, Homology of Classical Groups Over Finite Fields and Their Associated Infinite Loop Spaces. VI, 434 pages. 1978.

Vol. 675: J. Galambos and S. Kotz, Characterizations of Probability Distributions. VIII, 169 pages. 1978.

Vol. 676: Differential Geometrical Methods in Mathematical Physics II, Proceedings, 1977. Edited by K. Bleuler, H. R. Petry and A. Reetz. VI, 626 pages. 1978.

Vol. 677: Séminaire Bourbaki, vol. 1976/77, Exposés 489–506. IV, 264 pages. 1978.

Vol. 678: D. Dacunha-Castelle, H. Heyer et B. Roynette. Ecole d'Eté de Probabilités de Saint-Flour. VII-1977. Edité par P. L. Hennequin. IX, 379 pages. 1978.

Vol. 679: Numerical Treatment of Differential Equations in Applications, Proceedings, 1977. Edited by R. Ansorge and W. Törnig. IX, 163 pages. 1978.

Vol. 680: Mathematical Control Theory, Proceedings, 1977. Edited by W. A. Coppel. IX, 257 pages. 1978.

Vol. 681: Séminaire de Théorie du Potentiel Paris, No. 3, Directeurs: M. Brelot, G. Choquet et J. Deny. Rédacteurs: F. Hirsch et G. Mokobodzki. VII, 294 pages. 1978.

Vol. 682: G. D. James, The Representation Theory of the Symmetric Groups. V, 156 pages. 1978.

Vol. 683: Variétés Analytiques Compactes, Proceedings, 1977. Edité par Y. Hervier et A. Hirschowitz. V, 248 pages. 1978.

Vol. 684: E. E. Rosinger, Distributions and Nonlinéar Partial Differential Equations. XI, 146 pages. 1978.

Vol. 685: Knot Theory, Proceedings, 1977. Edited by J. C. Hausmann. VII, 311 pages. 1978.

Vol. 686: Combinatorial Mathematics, Proceedings, 1977. Edited by D. A. Holton and J. Seberry. IX, 353 pages. 1978.

Vol. 687: Algebraic Geometry, Proceedings, 1977. Edited by L. D. Olson. V, 244 pages. 1978.

Vol. 688: J. Dydak and J. Segal, Shape Theory. VI, 150 pages. 1978.

Vol. 689: Cabal Seminar 76–77, Proceedings, 1976–77. Edited by A.S. Kechris and Y. N. Moschovakis. V, 282 pages. 1978.

Vol. 690: W. J. J. Rey, Robust Statistical Methods. VI, 128 pages. 1978.

Vol. 691: G. Viennot, Algèbres de Lie Libres et Monoïdes Libres. III, 124 pages. 1978.

Vol. 692: T. Husain and S. M. Khaleelulla, Barrelledness in Topological and Ordered Vector Spaces. IX, 258 pages. 1978.

Vol. 693: Hilbert Space Operators, Proceedings, 1977. Edited by J. M. Bachar Jr. and D. W. Hadwin. VIII, 184 pages. 1978.

Vol. 694: Séminaire Pierre Lelong – Henri Skoda (Analyse) Année 1976/77. VII, 334 pages. 1978.

Vol. 695: Measure Theory Applications to Stochastic Analysis, Proceedings, 1977. Edited by G. Kallianpur and D. Kölzow. XII, 261 pages. 1978.

Vol. 696: P. J. Feinsilver, Special Functions, Probability Semigroups, and Hamiltonian Flows. VI, 112 pages. 1978.

Vol. 697: Topics in Algebra, Proceedings, 1978. Edited by M. F. Newman. XI, 229 pages. 1978.

Vol. 698: E. Grosswald, Bessel Polynomials. XIV, 182 pages. 1978.

Vol. 699: R. E. Greene and H.-H. Wu, Function Theory on Manifolds Which Possess a Pole. III, 215 pages. 1979.